T0300720

Springer Undergraduate Texts in Mathematics and Technology

Springer Undergraduate Texts in Mathematics and Technology (SUMAT) publishes textbooks aimed primarily at the undergraduate. Each text is designed principally for students who are considering careers either in the mathematical sciences or in technology-based areas such as engineering, finance, information technology and computer science, bioscience and medicine, optimization or industry. Texts aim to be accessible introductions to a wide range of core mathematical disciplines and their practical, real-world applications; and are fashioned both for course use and for independent study.

Timothy G. Feeman

Applied Linear Algebra and Matrix Methods

Timothy G. Feeman
Department of Mathematics and Statistics
Villanova University
Villanova, PA, USA

ISSN 1867-5506 ISSN 1867-5514 (electronic)
Springer Undergraduate Texts in Mathematics and Technology
ISBN 978-3-031-39561-1 ISBN 978-3-031-39562-8 (eBook)
https://doi.org/10.1007/978-3-031-39562-8

Mathematics Subject Classification: 15-01, 15AXX, 15A06, 15A18, 15A23

This Springer imprint is published by the registered company Springer Nature Switzerland AG
The registered company address is: Gewerbestrasse 11, 6330 Cham, Switzerland

Paper in this product is recyclable.

For Greta, the best bunny

Introduction

The recent Covid-19 pandemic brought with it widespread uncertainty. For over two and a half years, all around the world, our lives were thrown into disarray. We reorganized and adapted, stayed at home, worked and learned remotely, and socialized at a distance. Our mental and emotional health continue to suffer from the loss of in-person human connection. Vulnerable communities have felt the brunt of the pandemic. Many people have died. Life expectancy has declined. New vaccines, developed in record time, provided a measure of relief and hope. Yet, repeatedly, we were confronted with the challenge of new variants of the virus that were better adapted to living in humans. The term "long Covid" has entered our everyday vocabulary. Through it all, we have come face to face with what it means to *not know*. With this uncertainty has come an increased appreciation and understanding of the need for good data, and even better analysis and modeling of that data. Without that, every response to the crisis, and to the next crisis, is a shot in the dark.

Data-driven research necessarily rests on mathematical foundations, with *linear algebra* as one of the cornerstones. More than ever, linear algebra and the matrix methods at its core are essential tools for students and practitioners of statistics, data science, finance, computing science, and more. Yet the standard college curriculum is designed largely to serve traditional constituents such as mathematicians and electrical engineers.

At my university, the introduction to basic linear algebra has long been embedded in a course called *Differential Equations with Linear Algebra*, with the latter making up about 35% of the content. The applications are weighted toward physics and engineering. A second course, called simply *Linear Algebra*, is addressed primarily to majors in mathematics and takes a more abstract approach to the subject, helping students to hone their proof-writing skills and build a firm theoretical foundation for future study. In recent years, however, increasing numbers of students from non-mathematics majors have enrolled in this course and brought with them a growing hunger for applications. More and more, the students want to know how linear algebra can help them to understand and solve contemporary real-world problems in a host of disciplines. The mathematics majors want this, too.

The book you are now holding in your hands or viewing on your screen aims to provide this growing and diverse group of students with an applied linear algebra toolkit they can use to successfully grapple with the complex and challenging problems that lie ahead. Applications such as least squares problems, information retrieval, linear regression, Markov processes, finding connections in networks, and more are introduced in the small scale as early as possible and then explored in more generality as projects. There are some theorems. It is hard to do mathematics without them. As much as possible, I have drawn on the geometry of vectors in real Euclidean space as the basis for the mathematics, with the concept of orthogonality taking center stage. I find it useful to think of a theorem as a way of highlighting a particular picture, pattern, or calculation that we will end up using over and over again. Important matrix factorizations, as well as the concepts of eigenvalues and eigenvectors, emerge organically from the interplay between matrix computations and geometry. In this way, I hope to show that the theory of matrices and linear algebra arises from applications and geometric considerations, and, in turn, fosters a deeper understanding that leads to new applications.

Twenty-five years ago, I mentioned to a scientist at Kodak, who worked in digital photography, that some of my linear algebra students were chafing at the amount of theory involved and were demanding to know what it was all meant to be used for. His response was emphatic: **"Linear algebra is for everything!"** That advice presents a challenge to mathematics education that is even more pressing today. I hope this book is a step in the right direction.

—Tim Feeman, Villanova University, May 2023.

Advice for Instructors

First of all, if you are an instructor using this text for your course—Thank you! You and your students are the reason I wrote the book in the first place. I hope your course is a success and that the students come away from it feeling well-equipped to tackle a host of real-world problems in their primary fields of study.

This book is designed for an introductory matrix-based linear algebra course. The emphasis is on applications of current interest. Students taking the course should have completed one or two university-level mathematics courses, which would typically be calculus and/or statistics courses (but need not be). The experience gained working on complex multi-step problems will be especially beneficial in this course.

Whatever the prior knowledge and backgrounds of your students, a thorough grounding in Chapters 1–3 of the book is essential. One of my main premises in this book is that interesting applications become accessible as soon as we grasp the geometry of vectors in real Euclidean space. The inner product is the main tool. Thus, both a basic scheme for information retrieval and the concept of statistical correlation appear right away in Chapter 1. Matrices and matrix multiplication, introduced in Chapter 2, provide a way to organize larger data sets and to handle many vector-based computations at once. Chapter 3 presents a broad range of problems, applications, and contexts that can be tackled on a practical level without getting lost in theoretical details about subspaces, linear independence, or abstract linear transformations. It is probably not realistic to cover all of these contexts. Sections 3.1–3.6 include applications that are likely to be relevant to most students. In any case, make a selection that reflects your interests and those of the class. The exercises in these chapters are designed to reinforce the basic concepts and computations in low-dimensional settings. The projects give students the opportunity to explore higher-dimensional problems using real-world data. The projects are the cornerstones of learning around which the book is built.

Solving systems of linear equations is where it all began for linear algebra as a subject. So, in Chapter 4, Sections 4.1–4.5 are essential to a first course. Students will probably find the application to Leontief input–output matrices more exciting than cubic splines. I, on the other hand, have been intrigued by cubic splines ever since I saw wooden models made with actual splines in a big display case in the old engineering building at The University of Michigan, when I was in graduate school. The sections of Chapter 4 on the LU decomposition and affine projections can be omitted, especially if time is an issue.

The general least squares problem, with applications to multiple regression and curve fitting, is the subject of Chapter 5, Sections 5.1–5.3. Most of the exercises and projects in the chapter are based on these sections. The rest of the chapter is important for a deeper theoretical understanding, but you may wish to move on to get to more applications.

In fact, one could jump from Section 5.3 all the way to Chapter 7, on eigenvalues. The applications of QR factorization in Chapter 6 are certainly powerful. The idea that projection into the column space of a matrix is simpler when the columns form an orthonormal set is especially useful. However, my classroom experience suggests that students may not see much payoff in the computational advantages of using QR given that we are using computers to solve all but the lowest dimensional problems. If you have time, of course, then Chapter 6 has some great mathematics in it. It may make sense to come back to some concepts from Chapter 6 as they arise in later chapters.

The main reason to jump to Chapter 7 is that eigenvalues and eigenvectors are such important concepts with huge applications. Section 7.6 looks at some population models, including the Fibonacci numbers. Section 7.7, on rotations in three-dimensional space, is relevant to computer graphics.

From Chapter 7, one can move to either Chapter 8 or Chapter 9, though you should make every effort to discuss both. Virtually every student I have had has been excited by the applications in these two chapters. Chapter 8 features applications of eigenvalues to Markov processes. This includes web page ranking and ranking of sports teams. Chapter 9 looks at the eigenvalue decomposition for symmetric matrices. Here, it is worth looking at the concept of an orthogonal matrix, introduced in Chapter 6, if one has not done that already. The main application in Chapter 9 is an idea from spectral graph theory that uses the graph Laplacian matrix to help identify well-connected clusters of vertices in a network.

The main concept in Chapter 10, that using a reduced rank approximation to a matrix can reveal the essential information hidden in the data while also dramatically reducing storage requirements, is tremendously powerful. If time is short, at least try to discuss the basic concept of the singular value decomposition. Have the students explore one of the applications in Sections 10.3–10.5, depending on their interests, and present in-class demonstrations of one or two others. In my experience, image compression makes a great ending for the course, particularly as many use it every day without thinking much about what's really going on.

Whatever you do, I hope you have time for the students to work on at least five or six projects during a one-semester course. In addition to addressing real-world problems, the projects give students the opportunity to enhance their programming skills. In class, I typically use base *R* to keep the coding as transparent as possible. I make templates available to the class for them to imitate. For project reports, using *R* Studio makes it relatively easy to incorporate narrative, useful chunks of code, and plots all in the same place. I always seem to have a few students who know how to do this and are willing to help others, including me! Some students may already be familiar with Python or MATLAB. Those work fine, too, but I explain to those students that I may not be able to help them with coding details. Your expertise may be different, so go with what you know.

I hope this brief discussion helps you in planning your course. Thank you!

Acknowledgments

First and foremost, I am grateful for the opportunity to work with two fine editors at Springer. The extraordinary Elizabeth Loew has been a constant source of encouragement for over a decade now, sharing her wealth of experience and advice, including even how to French braid my daughter's hair! Loretta Bartolini patiently shepherded this project along for about a year during the pandemic before handing it back to Elizabeth. Without their talents, know-how, and support, this book would not be finding its way into the world. I also thank an unknown reviewer of an early draft for taking the time to read closely and provide a host of insightful comments, observations, and suggestions. I especially thank the Villanova Institute for Teaching and Learning (VITAL) for a summer mini-grant that supported the development of a new undergraduate course in applied linear algebra. Numerous conversations with colleagues at Villanova, including Katie Haymaker, Mike Tait, and Katie Muller, have enriched my sense both of how to teach linear algebra and what to teach when we teach linear algebra. The 150 or so students who have studied applied linear algebra with me over the past five years have taught me a lot and made the course a richer experience for everyone. All this advice and discussion has made this book better, though any and all shortcomings are entirely my responsibility. Finally, I thank my six-year-old daughter, to whom this volume is dedicated, for keeping me on my toes, filling my life with joy, and already getting it that math is a thing.

Contents

Chapter 1
Vectors

Most of the ideas and applications found in this book involve just two types of mathematical objects: vectors and matrices. In the first two chapters of this book, we define these concepts and begin to explore their applications.

1.1 Coordinates and Vectors

Classical Greek geometry takes place in a flat, 2-dimensional plane or in a 3-dimensional space, like the one we live in. Shapes like rectangles, triangles, and circles lie in a plane, while rectangular boxes, pyramids, and spheres require three dimensions. The insight that we can assign numerical coordinates to points in space was developed by Descartes, among others, in the early phase of modern mathematics. In a plane, we do this by drawing a pair of perpendicular copies of the real number line, usually called the x-axis and the y-axis. Typically, the x-axis is drawn as a horizontal line with the positive numbers to the right, while the y-axis is drawn vertically with the positive numbers going up. The point where these two axes cross is called the *origin* and occurs at the 0 mark on both lines. Thus, the origin has coordinates $(0, 0)$. For any arbitrary point P in the plane, we can draw a rectangle that has P at one corner and the origin at the diagonally opposite corner. One vertical side of this rectangle connects P to the x-axis, which gives us the x-coordinate of P. Similarly, a horizontal edge of the rectangle connects P to the y-axis at the y-coordinate of P. In this way, every point in the plane is associated to an ordered pair of numbers (x, y), called the coordinates of the point. This insight allows us to

---

Supplementary Information The online version contains supplementary material available at https://doi.org/10.1007/978-3-031-39562-8_1.

T. G. Feeman, *Applied Linear Algebra and Matrix Methods*,
Springer Undergraduate Texts in Mathematics and Technology,
https://doi.org/10.1007/978-3-031-39562-8_1

make a connection between *geometric* objects, like circles, and *algebraic* objects, like equations. For instance, a picture of all points in the plane whose coordinates (x, y) satisfy the requirement that $x^2 + y^2 = 1$ is a circle of radius 1 with center at the origin.

For a 3-dimensional space, we include a third coordinate axis, typically called the z-axis, that is perpendicular to both the x- and y-axes that formed the plane. It is customary to draw the three axes with a *right-hand* orientation, meaning that, if we point the fingers of our right hand along the positive direction of the x-axis and curl our fingers towards the positive direction of the y-axis, then our thumb will point in the positive direction of the z-axis. The origin now has coordinates $(0, 0, 0)$ and an arbitrary point lies at one corner of a 3-dimensional rectangular box that has the origin at the diametrically opposite corner. The coordinates of the point are determined by where the edges of the box cross the x-axis, y-axis, and z-axis. In three dimensions, a picture of all points whose coordinates (x, y, z) satisfy the requirement that $x^2 + y^2 + z^2 = 1$ is a sphere of radius 1 centered at the origin.

We can use our imagination to increase the number of dimensions to any positive whole number we want. For each dimension, we have to imagine an additional axis placed perpendicular to all of the other axes. For an N-dimensional space, we need N different axes. Admittedly, this is difficult, maybe even impossible, to picture. Nonetheless, we can represent every point in an N-dimensional space by listing its coordinates relative to the N different axes. This gives us an ordered list of N numbers to describe each point.

Remark With two or three dimensions, we often label the different coordinate directions as x, y, and z. However, many applications will require dozens or even hundreds of coordinates. To distinguish them in writing, it is simpler to adopt one choice of letter, say x, and then use the numerical subscripts $1, 2, \ldots, N$ for the different directions. For example, in a 5-dimensional space, we may refer to the x_1-axis, the x_2-axis, the x_3-axis, the x_4-axis, and the x_5-axis. Subscript notation will also come in handy when we work with matrices.

To find the coordinates of a point in an N-dimensional space, as just imagined, we consider an N-dimensional "box" that has the desired point at one corner, diagonally opposite to the origin. Now draw the diagonal of the box as an arrow that starts at the origin and ends at the given point. That arrow is called the **vector** associated to the given point. In order to preserve the distinction between vectors (arrows) and points, we express a vector numerically by listing the coordinates of its terminal point in a column. For each positive integer N, the set of all such columns of N real numbers forms the vector space $\mathbb{R}^N$. (We read this out loud as "R N").

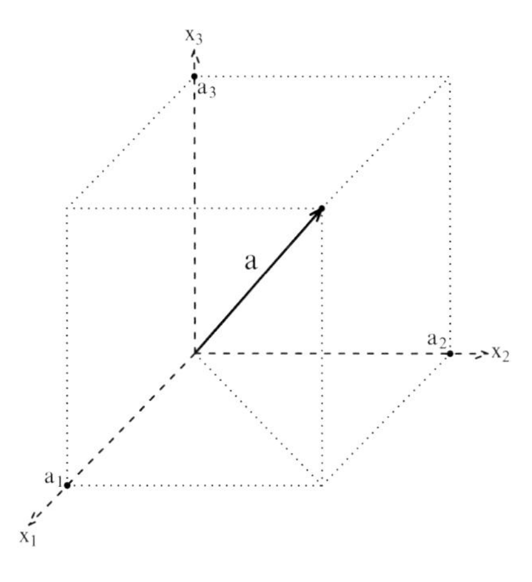

Fig. 1.1 In $\mathbb{R}^3$, a vector **a** can be represented by an arrow that starts at the origin and ends at the point with the same coordinates as the vector

Definition 1.1 For each positive integer N, the space $\mathbb{R}^N$ is the set

$$\mathbb{R}^N = \left\{ \begin{bmatrix} a_1 \\ a_2 \\ \vdots \\ a_N \end{bmatrix} : a_1, a_2, \ldots, a_N \text{ in } \mathbb{R} \right\}. \tag{1.1}$$

Each element of $\mathbb{R}^N$ is called a **vector**. The individual numbers a_1, a_2, and so on are called the *coordinates* of the vector they define.

Example 1.2 The space $\mathbb{R}^2$ corresponds to the set of all origin-based arrows in a plane. The vector $\begin{bmatrix} 3 \\ 1 \end{bmatrix}$ represents an arrow that starts at the origin $(0, 0)$ and ends at the terminal point with coordinates $(3, 1)$. Similarly, a vector $\begin{bmatrix} a_1 \\ a_2 \\ a_3 \end{bmatrix}$ in $\mathbb{R}^3$ corresponds to an arrow in $x_1x_2x_3$-space that starts at the origin $(0, 0, 0)$ and terminates at the point (a_1, a_2, a_3). This is illustrated in Figure 1.1. For $N > 3$, we can *imagine* such an arrow even if we are not able to actually draw it.

Just as each numerical vector in $\mathbb{R}^N$ can be identified with an arrow, the geometry of arrows gives rise to an arithmetic for vectors, like so. Given two vectors that start at the origin, they form a parallelogram by attaching a parallel copy of each arrow onto the end of the other arrow. We think of this as adding one arrow to the other. The sum of the vectors is the arrow that starts at the origin and forms a diagonal of this parallelogram. The coordinates of the terminal point of the diagonal of the parallelogram are obtained by adding the corresponding coordinates of the two vectors we started with. Figure 1.2 illustrates this in $\mathbb{R}^2$.

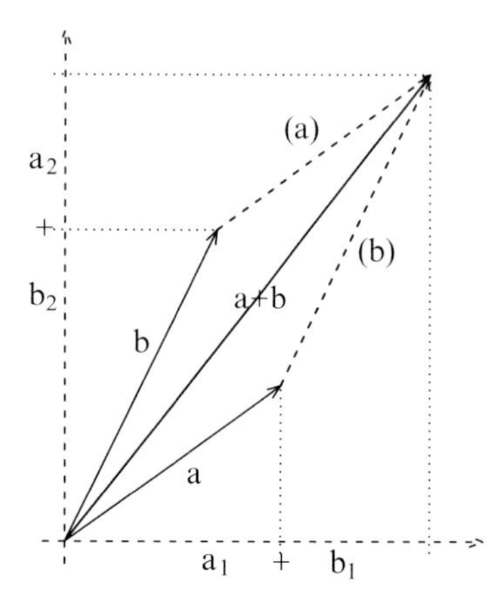

Fig. 1.2 The vectors **a** and **b**, based at the origin, form a parallelogram when copies (**a**) and (**b**) are included. One diagonal of the parallelogram is the origin-based vector **a** + **b**. Each coordinate of **a** + **b** is the sum of the corresponding coordinates of **a** and **b** individually

A single arrow generates an entire straight line that extends infinitely in both directions. Each point on this line is the terminal point of an arrow that starts at the origin. This new arrow is a scale model of the original arrow that has been stretched or shrunk by some scale factor and possibly flipped over to point in the opposite direction. The coordinates of the terminal point of the new arrow are obtained by multiplying the original coordinates by the scale factor and also by -1 if the direction has been changed. Figure 1.3 illustrates this in $\mathbb{R}^2$.

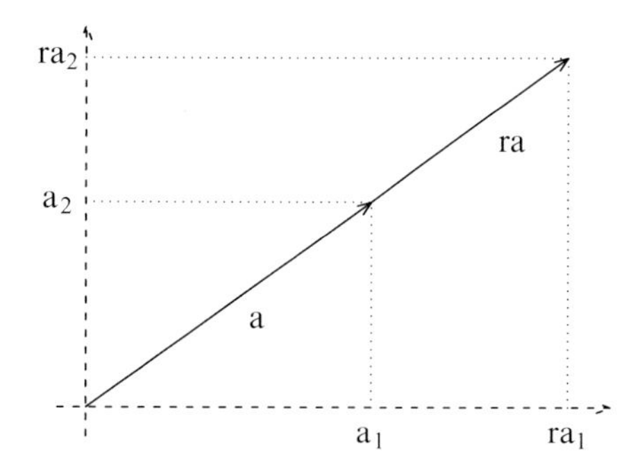

Fig. 1.3 For a vector **a** and a number r, the rectangles formed by **a** and r**a** have the same aspect ratio. That is, every dimension of one rectangle is r times the corresponding dimension of the other rectangle

We sum up this discussion by expanding our definition of $\mathbb{R}^N$.

Definition 1.3 The **vector space** $\mathbb{R}^N$ consists of all column vectors of N coordinates together with two arithmetic operations, called **vector addition** and **scalar multiplication**, defined like so.

- **Vector addition.** We add elements of $\mathbb{R}^N$ coordinate by coordinate. Thus, for arbitrary vectors $\begin{bmatrix} a_1 \\ a_2 \\ \vdots \\ a_N \end{bmatrix}$ and $\begin{bmatrix} b_1 \\ b_2 \\ \vdots \\ b_N \end{bmatrix}$ in $\mathbb{R}^N$,

$$\begin{bmatrix} a_1 \\ a_2 \\ \vdots \\ a_N \end{bmatrix} + \begin{bmatrix} b_1 \\ b_2 \\ \vdots \\ b_N \end{bmatrix} = \begin{bmatrix} a_1 + b_1 \\ a_2 + b_2 \\ \vdots \\ a_N + b_N \end{bmatrix}. \tag{1.2}$$

  The sum is also an element of $\mathbb{R}^N$.
- **Scalar multiplication.** The term *scalar* is another word for *real number*. Scalar multiplication is performed coordinate by coordinate. That is, for $\begin{bmatrix} a_1 \\ a_2 \\ \vdots \\ a_N \end{bmatrix}$ in $\mathbb{R}^N$ and any real number r,

$$r \cdot \begin{bmatrix} a_1 \\ a_2 \\ \vdots \\ a_N \end{bmatrix} = \begin{bmatrix} r \cdot a_1 \\ r \cdot a_2 \\ \vdots \\ r \cdot a_N \end{bmatrix}. \tag{1.3}$$

Remark When $N = 1$, we write $\mathbb{R}$ instead of $\mathbb{R}^1$. This is the set of all real numbers. Technically, there is a distinction between a single number, say -3, and the corresponding column vector, in this case $[-3]$. It will often be convenient to gloss over this distinction, but it is always there and sometimes it will matter.

Remark The distinction between a point and the vector that ends at that point can be fuzzy. In this book, we write $(a_1, \ldots, a_N)$ to denote the point and use column vector notation $\begin{bmatrix} a_1 \\ \vdots \\ a_N \end{bmatrix}$ for the vector.

Remark We have chosen to write each vector in $\mathbb{R}^N$ as a column of numbers. When we are writing or typing, though, it may be convenient to write the vector in a row instead, so it fits within a line of text. Thus, each vector is associated to a corresponding row vector that has the same coordinates. This row vector is called the **transpose** of the column vector. We also say that the column vector is the transpose of the row vector. The transpose of a vector is indicated symbolically with a superscript ^{T}. Thus,

$$\begin{bmatrix} a_1 \\ a_2 \\ \vdots \\ a_N \end{bmatrix}^T = \begin{bmatrix} a_1 \; a_2 \; \cdots \; a_N \end{bmatrix} \text{ and}$$

$$\begin{bmatrix} a_1 \; a_2 \; \cdots \; a_N \end{bmatrix}^T = \begin{bmatrix} a_1 \\ a_2 \\ \vdots \\ a_N \end{bmatrix} . \tag{1.4}$$

For the arithmetic operations (1.2) and (1.3), we apply the arithmetic of real numbers to each coordinate of a vector. Not surprisingly, this arithmetic for vectors shares other properties with basic numerical arithmetic.

- The addition operation is both **associative** and **commutative**. That is, for any vectors $\mathbf{a} = \begin{bmatrix} a_1 \\ a_2 \\ \vdots \\ a_N \end{bmatrix}$, $\mathbf{b} = \begin{bmatrix} b_1 \\ b_2 \\ \vdots \\ b_N \end{bmatrix}$, and $\mathbf{c} = \begin{bmatrix} c_1 \\ c_2 \\ \vdots \\ c_N \end{bmatrix}$ in $\mathbb{R}^N$, we have

$$(\mathbf{a} + \mathbf{b}) + \mathbf{c} = \mathbf{a} + (\mathbf{b} + \mathbf{c}) \; ;$$

  and

$$\mathbf{a} + \mathbf{b} = \mathbf{b} + \mathbf{a} \, .$$

- Addition and scalar multiplication are **distributive**. That is, for all scalars r and all vectors $\mathbf{a}$ and $\mathbf{b}$, we have $r \cdot (\mathbf{a} + \mathbf{b}) = r \cdot \mathbf{a} + r \cdot \mathbf{b}$, and for all scalars r and s and every vector $\mathbf{a}$, we have $(r + s) \cdot \mathbf{a} = r \cdot \mathbf{a} + s \cdot \mathbf{a}$.
- Scalar multiplication is **associative**. That is, for all scalars r and s and every vector $\mathbf{a}$, we have $(r \cdot s) \cdot \mathbf{a} = r \cdot (s \cdot \mathbf{a})$.
- The vector $\begin{bmatrix} 0 \\ 0 \\ \vdots \\ 0 \end{bmatrix}$ with all of its coordinates equal to 0 is called the **zero vector**, denoted by $\mathbf{0}$. The zero vector acts as an **additive identity:** For every vector $\mathbf{a}$, we have $\mathbf{a} + \mathbf{0} = \mathbf{0} + \mathbf{a} = \mathbf{a}$. Technically, there is a different zero vector for each value of N, but they all have the same arithmetic properties.

- Every vector in $\mathbb{R}^N$ has an **additive inverse**. Specifically, we have $\mathbf{a} + (-1) \cdot \mathbf{a} = \mathbf{0}$, by the definition of the arithmetic operations. The vector $(-1) \cdot \mathbf{a}$ is also denoted by $-\mathbf{a}$ and is the additive inverse of $\mathbf{a}$.
- The scalar 1 is a **multiplicative identity**, since, for every vector $\mathbf{a}$, we have $1 \cdot \mathbf{a} = \mathbf{a}$.

Everyone loves to play with building blocks, right? Well, suppose we have a set of vectors in $\mathbb{R}^N$, say $\{\mathbf{a}_1, \mathbf{a}_2, \ldots, \mathbf{a}_K\}$. What can we build from these blocks? Our only tools right now are vector addition and scalar multiplication. With these, we can construct every vector of the form

$$r_1 \cdot \mathbf{a}_1 + r_2 \cdot \mathbf{a}_2 + \cdots + r_K \cdot \mathbf{a}_K \,, \tag{1.5}$$

for scalars r_1, r_2, up to r_K. That is, we can obtain every possible sum of scalar multiples of the given vectors.

Definition 1.4 We adopt the following terminology to facilitate our discussion of collections of vectors.

(i) Given a collection $\mathcal{S}$ of vectors in some space $\mathbb{R}^N$, every vector in $\mathbb{R}^N$ that can be expressed as a sum of numerical multiples of vectors in $\mathcal{S}$ is called a **linear combination** of these vectors. Thus, the phrase *is a linear combination of* means the same thing as *is a sum of numerical multiples of*. For example, every vector of the form shown in (1.5) is a linear combination of the vectors $\{\mathbf{a}_1, \mathbf{a}_2, \ldots, \mathbf{a}_K\}$.

(ii) Given a collection $\mathcal{S}$ of vectors in some space $\mathbb{R}^N$, the **span** of $\mathcal{S}$, denoted by $span(\mathcal{S})$, is the set of all possible linear combinations of the vectors in $\mathcal{S}$. For example, the set of *all* vectors of the form shown in (1.5) is equal to $span\,(\{\mathbf{a}_1, \mathbf{a}_2, \ldots, \mathbf{a}_K\})$.

(iii) A nonempty set $\mathcal{U}$ of vectors in some space $\mathbb{R}^N$ is called a **subspace** of $\mathbb{R}^N$ if, and only if, it contains all possible linear combinations of its elements. For example, if $\mathcal{U} = span(\mathcal{S})$ for some collection $\mathcal{S}$ of vectors in $\mathbb{R}^N$, then $\mathcal{U}$ is a subspace of $\mathbb{R}^N$. Notice also that both $\mathcal{U} = \{\mathbf{0}\}$ and $\mathcal{U} = \mathbb{R}^N$ are subspaces of $\mathbb{R}^N$.

Going back to playing with building blocks, suppose we have a set of vectors $\mathcal{S}$ and we start to build the subspace $span(\mathcal{S})$. We may find that some of the vectors in $\mathcal{S}$ are not adding anything new because they are themselves linear combinations of other vectors in $\mathcal{S}$. If we can somehow identify and throw away all such redundant elements of $\mathcal{S}$, we will be left with a new set $\mathcal{S}'$, where $span(\mathcal{S}') = span(\mathcal{S})$, and none of the vectors in $\mathcal{S}'$ is a linear combination of the others. This concept also deserves its own terminology.

Definition 1.5 (i) A collection $\mathcal{S}$ of vectors in some space $\mathbb{R}^N$ is said to be **linearly independent** provided that none of the vectors in $\mathcal{S}$ can be expressed as a linear combination of the other vectors in $\mathcal{S}$.

(ii) A linearly independent collection $\mathcal{S}$ of vectors in some space $\mathbb{R}^N$ is called a **basis** for the subspace *span* $(\mathcal{S})$ that it spans.

(iii) The number of individual vectors in a basis for a given subspace of $\mathbb{R}^N$ is called the **dimension** of the subspace. In general, there are many possible choices of a basis for the same subspace. We will not attempt to prove here that every such basis has the same number of vectors in it, so that the concept of *dimension* is well-defined.

1.2 The Vector Norm

Every vector in $\mathbb{R}^N$ can be identified with an arrow. The direction in which the arrow points is the direction of the vector. Every positive scalar multiple of the vector points in the same direction. A negative scalar multiple points in the opposite direction. We can distinguish among all vectors that point in the same direction by computing their **lengths**.

In $\mathbb{R}^2$, the vector $\mathbf{a} = \begin{bmatrix} a_1 \\ a_2 \end{bmatrix}$ forms the diagonal of a rectangle with dimensions $|a_1|$ and $|a_2|$. By the Pythagorean theorem, the length of $\mathbf{a}$ is equal to $\sqrt{a_1^2 + a_2^2}$. In $\mathbb{R}^3$, we see in Figure 1.1 that the vector $\mathbf{a}$, with coordinates a_1, a_2, and a_3, forms the hypotenuse of a right triangle where one adjacent side has length $\sqrt{a_1^2 + a_2^2}$ and the other has length $|a_3|$. By the Pythagorean theorem again, the length of $\mathbf{a}$ is equal to

$$\sqrt{\left(\sqrt{a_1^2 + a_2^2}\right)^2 + a_3^2} = \sqrt{a_1^2 + a_2^2 + a_3^2}\,.$$

This computation extends, one dimension at a time, to give us a formula for the length of any vector in $\mathbb{R}^N$, for any N.

Formula 1.6 The **length**, also called the **norm**, of a vector $\mathbf{a} = \begin{bmatrix} a_1 \\ a_2 \\ \vdots \\ a_N \end{bmatrix}$ in $\mathbb{R}^N$ is denoted by $||\mathbf{a}||$ and defined by

$$||\mathbf{a}|| = \sqrt{a_1^2 + a_2^2 + \cdots + a_N^2}\,. \tag{1.6}$$

Remark Note that $||\mathbf{a}||^2$ is equal to the sum of the squares of the coordinates of $\mathbf{a}$. In many computations, it is simpler to work with $||\mathbf{a}||^2$ in order to avoid square roots.

The square of a real number is never negative. Formula 1.6, therefore, guarantees that a vector cannot have a negative length: $||\mathbf{a}|| \geq 0$, for all $\mathbf{a}$ in $\mathbb{R}^N$. Moreover, only the all-0s vector $\mathbf{0}$ has norm equal to 0. That is, $||\mathbf{0}|| = 0$, and if $\mathbf{a} \neq \mathbf{0}$, then $||\mathbf{a}|| > 0$.

To see how vector arithmetic affects the norm, start with scalar multiplication. Suppose $\mathbf{a}$ has coordinates $a_1, a_2, \ldots, a_N$. Then $r \cdot \mathbf{a}$ has coordinates $r \cdot a_1, r \cdot a_2, \ldots, r \cdot a_N$. Applying Formula 1.6, we compute

$$\begin{aligned} ||r \cdot \mathbf{a}|| &= \sqrt{(r \cdot a_1)^2 + (r \cdot a_2)^2 + \cdots + (r \cdot a_N)^2} \\ &= \sqrt{r^2 \cdot \left(a_1^2 + a_2^2 + \cdots + a_N^2\right)} \\ &= \sqrt{r^2} \cdot \sqrt{a_1^2 + a_2^2 + \cdots + a_N^2} \\ &= |r| \cdot ||\mathbf{a}|| \,. \end{aligned} \tag{1.7}$$

In other words, multiplying a vector by the scalar r changes the length of the vector by the factor $|r|$. We are using the fact that $\sqrt{r^2} = |r|$.

Example 1.7 For any nonzero vector $\mathbf{a}$, both $\pm\frac{1}{||\mathbf{a}||} \cdot \mathbf{a}$ have norm 1. Moreover, for any desired length ℓ, the vectors $\pm\frac{\ell}{||\mathbf{a}||} \cdot \mathbf{a}$ are parallel to $\mathbf{a}$ and have length ℓ.

A vector of length 1 is called a **unit vector.** If $||\mathbf{a}|| = 1$, then $||\mathbf{a}||^2 = 1$, as well. Therefore, $\mathbf{a}$ is a unit vector if, and only if,

$$||\mathbf{a}||^2 = a_1^2 + a_2^2 + \cdots + a_N^2 = 1 \,. \tag{1.8}$$

Next, consider two nonzero vectors $\mathbf{a}$ and $\mathbf{b}$ and the triangle they form with their sum $\mathbf{a} + \mathbf{b}$, as shown in Figure 1.4. In geometry class in school, we learn that the length of any one side of a triangle cannot be greater than the sum of the lengths of the other two sides. This makes sense—if two sides put together were shorter than the third side, they would not be able to join up to create a triangle. Applying this principle in our current context, we get the inequality

$$||\mathbf{a} + \mathbf{b}|| \leq ||\mathbf{a}|| + ||\mathbf{b}|| \,. \tag{1.9}$$

This is aptly known as the **triangle inequality** for vectors.

The sum $\mathbf{a} + \mathbf{b}$ forms one diagonal of the parallelogram generated by $\mathbf{a}$ and $\mathbf{b}$. The other diagonal is a parallel copy of $\mathbf{a} - \mathbf{b}$ (or of $\mathbf{b} - \mathbf{a}$ if it points the other way). This difference vector forms a triangle together with $\mathbf{a}$ and $\mathbf{b}$. The sides of this triangle have lengths $||\mathbf{a}||$, $||\mathbf{b}||$, and $||\mathbf{a} - \mathbf{b}||$. Each of these lengths must be less than or equal to the sum of the other two. This gives us two new inequalities.

$$\begin{aligned} &||\mathbf{a}|| \leq ||\mathbf{a} - \mathbf{b}|| + ||\mathbf{b}|| \Rightarrow ||\mathbf{a}|| - ||\mathbf{b}|| \leq ||\mathbf{a} - \mathbf{b}|| \,, \text{ and} \\ &||\mathbf{b}|| \leq ||\mathbf{a} - \mathbf{b}|| + ||\mathbf{a}|| \Rightarrow ||\mathbf{b}|| - ||\mathbf{a}|| \leq ||\mathbf{a} - \mathbf{b}|| \,. \end{aligned}$$

By definition, the absolute value $\Big| \, ||\mathbf{a}|| - ||\mathbf{b}|| \, \Big|$ is equal to whichever is the greater of $||\mathbf{a}|| - ||\mathbf{b}||$ and $||\mathbf{b}|| - ||\mathbf{a}||$. Therefore, we can combine these two inequalities into the following variant of the triangle inequality.

$$\Big| \, ||\mathbf{a}|| - ||\mathbf{b}|| \, \Big| \leq ||\mathbf{a} - \mathbf{b}|| \, . \tag{1.10}$$

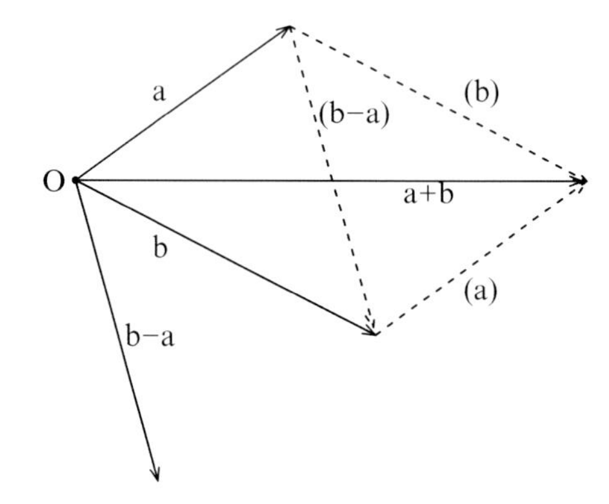

Fig. 1.4 The vectors **a** and **b**, based at the origin **O**, form a parallelogram when copies (**a**) and (**b**) are included. The vector $\mathbf{a} + \mathbf{b}$ is one diagonal of the parallelogram. The other diagonal is a copy of the difference vector $\mathbf{b} - \mathbf{a}$. Notice that **b** is a diagonal of the parallelogram formed by **a** and $\mathbf{b} - \mathbf{a}$

Remark There are reasonable notions of the norm of a vector other than the one in (1.6). For instance, the (*1-norm*) of a vector **a** is defined by $||\mathbf{a}||_1 = \sum_{i=1}^{N} |a_i|$, and the ∞-*norm* is defined by $||\mathbf{a}||_\infty = \max\{|a_i| \, : \, 1 \leq i \leq N\}$. For both of these, only the all-0s vector has norm 0, and the norm of $r\mathbf{a}$ is the norm of **a** multiplied by $|r|$. Both norms satisfy the triangle inequality (1.9). In this broader context, the norm defined in (1.6) is called the (*Euclidean norm*) or the (*2-norm*) and is denoted by $||\mathbf{a}||_2$.

1.3 Angles and the Inner Product

We now look at angles between vectors. Technically, there are two angles between any two nonzero vectors in $\mathbb{R}^N$. These angles are supplementary, meaning that they add up to a full 2π radians. By convention, when we say *the* angle between the vectors, we mean the smaller of the two, or π if they are equal. So now, take two nonzero vectors **a** and **b** in $\mathbb{R}^N$, and denote the angle between them by θ. If **a** and **b** lie along the same line, so that $\mathbf{b} = r \cdot \mathbf{a}$, for some real number $r \neq 0$, then either $\theta = 0$, if $r > 0$, or $\theta = \pi$, if $r < 0$. If **a** and **b** are not collinear, then we can form the triangle with sides **a**, **b**, and $\mathbf{a} - \mathbf{b}$. See Figure 1.5. The **law of cosines** asserts that

$$||\mathbf{a} - \mathbf{b}||^2 = ||\mathbf{a}||^2 + ||\mathbf{b}||^2 - 2 \cdot ||\mathbf{a}|| \cdot ||\mathbf{b}|| \cdot \cos(\theta) \, . \tag{1.11}$$

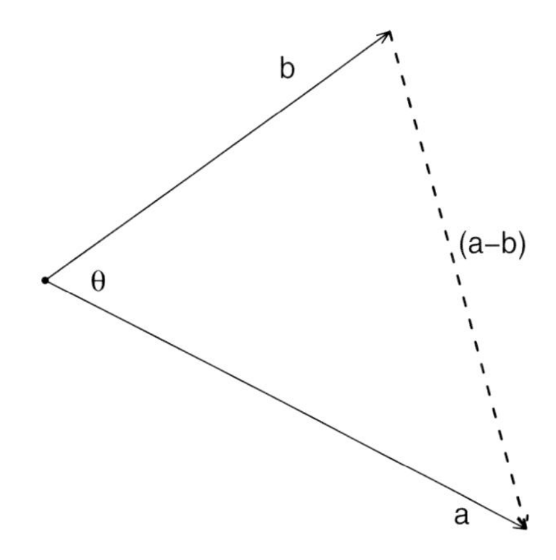

Fig. 1.5 The law of cosines (1.11) describes the relationship between the lengths of the three sides of a triangle together with any one angle. It is a generalization of the Pythagorean theorem, where the angle must be a right angle

We can rewrite Formula (1.11) in terms of the coordinates of the vectors involved. To do this, observe that

$$\begin{aligned} ||\mathbf{a}-\mathbf{b}||^2 &= \left((a_1-b_1)^2+(a_2-b_2)^2+\cdots+(a_N-b_N)^2\right) \\ &= (a_1^2+\cdots+a_N^2)+(b_1^2+\cdots+b_N^2)-2\cdot(a_1b_1+\cdots+a_Nb_N) \\ &= ||\mathbf{a}||^2+||\mathbf{b}||^2-2\cdot(a_1b_1+a_2b_2+\cdots+a_Nb_N)\,. \end{aligned}$$

Comparing this to the right-hand side of (1.11), we see that

$$||\mathbf{a}||\cdot||\mathbf{b}||\cdot\cos(\theta)=a_1b_1+a_2b_2+\cdots+a_Nb_N\,.$$

Solving this for $\cos(\theta)$ yields the following formula (Figure 1.5).

Formula 1.8 For two nonzero vectors **a** and **b** in $\mathbb{R}^N$, the angle θ between them is the angle between 0 and π radians whose *cosine* is given by

$$\cos(\theta)=\frac{a_1b_1+a_2b_2+\cdots+a_Nb_N}{||\mathbf{a}||\,||\mathbf{b}||}\,. \tag{1.12}$$

Notice that this formula is correct even if **a** and **b** are collinear. The expression in the numerator of (1.12) is a single number formed by first multiplying corresponding coordinates of **a** and **b** and then adding the results. This quantity is called the **inner product** of the two vectors. Anticipating our work with matrices, we denote the inner product of **a** and **b** as $\mathbf{a}^T\mathbf{b}$, where $\mathbf{a}^T=\begin{bmatrix}a_1 & \cdots & a_N\end{bmatrix}$, as in (1.4). Technically, we are *defining* $\mathbf{a}^T\mathbf{b}$ to equal the inner product of **a** and **b**.

Definition 1.9 For two vectors **a** and **b**, both in $\mathbb{R}^N$, the expression $\mathbf{a}^T\mathbf{b}$ is defined to be

$$\mathbf{a}^T\mathbf{b} = \begin{bmatrix} a_1 & \cdots & a_N \end{bmatrix} \cdot \begin{bmatrix} b_1 \\ \vdots \\ b_N \end{bmatrix}$$
$$= a_1b_1 + a_2b_2 + \cdots + a_Nb_N\,. \tag{1.13}$$

This number is called the **inner product** of **a** and **b**.

Remark To visualize the inner product $\mathbf{a}^T\mathbf{b}$, as in Definition 1.9, think of the row vector $\mathbf{a}^T$ and the column vector **b** as two sides of a zipper, each with N teeth. We zip these together by matching up the teeth: a_1 with b_1, a_2 with b_2, etc., adding up as we go. When everything is all zipped up, we get the inner product $a_1b_1+\cdots+a_Nb_N$. Notice that $\mathbf{b}^T\mathbf{a} = \mathbf{a}^T\mathbf{b}$.

In light of Definition 1.9, Formula 1.8 asserts that the *cosine* of the angle between two nonzero vectors is given by their inner product divided by the product of their lengths.

$$\cos(\theta) = \frac{\mathbf{a}^T\mathbf{b}}{||\mathbf{a}||\,||\mathbf{b}||}\,. \tag{1.14}$$

We can also compute the *sine* of the angle between any two nonzero vectors. Since $\cos^2(\theta) + \sin^2(\theta) = 1$, for any angle θ, we get the following.

$$\begin{aligned}\sin^2(\theta) &= 1 - \cos^2(\theta) \\ &= 1 - \frac{\left(\mathbf{a}^T\mathbf{b}\right)^2}{||\mathbf{a}||^2 \cdot ||\mathbf{b}||^2} \\ &= \frac{||\mathbf{a}||^2 \cdot ||\mathbf{b}||^2 - \left(\mathbf{a}^T\mathbf{b}\right)^2}{||\mathbf{a}||^2 \cdot ||\mathbf{b}||^2}\,.\end{aligned} \tag{1.15}$$

Since $\sin^2(\theta)$ is nonnegative, it follows also that

$$0 \le ||\mathbf{a}||^2 \cdot ||\mathbf{b}||^2 - (\mathbf{a}^T\mathbf{b})^2\,, \tag{1.16}$$

for **a** and **b** nonzero. Taking square roots in (1.15), we get

$$\sin(\theta) = \frac{\left(||\mathbf{a}||^2 \cdot ||\mathbf{b}||^2 - \left(\mathbf{a}^T\mathbf{b}\right)^2\right)^{1/2}}{||\mathbf{a}||\,||\mathbf{b}||}\,. \tag{1.17}$$

To express $\sin(\theta)$ using coordinates, consider the following chain of equalities.

$$
\begin{aligned}
&||\mathbf{a}||^2 \cdot ||\mathbf{b}||^2 - \left(\mathbf{a}^T\mathbf{b}\right)^2 \\
&= \left({a_1}^2 + \cdots + {a_N}^2\right) \cdot \left({b_1}^2 + \cdots + {b_N}^2\right) - (a_1 b_1 + \cdots + a_N b_N)^2 \\
&= \left(\sum_{j=1}^{N} {a_j}^2 {b_j}^2 + \sum_{\substack{j,k=1 \\ j \neq k}}^{N} {a_j}^2 {b_k}^2 \right) - \left(\sum_{j=1}^{N} {a_j}^2 {b_j}^2 + \sum_{\substack{j,k=1 \\ j \neq k}}^{N} a_j b_j a_k b_k \right) \\
&= \sum_{\substack{j,k=1 \\ j \neq k}}^{N} {a_j}^2 {b_k}^2 - \sum_{\substack{j,k=1 \\ j \neq k}}^{N} a_j b_j a_k b_k \\
&= \sum_{j=1}^{N-1} \sum_{k=j+1}^{N} \left[\left({a_j}^2 {b_k}^2 - a_j b_j a_k b_k \right) + \left({a_k}^2 {b_j}^2 - a_k b_k a_j b_j \right) \right] \\
&= \sum_{j=1}^{N-1} \sum_{k=j+1}^{N} \left(a_j b_k - a_k b_j \right)^2 \qquad (1.18)
\end{aligned}
$$

Thus, for the angle θ between two nonzero vectors **a** and **b**, we have

$$
\sin(\theta) = \frac{\left(\sum_{j=1}^{N-1} \sum_{k=j+1}^{N} \left(a_j b_k - a_k b_j \right)^2 \right)^{1/2}}{||\mathbf{a}|| \cdot ||\mathbf{b}||}. \qquad (1.19)
$$

The sum in the numerator of (1.19) is a sum of squares, so it is never negative. This makes sense since the angle between two vectors is always between 0 and π radians and, hence, has a *sine* value that is nonnegative.

One of the most famous inequalities in mathematics is a consequence of (1.16).

Theorem 1.10 (The Cauchy–Schwarz Inequality) *For all vectors* **a** *and* **b** *in* $\mathbb{R}^N$,

$$
|\mathbf{a}^T\mathbf{b}| \leq ||\mathbf{a}|| \cdot ||\mathbf{b}||\,. \qquad (1.20)
$$

Proof For nonzero vectors **a** and **b**, it follows from (1.16) that $(\mathbf{a}^T\mathbf{b})^2 \leq ||\mathbf{a}||^2 \cdot ||\mathbf{b}||^2$. Taking square roots on both sides yields $|\mathbf{a}^T\mathbf{b}| \leq ||\mathbf{a}||\,||\mathbf{b}||$, as claimed. If either of **a** or **b** is the zero vector **0**, then both sides of (1.20) have the value 0 and we have proved our result. □

Remark Equality occurs in the Cauchy–Schwarz inequality if, and only if, one of the vectors is the zero vector, or the *sine* of the angle between the vectors has the value 0. In that case, the angle between the vectors is either 0 or π, which means they lie along the same line and, thus, are scalar multiples of each other.

1.4 Inner Product and Vector Arithmetic

It is useful to know how the inner product, given in Definition 1.9, interacts with other vector arithmetic operations.

- For all $\mathbf{a}$ and $\mathbf{b}$ in $\mathbb{R}^N$,

$$\mathbf{a}^T \mathbf{b} = \mathbf{b}^T \mathbf{a}. \tag{1.21}$$

  This property of the inner product is called **symmetry**.
- For $\mathbf{a}$ and $\mathbf{b}$ in $\mathbb{R}^N$ and any real number r,

$$(r\,\mathbf{a})^T \mathbf{b} = r \cdot (\mathbf{a}^T \mathbf{b}) = \mathbf{a}^T (r\mathbf{b}). \tag{1.22}$$

  Thus, a scalar multiple can be factored out of either the first vector or the second vector. This property of the inner product is called **homogeneity**.
- The property called **additivity** works like so. Let $\mathbf{a}$, $\mathbf{b}$, and $\mathbf{c}$ all belong to $\mathbb{R}^N$. Then

$$(\mathbf{a} + \mathbf{b})^T \mathbf{c} = \mathbf{a}^T \mathbf{c} + \mathbf{b}^T \mathbf{c}\,; \text{ and} \tag{1.23}$$

$$\mathbf{a}^T (\mathbf{b} + \mathbf{c}) = \mathbf{a}^T \mathbf{b} + \mathbf{a}^T \mathbf{c}. \tag{1.24}$$

  These statements look a lot like the distributive property for arithmetic with numbers.
- For all $\mathbf{a}$ in $\mathbb{R}^N$,

$$\mathbf{a}^T \mathbf{a} = a_1^2 + a_2^2 + \cdots + a_N^2 = ||\mathbf{a}||^2. \tag{1.25}$$

  Thus, $\mathbf{a}^T \mathbf{a} \geq 0$, for every $\mathbf{a}$, and $\mathbf{a}^T \mathbf{a} = 0$ if, and only if, $\mathbf{a} = \mathbf{0}$, the zero vector. To express this property, we say the inner product is **positive definite.**

Remark The inner product of two vectors is sometimes called the *dot product* or *scalar product* and written as $\mathbf{a} \bullet \mathbf{b}$ or $\langle \mathbf{a} | \mathbf{b} \rangle$.

1.5 Statistical Correlation

In statistics, we may be interested in knowing whether there is a correlation between two different measurements made on each of a set of objects. For example, an above average value of one measurement might tend to go with an above average value of the other. Or perhaps when one value is above average, the other tends to be below average, and vice versa. Of course, there may be no significant relationship between the measurements.

To get a feel for this concept, try to decide which of the three possibilities might apply to each of the following pairs of measurements:

- A student's grade point average and the number of hours per week they spend studying
- The average hours of sleep a student gets each day and the number of milligrams of caffeine they consume per day
- The number of hours a student spends exercising each week and their height

To analyze these ideas, suppose we have N pairs of observations

$$(x_1,\ y_1),\ (x_2,\ y_2),\ \dots,\ (x_N,\ y_N)$$

of the quantities X and Y. If possible, create a picture by plotting each pair as a point in the xy-plane. This is called a **scatter plot**. The scatter plot can give us a qualitative sense of the relationship between the two quantities. If the points in the scatter plot more or less lie along a path ascending from the lower left corner to the upper right corner of the picture, that suggests the quantities are positively correlated. For negatively correlated quantities, the points lie roughly along a path from the top left corner to the lower right corner. A clearer trend in the scatter plot suggests a stronger correlation. We really need to numerically quantify this concept.

For each quantity, we compute the average of the observed values, also called the **sample mean**. These averages are denoted by $\bar{x}$ and $\bar{y}$. That is,

$$\bar{x} = \frac{x_1 + x_2 + \cdots + x_N}{N} \text{ and } \bar{y} = \frac{y_1 + y_2 + \cdots + y_N}{N}.$$

In the first pair of observations, if the measurements x_1 and y_1 are both above average or both below average, then the product $(x_1 - \bar{x}) \cdot (y_1 - \bar{y})$ is positive. If one of the measurements is above average and the other is below average, then the product $(x_1 - \bar{x}) \cdot (y_1 - \bar{y})$ is negative. The same situation holds for every pair of observations. Now look at the sum

$$(x_1 - \bar{x}) \cdot (y_1 - \bar{y}) + (x_2 - \bar{x}) \cdot (y_2 - \bar{y}) + \cdots + (x_N - \bar{x}) \cdot (y_N - \bar{y})\,. \qquad (1.26)$$

Intuitively, this sum ought to give us a sense of whether there is a correlation between paired measurements of X and Y. If the sum is positive, then the positive summands outweigh the negative ones, suggesting that x_j and y_j are usually either both above average or both below average. If the sum is negative, that suggests that x_j and y_j are usually on opposite sides of their respective averages. If there is a good mix of positive and negative summands, then the sum would probably be near 0 due to cancellation, suggesting that there is no real connection between the two different quantities. Unfortunately, the situation is not so simple. Examples 1.11 and 1.12 show that we can get the same value for (1.26) in very different contexts.

Example 1.11 Suppose we have the paired observations

$$(0.15,\ 4.5),\ (0.25,\ 4.5),\ (0.45,\ 8.0),\ \text{and}\ (0.75,\ 3.0)\,.$$

Then

$$\bar{x} = \frac{0.15 + 0.25 + 0.45 + 0.75}{4} = 0.4 \text{ and}$$

$$\bar{y} = \frac{4.5 + 4.5 + 8.0 + 3.0}{4} = 5.0\,.$$

The sum of the terms $(x_j - \bar{x}) \cdot (y_j - \bar{y})$ is

$$(-0.25)(-0.5) + (-0.15)(-0.5) + (0.05)(3.0) + (0.35)(-2.0) = -0.35\,.$$

The total is negative even though only one of the four summands is negative. What exactly does the -0.35 mean in context? Is there a correlation between the two quantities being measured? It is hard to say without more information.

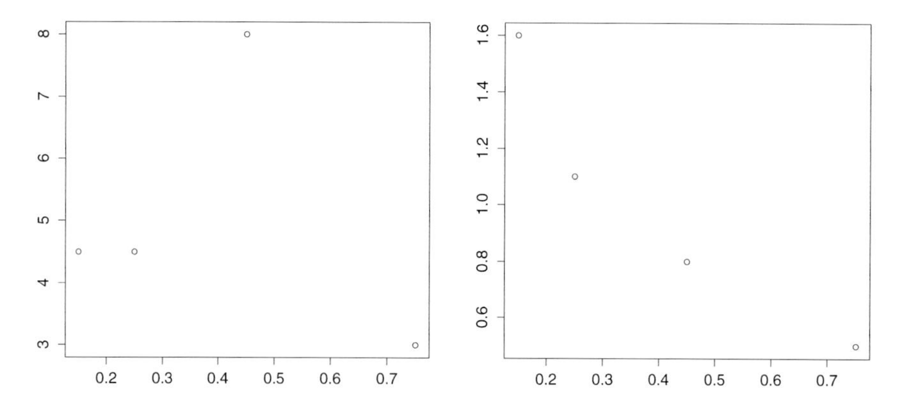

Fig. 1.6 Scatter plots suggest that the quantities measured in Example 1.11 (left) are not closely correlated, while those in Example 1.12 (right) seem to be negatively correlated

Example 1.12 Now consider the observations

$$(0.15,\ 1.6),\ (0.25,\ 1.1),\ (0.45,\ 0.8),\ \text{and}\ (0.75,\ 0.5)\,.$$

Thus, $\bar{x} = 0.4$ and $\bar{y} = 1.0$. The sum of the products $(x_j - \bar{x}) \cdot (y_j - \bar{y})$ is

$$(-0.25)(0.6) + (-0.15)(0.1) + (0.05)(-0.2) + (0.35)(-0.5) = -0.35\,.$$

All four summands are negative, suggesting the two quantities are negatively correlated. The scatter plot, in Figure 1.6, reinforces this belief. Still, it is not clear what exactly the value -0.35 means in context.

In general, simply making more observations will change the value of (1.26) without changing the quality of the relationship between X and Y. Also, if one of the quantities has a much wider range of possible values than the other, then the same difference away from the average is less significant for one variable than for the other. To adjust for these problems, we have to take into account the overall variability in the observed measurements of each quantity. For this, we compute the values

$$\sqrt{(x_1 - \bar{x})^2 + (x_2 - \bar{x})^2 + \cdots + (x_N - \bar{x})^2}$$

and

$$\sqrt{(y_1 - \bar{y})^2 + (y_2 - \bar{y})^2 + \cdots + (y_N - \bar{y})^2}\,.$$

Dividing the sum in (1.26) by these variability values allows us to adjust for both the number of pairs of observations and the range in values of the two quantities. This leads us to a statistic known as the **Pearson correlation coefficient**, denoted by the Greek letter $\boldsymbol{\rho}$ (pronounced 'rhō').

Formula 1.13 Given a set $\{(x_1,\ y_1),\ (x_2,\ y_2),\ \ldots,\ (x_N,\ y_N)\}$ of N paired observations of two quantities X and Y, the **Pearson correlation coefficient**, denoted $\boldsymbol{\rho}(X,\ Y)$, is defined by

$$\boldsymbol{\rho}(X,\ Y) = \frac{\sum_{j=1}^{N}(x_j - \bar{x})(y_j - \bar{y})}{\sqrt{\sum_{j=1}^{N}(x_j - \bar{x})^2} \cdot \sqrt{\sum_{j=1}^{N}(y_j - \bar{y})^2}}\,. \tag{1.27}$$

When X and Y are clear from the context, we will write simply ρ in place of $\rho(X,\ Y)$.

In fact, the Pearson correlation coefficient ρ is exactly the value of the *cosine* of the angle between two specific vectors. One vector has the differences $(x_i - \bar{x})$ as its coordinates, and the other has coordinates $(y_i - \bar{y})$. Formula (1.27) is the result of applying formula (1.14) to these two vectors. The Cauchy–Schwarz inequality assures us that $|\boldsymbol{\rho}| \leq 1$. A value of ρ close to 1 suggests a strong positive correlation between X and Y, while a value of ρ near -1 suggests a strong negative correlation. If ρ is near 0, then the variables X and Y are probably not closely correlated. For instance, if X and Y are independent random variables, then we would expect $\rho \approx 0$.

Example 1.14 Returning to Example 1.11, we have

$$\sqrt{(x_1 - \bar{x})^2 + (x_2 - \bar{x})^2 + (x_3 - \bar{x})^2 + (x_4 - \bar{x})^2} \approx 0.4583\,,$$

and

$$\sqrt{(y_1 - \bar{y})^2 + (y_2 - \bar{y})^2 + \cdots + (y_N - \bar{y})^2} \approx 3.6742\,.$$

The Y quantity exhibits much more variability than the X quantity. The Pearson correlation coefficient is

$$\rho(X,\,Y) \approx \frac{-0.35}{(0.4583)\cdot(3.6742)} \approx -0.2079\,.$$

This value of ρ is fairly close to 0 on the scale from -1 to $+1$. The angle between the corresponding mean-adjusted vectors of the two quantities is about 102 degrees, meaning that the vectors are not close to lying on the same line. This suggests that the two quantities are not correlated.

For the observations in Example 1.12, we again have

$$\sqrt{(x_1 - \bar{x})^2 + (x_2 - \bar{x})^2 + (x_3 - \bar{x})^2 + (x_4 - \bar{x})^2} \approx 0.4583\,.$$

But now,

$$\sqrt{(y_1 - \bar{y})^2 + (y_2 - \bar{y})^2 + \cdots + (y_N - \bar{y})^2} \approx 0.8124\,.$$

The Pearson correlation coefficient is

$$\rho \approx \frac{-0.35}{(0.4583)\cdot(0.8124)} \approx -0.9400\,.$$

This value of ρ is very close to -1. The angle between the corresponding mean-adjusted vectors of the two quantities is about 160 degrees. These vectors almost lie along the same line, pointing in opposite directions. This suggests a strong negative correlation between the two quantities.

Example 1.15 Crickets make their chirping sound by rapidly rubbing their wings against each other. The rate at which crickets produce chirps increases when the surrounding air temperature is higher. To examine whether there is a correlation between the chirp rate of a cricket and the air temperature, consider the five observations listed in Table 1.1.

A plot of these data points, shown in Figure 1.7, suggests a strong relationship, possibly even a linear relationship, between these two quantities. Let us compute the Pearson correlation coefficient.

The average value of the sampled temperatures is $\bar{x} = 84.06$, while the sample mean of the chirp rates is $\bar{y} = 17.92$. Plugging these numbers in to the formula (1.27), we get $\rho \approx 0.958$. This is close to 1, so we conclude that there is a strong correlation between the chirp rate of a cricket and the temperature, at least based on this small sample of observations.

Table 1.1 Five observations of air temperature and cricket chirp rate

Observation number	Temperature x_i (°F)	Chirps per second y_i
1	88.6	20.0
2	93.3	19.8
3	75.2	15.5
4	80.6	17.1
5	82.6	17.2

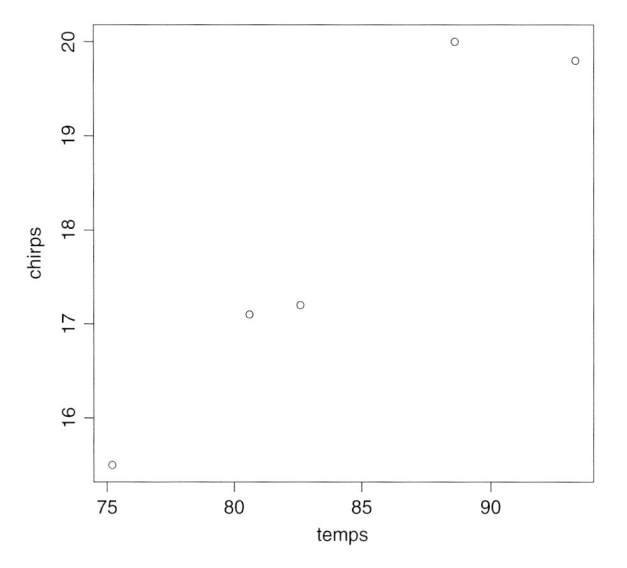

Fig. 1.7 A scatter plot of five observations of air temperature (x) and cricket chirp rate (y)

Example 1.16 Table 1.2 shows data from the year 1980 on child mortality rates, female literacy rates, and per capita gross national product for 57 developing countries. Looking at the scatter plots in Figure 1.8, child mortality and female literacy rates seem to be strongly negatively correlated. That means that the more literate the female population is, the less likely a child is to die before the age of 5. Child mortality rates and per capita GNP also have a negative correlation. However, the strength of the correlation is reduced due to considerable variability in the child mortality rates among the poorest countries. Lastly, there appears to be a positive correlation between female literacy rates and per capita GNP, though this is not as strong as one might expect. The data do not take into account income distribution, educational opportunities for girls and women, access to health care, or the level of participation of women in the workforce outside the home, among other factors.

Table 1.2 Data from 57 countries in 1980 on child mortality (CM) = number of deaths of children under age 5 per 1000 live births, female literacy rate percent (FLR), and per capita gross national product (pcGNP) (see Table 6.4 in [9])

n	CM	FLR	pcGNP	n	CM	FLR	pcGNP	n	CM	FLR	pcGNP
1	128	37	1870	20	118	47	1080	39	191	31	1010
2	204	22	130	21	269	17	290	40	182	19	300
3	202	16	310	22	189	35	270	41	37	88	1730
4	197	65	570	23	126	58	560	42	103	35	780
5	96	76	2050	24	167	29	240	43	67	85	1300
6	209	26	200	25	135	65	430	44	143	78	930
7	170	45	670	26	72	63	1420	45	83	85	690
8	240	29	300	27	128	49	420	46	223	33	200
9	241	11	120	28	152	84	420	47	240	19	450
10	55	55	290	29	224	23	530	48	312	21	280
11	75	87	1180	30	104	62	350	49	52	83	270
12	129	55	900	31	287	31	230	50	79	43	1340
13	24	93	1730	32	41	66	1620	51	61	88	670
14	165	31	1150	33	312	11	190	52	168	28	410
15	94	77	1160	34	77	88	2090	53	121	41	1310
16	96	80	1270	35	142	22	900	54	115	62	1470
17	148	30	580	36	262	22	230	55	186	45	300
18	98	69	660	37	215	12	140	56	178	45	220
19	161	43	420	38	246	9	330	57	142	67	560

The Pearson correlation coefficients corresponding to the different pairs of measured quantities are as follows.

$$\begin{cases} \rho(CM,\ FLR) \approx -0.8044\,, \\ \rho(CM,\ pcGNP) \approx -0.6572\,, \text{ and} \\ \rho(FLR,\ pcGNP) \approx 0.5645\,. \end{cases} \tag{1.28}$$

Example 1.17 A comparison of the distance versus the average gradient for the 51 categorized climbs of the 2018 Tour de France yields a Pearson correlation coefficient of 0.109. This is close to 0 and indicates that there is little correlation between these two quantities. Indeed, across all categories, two climbs of the same distance can have very different average gradients. Similarly, two climbs with the same gradient can be categorized differently depending on their lengths. Within each category, the correlation might be stronger. See Figure 1.9. We examine this more in Project 1.2.

Karl Pearson (1857–1936) was a British statistician who played a leading role in founding the modern study of statistics. Pearson was a protégé of Frances Galton and a leading proponent of the pseudo-science of eugenics, which purported to offer a scientific justification for white supremacy and the oppression of "inferior races".

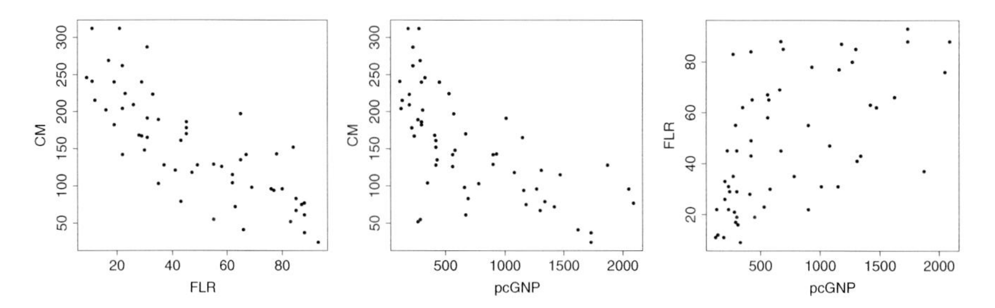

Fig. 1.8 The scatter plots show comparisons between (left) child mortality and female literacy rates, (center) child mortality rates and per capita GNP, and (right) female literacy rates and per capita GNP

He opposed allowing Jews to immigrate into Britain. So, perhaps we can strive to put his statistical notions to better use than he himself did.

1.6 Information Retrieval

A fundamental problem in information retrieval is to find documents that are most likely to contain useful information related to the terms provided by a user's search query. The recommended documents will come from some preexisting library of documents, while the search terms must be taken from an established dictionary of terms.

To facilitate the searches, we form a vector for each document in our library, called the **document vector**, that indicates which terms in the dictionary can be found in that document. A small example is shown in Table 1.3. The library consists of eight documents, labeled $B1$ through $B8$ in the table. These are the following mathematics books: $B1$, Allenby, *Rings, Fields, and Groups: An Introduction to Abstract Algebra*; $B2$, Axler, *Linear Algebra Done Right*; $B3$, Fridy, *Introductory Analysis: The Theory of Calculus*; $B4$, Gallian, *Contemporary Abstract Algebra*; $B5$, Hardy and Walker, *Applied Algebra: Codes, Ciphers, and Discrete Algorithms*; $B6$, Noble and Daniel, *Applied Linear Algebra*; $B7$, Ross, *Elementary Analysis: The Theory of Calculus*; and $B8$, Strang, *Linear Algebra and Its Applications*. The terms, shown in the leftmost column, were checked against the title and index of each document. Each column of the array shows a resulting document vector. In this example, an entry of 1 indicates that the given term is included in the corresponding document; an entry of 0 indicates that the given term is not found in that document. More sophisticated methods might use entries that vary according to the frequency with which each term appears in each document.

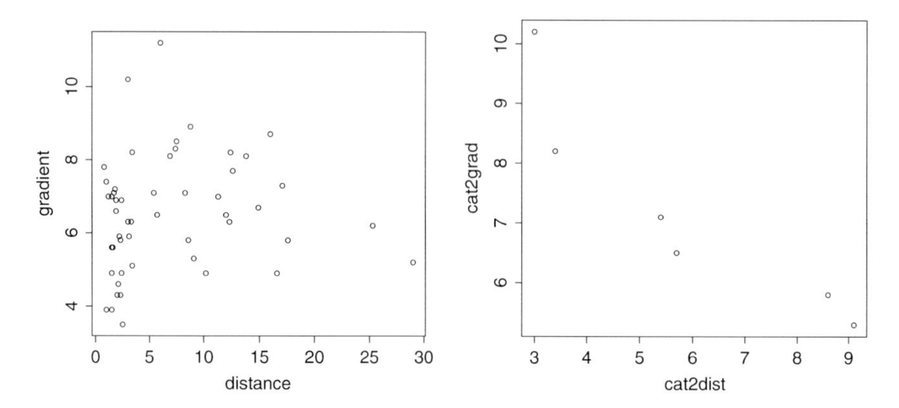

Fig. 1.9 The plot on the left suggests that there is little correlation between the distance and average % gradient for categorized climbs of the 2018 Tour de France. The plot on the right suggests a stronger correlation between distance and gradient for just the climbs in Category 2

When two documents use many of the same terms, their document vectors have many 1s in the same position. This yields a relatively large value for the inner product of the two document vectors. By contrast, when two documents tend to use different terms, the positions of the 1s do not coincide and the inner product of the document vectors is closer to 0. We should normalize these measurements to take into account the total number of terms used in each document. For instance, two shared terms out of four total terms represents a stronger similarity than two shared terms out of ten total terms. In this scenario, the length of each document vector is the square root of the total number of terms used in the document. Thus, the *cosine of the angle* between two document vectors gives a sense of how many overlapping terms there are in relation to the total number of terms present. We say that two documents are close if the cosine of the angle between their document vectors is close to 1.

To apply this to a search query by a user, we consider the query itself to be a document and construct the document vector for the query; we call this the **query vector**. Then we compute the cosine of the angle between the query vector and each document vector in the library. Finally, we choose a threshold value near 1 and recommend a particular document to the user if the cosine of the angle between the document vector and the query vector is greater than the threshold value.

Using a higher threshold results in higher **precision** in the search, in that the recommended documents will tend to be better matches for the query. However, a very precise search may omit some documents that could be of interest to the user. Lowering the threshold generates a larger **yield** of documents, but runs the risk of including more documents that are not of interest to the user. Finding algorithms that strike a good balance between precision and yield is an active field of research.

Table 1.3 A term–document array generated from selected terms in the titles and indices of eight mathematics books

	Documents							
Terms	B1	B2	B3	B4	B5	B6	B7	B8
Algebra	1	1	0	1	1	1	0	1
Analysis	0	0	1	0	0	0	1	0
Applied	0	0	0	0	1	1	0	1
Derivative	1	1	1	1	0	1	1	0
Eigenvalue	0	1	0	0	0	1	0	1
Factorization	0	0	0	1	0	0	0	1
Group	1	0	0	1	1	0	0	1
Inner product	0	1	1	0	0	1	0	1
Linear	1	1	1	1	1	1	0	1
Matrix	1	1	0	1	1	1	0	1
Nullspace	0	1	0	0	0	1	0	1
Orthogonal	0	1	0	1	0	1	0	1
Permutation	1	1	0	1	1	1	0	1
Riemann	0	0	1	0	0	0	1	0
Sequence	1	0	1	0	0	0	1	0
Symmetric	1	0	1	1	1	1	0	1
Transpose	0	1	0	0	0	1	0	1
Vector	0	1	0	1	1	1	0	1

Algorithm 1.18 (Term–Document Information Retrieval)

1. Form the *query vector* $\mathbf{q}$ from the terms in the search query.
2. For each document vector $\mathbf{a}$ in the library, compute the *cosine*:

$$\cos(\theta) = \frac{\mathbf{a}^T \mathbf{q}}{||\mathbf{a}||\,||\mathbf{q}||} .$$

3. Choose a suitable threshold value α and recommend each document for which $\cos(\theta) > \alpha$.

Example 1.19 Assessing the query with the terms **group** and **matrix** against the document library in Table 1.3, the computed cosine values are

$$\left[0.500\ 0.213\ 0.000\ 0.447\ 0.500\ 0.196\ 0.000\ 0.378\right] .$$

For the threshold $\alpha = 0.5$, we recommend documents B1 and B5, with B4 narrowly missing the cut. Adding the term **vector** to the query yields the cosine values

$$\left[0.408\ 0.348\ 0.000\ 0.548\ 0.612\ 0.320\ 0.000\ 0.463\right] .$$

Now documents B4 and B5 pass the 0.5 threshold, with B8 close behind.

To streamline the computation of the various cosine values, we can *normalize* the document vectors to be unit vectors. That is, replace each document vector $\mathbf{a}$ by $\widehat{\mathbf{a}} = \mathbf{a}/||\mathbf{a}||$. Also, let $\widehat{\mathbf{q}} = \mathbf{q}/||\mathbf{q}||$ be the normalized query vector. The inner products $\widehat{\mathbf{a}}^T\widehat{\mathbf{q}}$ now give us the cosine values.

Remark The cosine method for information retrieval is a useful starting point, but has its limitations. For instance, this method counts only the specific terms that two documents share. There is no attempt to find possible connections between different terms, such as *synonyms*, or to deal with words that have multiple meanings. Later on, we will see how we can use a matrix factorization called the singular value decomposition to improve our search engine design and get stronger responses to queries.

1.6.1 Comparing Movie Viewers

We can also apply the cosine method to compare movie ratings of different viewers of a movie streaming service. When a viewer submits their ratings of the movies in a specific library, they generate their own personal *ratings vector*. If the angle between two rating vectors is small, then those viewers might be considered to have a similar taste in movies. The streaming service can use this information to make movie recommendations to its users.

To get started, suppose we ask a set of viewers to rate the movies in a given collection according to the following simple scheme: $+1$ represents a favorable "thumbs up" rating; -1 represents an unfavorable "thumbs down"; 0 means the viewer did not watch or did not rate the film. For each viewer, we form a vector of their ratings of the movies in the collection. The array in Table 1.4 shows hypothetical ratings of five prominent films from 2018 by six fictitious viewers. Each column of the array shows the ratings vector for a particular viewer.

Table 1.4 An array of ratings of selected movies by six viewers. A 1 indicates a favorable "thumbs up" rating; a -1 indicates an unfavorable "thumbs down" rating; a 0 indicates that the viewer either did not watch the movie or did not rate it

	Viewers					
Movies	Amy	Ben	Chloe	Doug	Ebony	Fred
Call Me By Your Name	1	-1	0	1	1	0
Darkest Hour	1	1	0	1	0	-1
Dunkirk	-1	-1	1	1	-1	1
Get Out	0	1	1	0	1	1
Lady Bird	1	1	-1	-1	1	1

With this simple rating system, the inner product of the rating vectors for two viewers counts the number of films on which they agree (both rated $+1$ or both rated -1) minus the number of films on which they disagree (one $+1$ and one -1). If this inner product is positive, the two viewers agree more than they disagree. If the inner product is negative, the two viewers disagree more than they agree. For example, here are the inner products of the viewer rating vectors shown in Table 1.4. (The inner product of a viewer's vector with itself is the total number of films the viewer rated.)

	Amy	Ben	Chloe	Doug	Ebony	Fred
Amy	(4)	2	-2	0	3	-1
Ben		(5)	-1	-2	2	0
Chloe			(3)	2	-1	1
Doug				(4)	-1	-1
Ebony					(4)	1
Fred						(4)

This suggests that the most compatible pair of viewers is Amy and Ebony. The least compatible pairs are Amy and Chloe and Ben and Doug, at least as far as this analysis goes.

The inner product by itself is inadequate for determining compatibility because there are different ways to get the same result. For example, two viewers might agree on all of their ratings but only rate a small number of films. Two other viewers might rate lots of films but agree only slightly more often than they disagree. In both cases, the inner product will be a small, positive number. We can adjust for this by computing the *cosine* of the angle between two viewer's rating vectors, using formula (1.14). This will take the number of movies each viewer has rated into account. For two viewers who rate two movies and agree in their ratings, the inner product of the corresponding rating vectors is 2, while the angle between them is 0, since they are the same vector. For two viewers who rate ten films, agreeing on six of them but disagreeing on four, the inner product is also 2. This time, we get $\cos(\theta) = (6 - 4)/\sqrt{10}^2 = 0.2$, and $\theta \approx 78.5°$. These viewers do not have similar preferences.

1.7 Distance on a Sphere

The sphere of radius $R \geq 0$ centered at the origin in $\mathbb{R}^3$ consists of the terminal points of all origin-based vectors of length R. When the radius is $R = 1$, we call this the **unit sphere** in $\mathbb{R}^3$.

Any two noncollinear vectors $\mathbf{p}$ and $\mathbf{q}$ of length $R > 0$ generate a plane in $\mathbb{R}^3$. This plane intersects the sphere of radius R in a circle of radius R. This is called a **great circle** on the sphere. It has the same radius as the sphere itself. The arc of the great circle that connects the terminal points of $\mathbf{p}$ and $\mathbf{q}$ represents the shortest path *along the surface of the sphere* between these points. The circumference of the full great circle is $2\pi R$. Thus, the length of the arc is equal to $(\theta/2\pi)\cdot(2\pi R) = R\theta$, where θ is the angle between the vectors $\mathbf{p}$ and $\mathbf{q}$. We know that $\cos(\theta) = \left(\mathbf{p}^T\mathbf{q}\right)/R^2$. It follows that the distance, measured along the surface of the sphere, between the terminal points of $\mathbf{p}$ and $\mathbf{q}$ is given by the formula

$$R\theta = R\arccos\left(\left(\mathbf{p}^T\mathbf{q}\right)/R^2\right). \tag{1.29}$$

The distance formula in (1.29) can also be expressed in terms of longitude and latitude. The longitude of a point P on the sphere is the angle measured around the equator between the prime meridian and the meridian for P. All points on the same meridian as P have the same longitude. The latitude of P is the angle measured along a meridian between the equator and the circle on the sphere that is parallel to the equator and passes through P. Figure 1.10 illustrates these measurements. On the sphere of radius R centered at the origin in $\mathbb{R}^3$, the point with longitude λ and latitude ϕ has 3-dimensional Cartesian coordinates $x = R\cdot\cos(\phi)\cdot\cos(\lambda)$, $y = R\cdot\cos(\phi)\cdot\sin(\lambda)$, and $z = R\cdot\sin(\phi)$. Note that $x^2+y^2 = R^2\cdot\cos^2(\phi)$ and that $x^2+y^2+z^2 = R^2$, the Cartesian equation for the sphere.

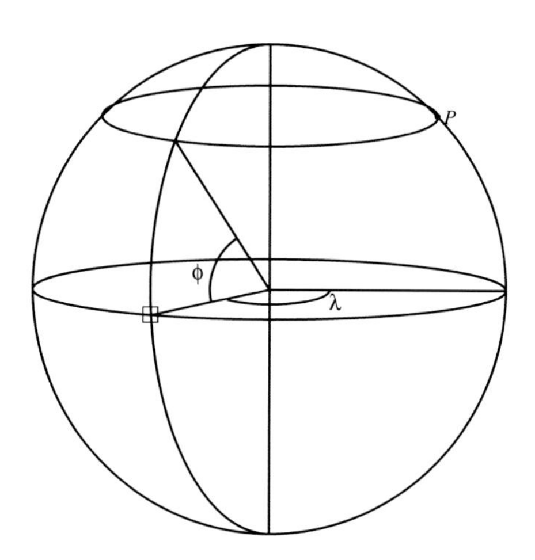

Fig. 1.10 The point P has latitude ϕ and longitude λ. These are measured from the reference point where the equator and prime meridian intersect, marked with a box □ in the figure

Suppose now that the vectors $\mathbf{p}$ and $\mathbf{q}$ have terminal points on the sphere of radius R with longitudes λ_1 and λ_2 and latitudes ϕ_1 and ϕ_2, respectively, measured in radians. In Cartesian coordinates,

$$\mathbf{p} = R\cdot\begin{bmatrix}\cos(\phi_1)\cdot\cos(\lambda_1)\\ \cos(\phi_1)\cdot\sin(\lambda_1)\\ \sin(\phi_1)\end{bmatrix} \text{ and } \mathbf{q} = R\cdot\begin{bmatrix}\cos(\phi_2)\cdot\cos(\lambda_2)\\ \cos(\phi_2)\cdot\sin(\lambda_2)\\ \sin(\phi_2)\end{bmatrix}.$$

Evaluating (1.29), the great circle distance, $R\theta$, becomes

$$R \arccos\left(\cos(\phi_1)\cos(\phi_2)\cos(\lambda_1 - \lambda_2) + \sin(\phi_1)\sin(\phi_2)\right). \tag{1.30}$$

Example 1.20 To compute the great circle distance from London, England, to Beijing, China, we start with the longitude and latitude coordinates of $\lambda_1 = 0$ and $\phi_1 = 51.5\pi/180$ for London and $\lambda_2 = 116.4\pi/180$ and $\phi_2 = 39.9\pi/180$ for Beijing. These are measured in radians. On Earth, the mean radius R is about 3958.8 miles or 6371.1 kilometers. Thus, the angle between London and Beijing is about 1.28 radians (73.2°) and the distance is approximately 5055 miles or 8135 kilometers. The great circle route is illustrated in Figure 1.11.

Fig. 1.11 The great circle route from London to Beijing

1.8 Bézier Curves

Two nonzero vectors **u** and **v** in $\mathbb{R}^N$ that do not lie on the same line form a triangle along with their difference vector $\mathbf{v} - \mathbf{u}$. For every real number t, the terminal point of the vector

$$\mathbf{u} + t \cdot (\mathbf{v} - \mathbf{u}) = (1 - t) \cdot \mathbf{u} + t \cdot \mathbf{v}$$

lies along the line through the terminal points of **u** and **v**. Restricting t so that $0 \leq t \leq 1$, we get the line segment between the terminal points of **u** and **v**. This is an example of a **parameterized curve**, or **vector function**. The most general definition is as follows.

Definition A **vector function** in $\mathbb{R}^N$ is a function **f** that assigns to each real number t in some interval $[a, b]$ a vector $\mathbf{f}(t)$ in $\mathbb{R}^N$. A **graph** of the vector function **f** consists of a plot of the terminal points of the vectors $\mathbf{f}(t)$ for $a \leq t \leq b$.

One way to think about the graph of a vector function is to imagine that the parameter t represents time and that the terminal point of $\mathbf{f}(t)$ represents the location at time t of a fly or a dust particle moving about in $\mathbb{R}^N$. The graph of $\mathbf{f}$ is the path traced out by the fly. Even in $\mathbb{R}^2$, the vector function point of view allows us to describe a huge array of curves, far more than can be represented by the traditional "$y = f(x)$" function construct. For example, the unit circle in $\mathbb{R}^2$, with equation $x^2 + y^2 = 1$, cannot be described using a single formula of the $y = f(x)$ type. But the circle is the graph of the vector function $\mathbf{f}$ given by

$$\mathbf{f}(t) = \begin{bmatrix} \cos(t) \\ \sin(t) \end{bmatrix}, \text{ for } 0 \leq t \leq 2\pi .$$

We need the tools of calculus to fully analyze the properties of vector functions and their graphs, including continuity, smoothness, how sharply curved the graph is, the length of the curve (since we may not be able to hold a straight ruler up to it), and more. We will not attempt this project here. Instead, let us look at one particular family of vector functions that is familiar to people who design video games and other computer graphics—**Bézier curves**.

Bézier developed these curves in the 1960s while working on auto body design for Renault. As often happens in the history of mathematics, it turns out that these same curves had been described in a theoretical mathematical context years earlier, in this case by Sergei Bernstein, in a famous (to mathematicians, that is) paper published in 1912. A common feature of lags like this that occurred in the twentieth century is that, by the 1960s, there was enough computer power available to quickly generate thousands of examples, where mathematicians from fifty years earlier or more often worked with thought experiments. Two other dramatic examples are the development of fractals and the invention of the CT scan in medical imaging. In both cases, the core mathematical groundwork was laid in the 1910s, by Julia and Fatou in the case of fractals and by Radon in the case of the imaging problem. But without computers, not even one exciting fractal or one remotely realistic imaging problem can be pictured or solved. Thus, in the 1960s and 1970s, the mathematics for both of these problems was freshly discovered and put to use on a whole new level, by Mandelbrot in the case of fractals and Cormack and Hounsfield for CT imaging.

Bézier curves start with a set of four points in the plane, which we will view as being the terminal points of four vectors in $\mathbb{R}^2$. Label these vectors as $\mathbf{p}_0$, $\mathbf{p}_1$, $\mathbf{p}_2$, and $\mathbf{p}_3$. The first and last points are the starting and ending points of the curve, respectively. The other two points function as control points for the curve. The idea is that the curve will start at the terminal point of $\mathbf{p}_0$ and head towards the terminal point of $\mathbf{p}_1$. As it does this, $\mathbf{p}_2$ will make its presence felt and pull the curve towards its terminal point. All the while, the influence of $\mathbf{p}_3$ gradually increases until the curve finally ends at the terminal point of $\mathbf{p}_3$.

To implement this mathematically, our main tool is the vector function formula for the line segment between two points, described above. Borrowing the notation used by the computer graphics designer Freya Holmér, we will call this function

lerp, for **linear interpolation**. That is, for two vectors $\mathbf{u}$ and $\mathbf{v}$ and a real number t, we define

$$\mathbf{lerp}(\mathbf{u},\ \mathbf{v},\ t) = \mathbf{u} + t \cdot (\mathbf{v} - \mathbf{u}) = (1 - t) \cdot \mathbf{u} + t \cdot \mathbf{v}\,. \tag{1.31}$$

Observe that $\mathbf{lerp}(\mathbf{u},\ \mathbf{v},\ 0) = \mathbf{u}$ and $\mathbf{lerp}(\mathbf{u},\ \mathbf{v},\ 1) = \mathbf{v}$. Now imagine that we have two particles moving at the same time, one along the line segment between the endpoints of $\mathbf{p}_0$ and $\mathbf{p}_1$ and the other traversing the line segment between the terminal points of $\mathbf{p}_1$ and $\mathbf{p}_2$. At any time t with $0 \le t \le 1$, the locations of these particles are $\mathbf{lerp}(\mathbf{p}_0,\ \mathbf{p}_1,\ t)$ and $\mathbf{lerp}(\mathbf{p}_1,\ \mathbf{p}_2,\ t)$. Next, imagine a third particle that continually adjusts its path to follow the line between the positions of the first two particles. That is, the position of this third particle at time t, labeled now as $\mathbf{f}_1(t)$, is given by

$$\begin{aligned}\mathbf{f}_1(t) &= \mathbf{lerp}\left(\mathbf{lerp}(\mathbf{p}_0,\ \mathbf{p}_1,\ t),\ \mathbf{lerp}(\mathbf{p}_1,\ \mathbf{p}_2,\ t),\ t\right)\\ &= (1-t)\cdot[(1-t)\cdot\mathbf{p}_0 + t\cdot\mathbf{p}_1] + t\cdot[(1-t)\cdot\mathbf{p}_1 + t\cdot\mathbf{p}_2]\\ &= (1-t)^2\cdot\mathbf{p}_0 + 2(1-t)\cdot t\cdot\mathbf{p}_1 + t^2\cdot\mathbf{p}_2\,.\end{aligned}$$

This is the **quadratic Bézier curve** determined by the vectors $\mathbf{p}_0$, $\mathbf{p}_1$, and $\mathbf{p}_2$ and their terminal points. Now imagine a fourth particle traversing the quadratic Bézier curve determined by the vectors $\mathbf{p}_1$, $\mathbf{p}_2$, and $\mathbf{p}_3$. Label the position at time t of this particle as $\mathbf{f}_2(t)$. That is, for $0 \le t \le 1$, we have

$$\begin{aligned}\mathbf{f}_2(t) &= \mathbf{lerp}\left(\mathbf{lerp}(\mathbf{p}_1,\ \mathbf{p}_2,\ t),\ \mathbf{lerp}(\mathbf{p}_2,\ \mathbf{p}_3,\ t),\ t\right)\\ &= (1-t)^2\cdot\mathbf{p}_1 + 2(1-t)\cdot t\cdot\mathbf{p}_2 + t^2\cdot\mathbf{p}_3\,.\end{aligned}$$

Finally, to generate the full Bézier curve from all four vectors, imagine yet another particle that continually adjusts its path to follow the line between the two quadratic Bézier curves just defined. Thus, the position of this particle, for $0 \le t \le 1$, is given by

$$\begin{aligned}\mathbf{B}(t) &= \mathbf{lerp}(\mathbf{f}_1(t),\ \mathbf{f}_2(t),\ t)\\ &= (1-t)\cdot\mathbf{f}_1(t) + t\cdot\mathbf{f}_2(t)\\ &= (1-t)^3\cdot\mathbf{p}_0 + 3(1-t)^2\cdot t\cdot\mathbf{p}_1 + 3(1-t)\cdot t^2\cdot\mathbf{p}_2 + t^3\cdot\mathbf{p}_3\,.\end{aligned} \tag{1.32}$$

Figure 1.12 depicts some examples of Bézier curves. The coefficients on $\mathbf{B}(t)$ are interesting in their own right. Indeed, we can expand the number 1 like so.

$$\begin{aligned}1 &= 1^3\\ &= ((1-t)+t)^3\\ &= (1-t)^3 + 3(1-t)^2\cdot t + 3(1-t)\cdot t^2 + t^3\,.\end{aligned}$$

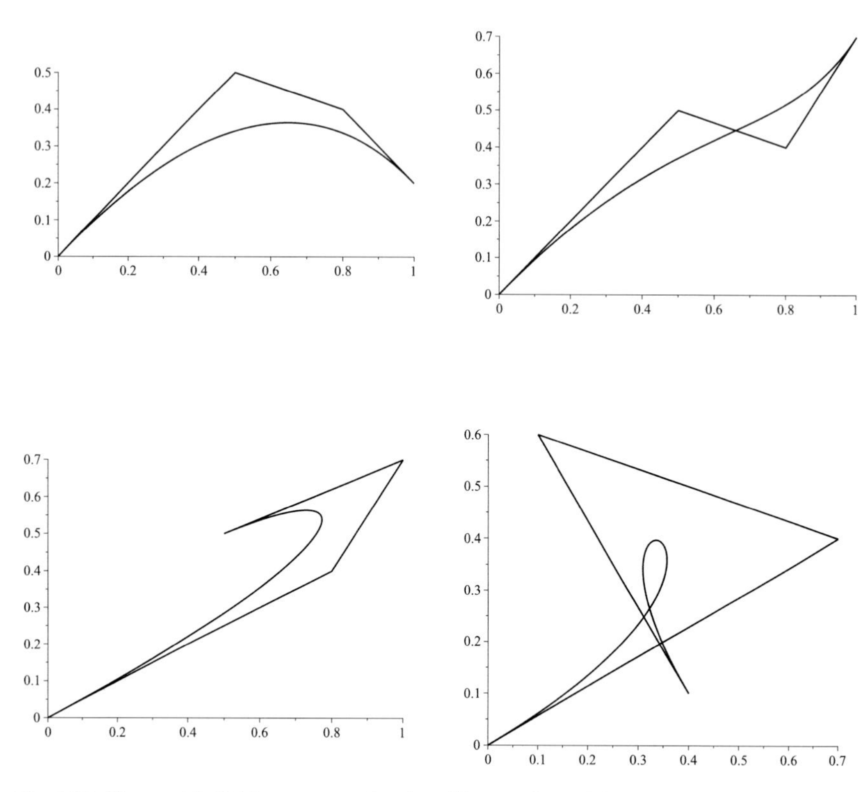

Fig. 1.12 These cubic Bézier curves in the plane illustrate four of the countless possible designs

For $0 \leq t \leq 1$, we also have $0 \leq (1-t) \leq 1$. Thus, the polynomials $(1-t)^3$, $3(1-t)^2 \cdot t$, $3(1-t)t^2$, and t^3 are nonnegative and add up to 1, for all $0 \leq t \leq 1$. Figure 1.13 shows the graphs of these polynomials for $0 \leq t \leq 1$.

A linear combination of vectors where the scalars are nonnegative and add up to 1 is called a **convex combination** of those vectors. The terminal point of a convex combination of vectors in $\mathbb{R}^2$ must lie inside the smallest convex polygon that contains the terminal points of all the vectors themselves. Thus, every point on the cubic Bézier curve $\mathbf{B}(t)$ lies inside a convex quadrilateral determined by the four guide points.

For any natural number n, we may expand $1 = 1^n = ((1-t)+t)^n$ and use the resulting summands to form a Bézier curve of degree n with $n+1$ control points. Bernstein used this to construct a set of polynomials of increasing degree that could approximate any continuous function to any desired accuracy. In the contemporary context of computer graphics, cubic Bézier curves have attracted the most attention. It is also possible to generate cubic Bézier curves in 3-dimensional space by choosing the four control points to be in $\mathbb{R}^3$.

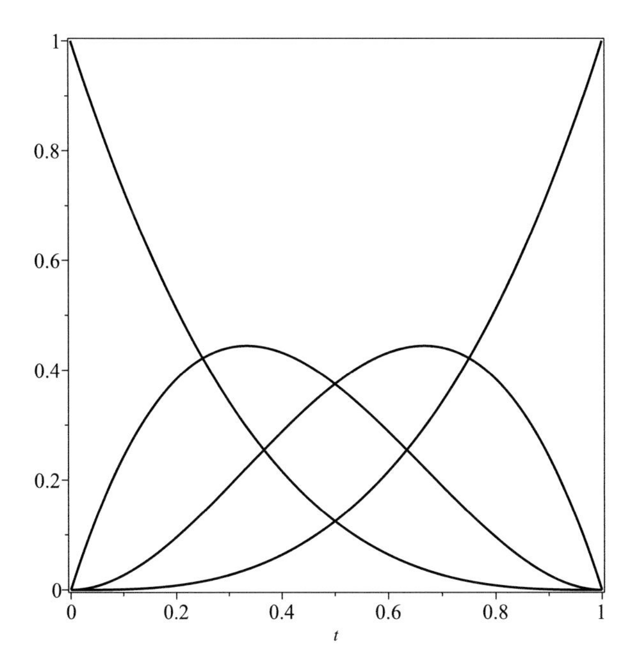

Fig. 1.13 This figure shows the graphs on the interval $0 \leq t \leq 1$ of the polynomials t^3, $3 \cdot t^2 \cdot (1-t)$, $3 \cdot t \cdot (1-t)^2$, and $(1-t)^3$. At each value of t, the values of these polynomials add up to 1

1.9 Orthogonal Vectors

Two nonzero vectors are perpendicular precisely when the angle between them is $\pi/2$ radians. Since $\cos(\pi/2) = 0$, it follows from Formula 1.8 and the computation (1.14) that *two nonzero vectors are perpendicular precisely when their inner product is equal to 0*. The inner product is also equal to 0 if either one of the vectors involved is the zero vector $\mathbf{0}$. To allow for this subtle distinction, we use the new term **orthogonal** whenever the inner product of two vectors, zero or otherwise, is equal to 0. For nonzero vectors, *orthogonal* and *perpendicular* are synonyms.

Definition 1.21 Two vectors $\mathbf{a}$ and $\mathbf{b}$ in $\mathbb{R}^N$ are said to be **orthogonal** provided that $\mathbf{a}^T\mathbf{b} = 0$. That is, two vectors are orthogonal if their inner product is 0.

Notice that *the zero vector* $\mathbf{0}$ *is orthogonal to every vector* in $\mathbb{R}^N$. Moreover, if some vector $\mathbf{a}$ happened to be orthogonal to every vector, then it would also be orthogonal to itself; that is, $\mathbf{a}$ would have to satisfy the condition $\mathbf{a}^T\mathbf{a} = ||\mathbf{a}||^2 = 0$. That can only happen if $\mathbf{a} = \mathbf{0}$. Thus, we have shown the following basic fact, which is more important than one might suspect at first.

Proposition 1.22 *The zero vector in* $\mathbb{R}^N$ *is orthogonal to every vector in* $\mathbb{R}^N$. *Moreover, the zero vector is the only vector in* $\mathbb{R}^N$ *that has this property.*

Example 1.23 Consider the parallelogram generated by two nonzero vectors $\mathbf{a}$ and $\mathbf{b}$. If these vectors are orthogonal to each other, then $\mathbf{a}^T\mathbf{b} = 0$ and the parallelogram is a rectangle. Each of the diagonals $\mathbf{a} + \mathbf{b}$ and $\mathbf{a} - \mathbf{b}$ is the hypotenuse of a right triangle having $\mathbf{a}$ and $\mathbf{b}$ (or copies of them) as its adjacent sides. Hence, the diagonals have the same length: $||\mathbf{a} + \mathbf{b}|| = ||\mathbf{a} - \mathbf{b}||$.

Conversely, suppose the diagonals of a parallelogram happen to have the same length. Must this be a rectangle? Well, if $\mathbf{a} + \mathbf{b}$ and $\mathbf{a} - \mathbf{b}$ have the same length, then we get the following chain of equalities.

$$\begin{aligned}
0 &= ||\mathbf{a} + \mathbf{b}||^2 - ||\mathbf{a} - \mathbf{b}||^2 \\
&= (\mathbf{a} + \mathbf{b})^T(\mathbf{a} + \mathbf{b}) - (\mathbf{a} - \mathbf{b})^T(\mathbf{a} - \mathbf{b}) \\
&= \left(\mathbf{a}^T\mathbf{a} + \mathbf{a}^T\mathbf{b} + \mathbf{b}^T\mathbf{a} + \mathbf{b}^T\mathbf{b}\right) - \left(\mathbf{a}^T\mathbf{a} - \mathbf{a}^T\mathbf{b} - \mathbf{b}^T\mathbf{a} + \mathbf{b}^T\mathbf{b}\right) \\
&= 4 \cdot (\mathbf{a}^T\mathbf{b})\,.
\end{aligned}$$

Thus, $\mathbf{a}^T\mathbf{b} = 0$. The adjacent sides of the parallelogram are, indeed, orthogonal and we have a rectangle. We have now shown that *a parallelogram is a rectangle if, and only if, its diagonals have the same length.*

Example 1.24 When the edges of the parallelogram generated by two nonzero vectors $\mathbf{a}$ and $\mathbf{b}$ all have the same length, that is, when $||\mathbf{a}|| = ||\mathbf{b}||$, then the parallelogram is called a *rhombus*. The condition $||\mathbf{a}|| = ||\mathbf{b}||$ is equivalent to

$$\begin{aligned}
0 &= \mathbf{a}^T\mathbf{a} - \mathbf{b}^T\mathbf{b} \\
&= \mathbf{a}^T\mathbf{a} - \mathbf{a}^T\mathbf{b} + \mathbf{b}^T\mathbf{a} - \mathbf{b}^T\mathbf{b} \text{ (since } \mathbf{a}^T\mathbf{b} = \mathbf{b}^T\mathbf{a}) \\
&= \mathbf{a}^T(\mathbf{a} - \mathbf{b}) + \mathbf{b}^T(\mathbf{a} - \mathbf{b}) \\
&= (\mathbf{a} + \mathbf{b})^T(\mathbf{a} - \mathbf{b})\,. \qquad (1.33)
\end{aligned}$$

Since the diagonals of the parallelogram are given by $\mathbf{a} + \mathbf{b}$ and $\mathbf{a} - \mathbf{b}$, the computation in (1.33) shows that *a parallelogram is a rhombus if, and only if, its diagonals are orthogonal to each other.* A square is both a rectangle and a rhombus and is the only type of parallelogram that has both orthogonal edges *and* orthogonal diagonals.

1.10 Area of a Parallelogram

We now compute the *area* of the parallelogram generated by two nonzero, non-collinear vectors $\mathbf{a}$ and $\mathbf{b}$ in $\mathbb{R}^N$. The basic formula is *area* = *base* × *altitude*. If we take the vector $\mathbf{a}$ as the base, then $||\mathbf{a}||$ is the length of the base. With θ denoting the angle between the two vectors, the altitude is given by $||\mathbf{b}|| \cdot \sin(\theta)$. See Figure 1.14.

The area is then $||\mathbf{a}|| \cdot ||\mathbf{b}|| \cdot \sin(\theta)$. Applying Formulas (1.19) and (1.17), we get two expressions for the area of the parallelogram.

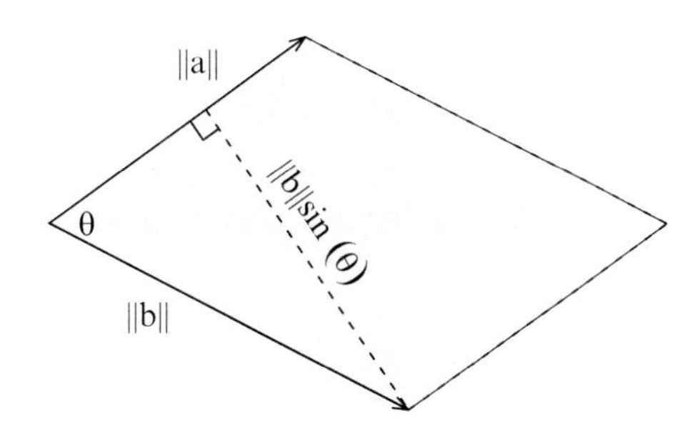

Fig. 1.14 The parallelogram formed by the vectors **a** and **b** has area $||\mathbf{a}|| \cdot ||\mathbf{b}|| \cdot \sin(\theta)$, where θ is the angle between the vectors

Formula 1.25 The area of the parallelogram generated by nonzero and non-collinear vectors **a** and **b** in $\mathbb{R}^N$ is

$$\begin{aligned}\text{Area} &= ||\mathbf{a}|| \cdot ||\mathbf{b}|| \cdot \frac{\left(||\mathbf{a}||^2 \cdot ||\mathbf{b}||^2 - (\mathbf{a}^T\mathbf{b})^2\right)^{1/2}}{||\mathbf{a}|| \cdot ||\mathbf{b}||} \\ &= \left(||\mathbf{a}||^2 \cdot ||\mathbf{b}||^2 - (\mathbf{a}^T\mathbf{b})^2\right)^{1/2}. \end{aligned} \tag{1.34}$$

Using coordinates, we have

$$\text{Area} = \left(\sum_{j=1}^{N-1} \sum_{k=j+1}^{N} \left(a_j b_k - a_k b_j\right)^2\right)^{1/2}. \tag{1.35}$$

Example 1.26 If two vectors **a** and **b** are orthogonal, so that $\mathbf{a}^T\mathbf{b} = 0$, then the parallelogram they generate is a rectangle. In this case, formula (1.34) reduces to the familiar fact that the area of a rectangle is equal to the product $||\mathbf{a}|| \cdot ||\mathbf{b}||$ of the lengths of its sides.

Example 1.27 The only way the parallelogram generated by **a** and **b** can have area 0 is to have equality in the Cauchy–Schwarz inequality: $|\mathbf{a}^T\mathbf{b}| = ||\mathbf{a}|| \cdot ||\mathbf{b}||$. That only happens when **a** and **b** are parallel, lying along the same line, in which case they do not generate a proper parallelogram. In $\mathbb{R}^2$, formula (1.35) has just one term, and the area of the parallelogram is $|a_1b_2 - a_2b_1|$. Thus, the vectors **a** and **b** lie on the same line if, and only if, $a_1b_2 - a_2b_1 = 0$. The number $a_1b_2 - a_2b_1$ is called the **determinant** of the 2×2 matrix $\begin{bmatrix} a_1 & b_1 \\ a_2 & b_2 \end{bmatrix}$ that has **a** and **b** as its columns. The condition that this determinant is nonzero is equivalent to the matrix having an inverse for matrix multiplication. More details of this are discussed later.

Example 1.28 For the parallelogram generated by two vectors in $\mathbb{R}^3$, Formula (1.35) has three terms. The area is equal to

$$A = \left((a_1b_2 - a_2b_1)^2 + (a_1b_3 - a_3b_1)^2 + (a_2b_3 - a_3b_2)^2\right)^{1/2}.$$

This area coincides with the norm of another vector in $\mathbb{R}^3$ called the **cross product** of $\mathbf{a}$ and $\mathbf{b}$ and denoted by $\mathbf{a} \times \mathbf{b}$. Specifically, for $\mathbf{a}$ and $\mathbf{b}$ in $\mathbb{R}^3$,

$$\mathbf{a} \times \mathbf{b} = \left[\, (a_2b_3 - a_3b_2)\ (a_3b_1 - a_1b_3)\ (a_1b_2 - a_2b_1) \,\right]^T. \tag{1.36}$$

The parallelogram formed by $\mathbf{a}$ and $\mathbf{b}$ has area equal to $||\mathbf{a} \times \mathbf{b}||$. The cross product has important applications in physics, but will not concern us much here. For more about the cross product and its properties, see Exercise #27 in this chapter and Project 6.1.

1.11 Projection and Reflection

When we developed the formula for the area of the parallelogram generated by two nonzero, noncollinear vectors $\mathbf{a}$ and $\mathbf{b}$ (see Formula 1.25 and (1.34)), we computed the altitude of the parallelogram by "dropping a perpendicular" from the end of the vector $\mathbf{b}$ to the line determined by the vector $\mathbf{a}$. In this way, we formed a right triangle, where $\mathbf{b}$ was the hypotenuse, one of the adjacent sides was parallel to $\mathbf{a}$, and the other adjacent side was orthogonal to $\mathbf{a}$. Using right-triangle trigonometry, we see that the length of the side parallel to $\mathbf{a}$ is given by $|\cos(\theta)| \cdot ||\mathbf{b}||$, where θ is the angle between $\mathbf{a}$ and $\mathbf{b}$. (We need the absolute value because the angle θ could be greater than $\pi/2$, in which case $\cos(\theta) < 0$ and the interior angle in the right triangle is $\pi - \theta$.) As a vector, this side of the triangle, parallel to $\mathbf{a}$, is obtained by multiplying $\cos(\theta)\, ||\mathbf{b}||$ by the unit vector $\mathbf{a}/||\mathbf{a}||$. See Example 1.7 above. In this way, we get the vector $(\cos(\theta)\, ||\mathbf{b}||/||\mathbf{a}||)\, \mathbf{a}$, which is a scalar multiple of $\mathbf{a}$. We know that $\cos(\theta) = \frac{\mathbf{a}^T\mathbf{b}}{||\mathbf{a}||\,||\mathbf{b}||}$. Thus, one adjacent side of the triangle is given by the vector $\left(\mathbf{a}^T\mathbf{b}/||\mathbf{a}||^2\right)\mathbf{a}$. This vector is called the **projection** of $\mathbf{b}$ along $\mathbf{a}$. It is shown on the left in Figure 1.15.

Definition 1.29 Let $\mathbf{a}$ and $\mathbf{b}$ be vectors in $\mathbb{R}^N$, with $\mathbf{a} \neq \mathbf{0}$.

(i) The **projection of b along a is defined to be the vector**

$$\text{proj}_{\mathbf{a}}\mathbf{b} = \frac{\mathbf{a}^T\mathbf{b}}{||\mathbf{a}||^2}\,\mathbf{a}. \tag{1.37}$$

(ii) The **component of b in the direction of a** is the number

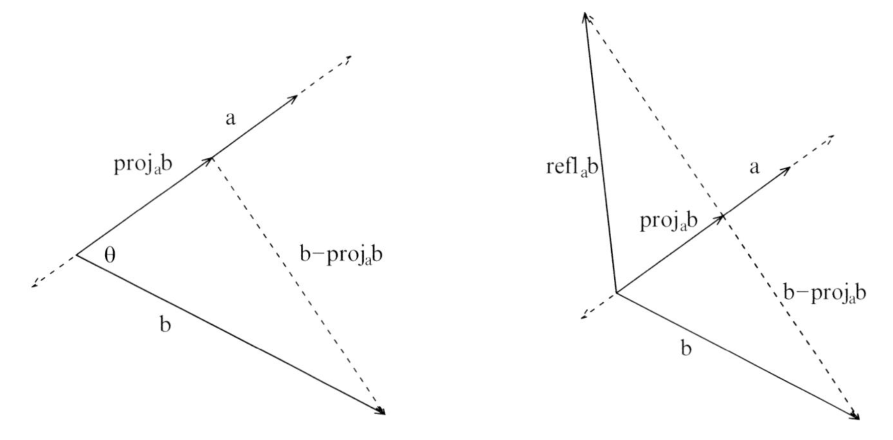

Fig. 1.15 *Left:* The projection of **b** along **a** lies along the line generated by **a**. The difference $\mathbf{b} - \text{proj}_{\mathbf{a}}\mathbf{b}$ is orthogonal to **a**. *Right:* The reflection of **b** across **a**

$$\text{comp}_{\mathbf{a}}\mathbf{b} = \cos(\theta)\, ||\mathbf{b}|| = \frac{\mathbf{a}^T\mathbf{b}}{||\mathbf{a}||}. \tag{1.38}$$

Notice that, if **b** is orthogonal to **a**, then $\text{proj}_{\mathbf{a}}\mathbf{b} = \mathbf{0}$. In general, if we defined the projection correctly, then the triangle with sides given by **b**, $\text{proj}_{\mathbf{a}}\mathbf{b}$, and $\left(\mathbf{b} - \text{proj}_{\mathbf{a}}\mathbf{b}\right)$ should be a right triangle with hypotenuse **b**. This will happen if the vector $\left(\mathbf{b} - \text{proj}_{\mathbf{a}}\mathbf{b}\right)$ is orthogonal to **a**. Let us check.

$$\begin{aligned}
\mathbf{a}^T\left(\mathbf{b} - \text{proj}_{\mathbf{a}}\mathbf{b}\right) &= \mathbf{a}^T\left(\mathbf{b} - \frac{\mathbf{a}^T\mathbf{b}}{||\mathbf{a}||^2}\,\mathbf{a}\right) \\
&= \left(\mathbf{a}^T\mathbf{b}\right) - \frac{\mathbf{a}^T\mathbf{b}}{||\mathbf{a}||^2}\left(\mathbf{a}^T\mathbf{a}\right) \\
&= \left(\mathbf{a}^T\mathbf{b}\right) - \frac{\mathbf{a}^T\mathbf{b}}{||\mathbf{a}||^2}\,||\mathbf{a}||^2 \\
&= \left(\mathbf{a}^T\mathbf{b}\right) - \mathbf{a}^T\mathbf{b} \\
&= 0\,, \text{ as we hoped.}
\end{aligned}$$

Another way to express this relationship is that the vector $\text{proj}_{\mathbf{a}}\mathbf{b}$ is the vector of the form $t \cdot \mathbf{a}$, for t a real number, that minimizes the norm $||\mathbf{b} - t \cdot \mathbf{a}||$. To confirm this, we compute

$$\begin{aligned}
||\mathbf{b} - t \cdot \mathbf{a}||^2 &= (\mathbf{b} - t \cdot \mathbf{a})^T\,(\mathbf{b} - t \cdot \mathbf{a}) \\
&= \mathbf{b}^T\mathbf{b} - 2t \cdot \mathbf{a}^T\mathbf{b} + t^2 \cdot \mathbf{a}^T\mathbf{a}
\end{aligned}$$

$$= ||\mathbf{b}||^2 - 2t \cdot \mathbf{a}^T\mathbf{b} + t^2 \cdot ||\mathbf{a}||^2 \tag{1.39}$$

This last expression is a quadratic in t that will be minimized at the vertex of the corresponding parabola. From the quadratic formula, the vertex occurs when $t = \frac{\mathbf{a}^T\mathbf{b}}{||\mathbf{a}||^2}$. This is exactly the value of t used in the definition of $\text{proj}_{\mathbf{a}}\mathbf{b}$. In this light, we see that the projection of a vector is a solution to a minimization problem. In the chapters ahead, we will explore a variety of applications that require us to solve minimization problems. Projections will play an important role in our analysis.

The word *projection* might evoke a picture of the vector **b** casting a shadow along the line formed by **a** when the rays from a distant light source shine perpendicularly to that line. Suppose, instead, we think of the line formed by **a** not as a projector screen, but as a mirror. Then the vector **b** will have a mirror image, or *reflection*, across that line. Temporarily label this reflection $\mathbf{b}'$. Since **b** is also the reflection of $\mathbf{b}'$, we should have $\text{proj}_{\mathbf{a}}\mathbf{b}' = \text{proj}_{\mathbf{a}}\mathbf{b}$. Then, since **b** and $\mathbf{b}'$ lie on opposite sides of the mirror, we must have $\mathbf{b}' - \text{proj}_{\mathbf{a}}\mathbf{b}' = -\left(\mathbf{b} - \text{proj}_{\mathbf{a}}\mathbf{b}\right)$. That is, $\mathbf{b}' - \text{proj}_{\mathbf{a}}\mathbf{b} = \text{proj}_{\mathbf{a}}\mathbf{b} - \mathbf{b}$. Solving for $\mathbf{b}'$ yields $\mathbf{b}' = 2 \cdot \text{proj}_{\mathbf{a}}\mathbf{b} - \mathbf{b}$. This vector is called the **reflection** of **b** across **a**. It is shown on the right in Figure 1.15.

Definition 1.30 Given two vectors **a** and **b** in $\mathbb{R}^N$, with $\mathbf{a} \neq \mathbf{0}$, the **reflection of b across a** is defined to be the vector

$$\text{refl}_{\mathbf{a}}\mathbf{b} = 2 \cdot \text{proj}_{\mathbf{a}}\mathbf{b} - \mathbf{b}\,. \tag{1.40}$$

1.12 The "All-1s" Vector

For a given value of N, denote by $\mathbf{1}_N$ the vector in $\mathbb{R}^N$ that has all of its coordinates equal to 1. When there is no possible confusion about the dimension, we write simply $\mathbf{1}$, rather than $\mathbf{1}_N$. The "all-1s" vector has the special property that the inner product of any vector **x** in $\mathbb{R}^N$ with $\mathbf{1}_N$ is the sum of the coordinates of **x**. That is,

$$\mathbf{1}^T\mathbf{x} = \mathbf{x}^T\mathbf{1} = x_1 + x_2 + \cdots + x_N\,.$$

In particular, any vector whose coordinates sum to 0 is orthogonal to $\mathbf{1}$. The norm is $||\mathbf{1}_N|| = (1 + \cdots + 1)^{1/2} = \sqrt{N}$. The projection along $\mathbf{1}_N$ of an arbitrary vector **x** is

$$\text{proj}_{\mathbf{1}}\mathbf{x} = \left(\frac{\mathbf{1}^T\mathbf{x}}{||\mathbf{1}||^2}\right)\mathbf{1} = \left(\frac{x_1 + x_2 + \cdots + x_N}{N}\right)\mathbf{1} = \bar{x}\mathbf{1}\,, \tag{1.41}$$

where $\bar{x} = (x_1 + x_2 + \cdots + x_N)/N$ is the average value, or **mean**, of the coordinates of **x**. It follows that the vector $\widehat{\mathbf{x}} = \mathbf{x} - \bar{x}\mathbf{1}$ is orthogonal to $\mathbf{1}$. The vectors **x**, $\bar{x}\mathbf{1}$, and $\widehat{\mathbf{x}} = \mathbf{x} - \bar{x}\mathbf{1}$ form a right triangle. By the Pythagorean theorem, it follows that

$$\sum_{i=1}^{N}(x_i-\bar{x})^2 = ||\mathbf{x}-\bar{x}\mathbf{1}||^2$$
$$= ||\mathbf{x}||^2 - ||\bar{x}\mathbf{1}||^2$$
$$= \sum_{i=1}^{N} x_i{}^2 - N(\bar{x})^2 . \tag{1.42}$$

In the context of statistical correlation introduced in Section 1.5 above, for two vectors $\mathbf{x}$ and $\mathbf{y}$ in $\mathbb{R}^N$ whose coordinates represent paired observations of two random variables, the Pearson correlation coefficient, defined in (1.27), is equal to the cosine of the angle between the mean-centered vectors $\widehat{\mathbf{x}} = \mathbf{x} - \bar{x}\mathbf{1}$ and $\widehat{\mathbf{y}} = \mathbf{y} - \bar{y}\mathbf{1}$. Equation (1.42) can simplify the computation of the correlation coefficient.

1.13 Exercises

1. For each given pair of vectors $\mathbf{a}$ and $\mathbf{b}$ in $\mathbb{R}^2$, do the following:

   (i) Create a sketch similar to Figure 1.4. Locate and label the vectors $\mathbf{a}+\mathbf{b}$ and $\mathbf{a}-\mathbf{b}$ in your sketch.
   (ii) Compute the sum $\mathbf{a}+\mathbf{b}$ and the difference $\mathbf{a}-\mathbf{b}$.
   (iii) Verify that

   $$||\mathbf{a}+\mathbf{b}||^2 + ||\mathbf{a}-\mathbf{b}||^2 = 2\cdot\left(||\mathbf{a}||^2 + ||\mathbf{b}||^2\right).$$

   (This illustrates the *parallelogram law*.)

   (a) $\mathbf{a} = \begin{bmatrix} 3 \\ -1 \end{bmatrix}$; $\mathbf{b} = \begin{bmatrix} -1 \\ 5 \end{bmatrix}$.
   (b) $\mathbf{a} = \begin{bmatrix} 2 \\ 4 \end{bmatrix}$; $\mathbf{b} = \begin{bmatrix} -2 \\ 3 \end{bmatrix}$.
   (c) $\mathbf{a} = \begin{bmatrix} 3 \\ 1 \end{bmatrix}$; $\mathbf{b} = \begin{bmatrix} 2 \\ 3 \end{bmatrix}$.
   (d) $\mathbf{a} = \begin{bmatrix} 3 \\ 1 \end{bmatrix}$; $\mathbf{b} = \begin{bmatrix} 1 \\ -3 \end{bmatrix}$.

2. For each vector given below: (i) Find a unit vector in the same direction, and (ii) find a unit vector in the opposite direction.

(a) $\begin{bmatrix} 2 \\ 3 \end{bmatrix}$; (b) $\begin{bmatrix} -1 \\ 2 \end{bmatrix}$; (c) $\begin{bmatrix} 1 \\ 2 \\ 3 \end{bmatrix}$; (d) $\begin{bmatrix} 2 \\ 1 \\ 5 \end{bmatrix}$; (e) $\begin{bmatrix} 1 \\ 1 \\ 1 \\ 1 \end{bmatrix}$.

3. Find a vector in $\mathbb{R}^3$ in the same direction as $\begin{bmatrix} 6 \\ 2 \\ -3 \end{bmatrix}$ with length 4.
4. For each given pair of vectors **a** and **b**:

   (i) Compute the inner product $\mathbf{a}^T\mathbf{b} = \mathbf{b}^T\mathbf{a}$.
   (ii) Compute the norms $||\mathbf{a}||$ and $||\mathbf{b}||$.
   (iii) Find the angle between the vectors **a** and **b**.

   (a) $\mathbf{a} = \begin{bmatrix} 6 \\ -2 \\ 3 \end{bmatrix}$; $\mathbf{b} = \begin{bmatrix} 2 \\ -1 \\ 2 \end{bmatrix}$.

   (b) $\mathbf{a} = \begin{bmatrix} 1 \\ 3 \\ 1 \end{bmatrix}$; $\mathbf{b} = \begin{bmatrix} 2 \\ 1 \\ -2 \end{bmatrix}$.

   (c) $\mathbf{a} = \begin{bmatrix} -1 \\ 3 \\ 0 \end{bmatrix}$; $\mathbf{b} = \begin{bmatrix} 2 \\ 1 \\ 5 \end{bmatrix}$.

   (d) $\mathbf{a} = \begin{bmatrix} 1 \\ 2 \\ 1 \\ 2 \end{bmatrix}$; $\mathbf{b} = \begin{bmatrix} 2 \\ 1 \\ -2 \\ 1 \end{bmatrix}$.

   (e) $\mathbf{a} = \begin{bmatrix} 1 \\ 2 \\ 1 \\ 2 \end{bmatrix}$; $\mathbf{b} = \begin{bmatrix} 3 \\ 2 \\ -1 \\ -3 \end{bmatrix}$.

5. Compute the inner product $\mathbf{a}^T\mathbf{b}$ if we know that $||\mathbf{a}|| = 80$, $||\mathbf{b}|| = 50$, and the angle between **a** and **b** is $3\pi/4$ radians (or 135°).
6. A street vendor charges \$4.50 for a burrito, \$4.00 for a taco, and \$1.50 for a soft drink. The vendor sells a burritos, b tacos, and c soft drinks on a given day. If $\mathbf{p} = \begin{bmatrix} 4.50 \\ 4.00 \\ 1.50 \end{bmatrix}$ and $\mathbf{v} = \begin{bmatrix} a \\ b \\ c \end{bmatrix}$, then what does the value of the inner product $\mathbf{p}^T\mathbf{v}$ represent?
7. The table shows the team home runs (HR) and team wins for the 2019 season for six teams in the American League of Major League Baseball: New York Yankees (NYY), Baltimore Orioles (Balt), Cleveland Indians (Cleve), Chicago

White Sox (CWS), Texas Rangers (TX), and Los Angeles Angels (LAA). Before you calculate anything, make a guess about the correlation coefficient. Then calculate the correlation coefficient of these two quantities based on this sample of observations. Did the result agree with your guess?

Team:	NYY	Balt	Cleve	CWS	TX	LAA
HR:	306	213	223	182	223	220
Wins:	103	54	93	72	78	72

8. The table shows the team earned run average (ERA; earned runs allowed per nine innings of play) and team wins for the 2019 season for six teams in the National League of Major League Baseball: Washington Nationals (Wash), Philadelphia Phillies (Phil), Milwaukee Brewers (Mil), Pittsburgh Pirates (Pitt), Los Angeles Dodgers (LAD), and San Francisco Giants (SFG). Before you calculate anything, make a guess about the correlation coefficient. Then calculate the correlation coefficient of these two quantities based on this sample of observations. Did the result agree with your guess?

Team	Wash	Phil	Mil	Pitt	LAD	SFG
ERA	4.27	4.53	4.40	5.18	3.37	4.38
Wins	93	81	89	69	106	77

9. Refer to the data in Table 1.2 and the discussion in Example 1.16.
   (a) Reproduce the scatter plots in Figure 1.8.
   (b) Compute the correlation coefficients
   $$\rho(CM, FLR),\ \rho(CM, pcGNP),\ \text{and}\ \rho(FLR, pcGNP)$$
   directly from the data in the table. (Use a computer.) Do your results agree with the values given in (1.28)?
   (c) Discuss what the correlation coefficients in (1.28) mean to you. Were they higher or lower than you expected? What, if anything, do they imply about causality? (E.g., Does a higher literacy rate by itself cause a lower child mortality rate?)
10. Table 1.5 shows a term–document array based on a library of five mathematical books: $B1$, Dunham, *Euler: The Master of Us All*; $B2$, Osserman, *Poetry of the Universe*; $B3$, Pappas, *The Joy of Mathematics*; $B4$, Pölya, *Patterns of Plausible Inference*; and $B5$, Yaglom, *Geometric Transformations I*. The dictionary of terms is shown in the leftmost column.

    For each query $\mathbf{q}$ given below, apply Algorithm 1.18 to do the following:

Table 1.5 A term–document array based on a library of five mathematical books

	Documents				
Terms	B1	B2	B3	B4	B5
Dimension	0	1	1	0	0
Geometry	1	1	1	0	1
Probability	1	1	1	1	0
Statistics	0	1	0	1	0
Symmetry	0	0	1	0	1

(i) Determine which documents are relevant to the query with a threshold of $\alpha = 0.5$.
(ii) Determine which documents are relevant to the query with a threshold of $\alpha = 0.8$.

(a) $\mathbf{q} = \begin{bmatrix} 0 & 1 & 1 & 0 & 0 \end{bmatrix}^T$ = geometry + probability.
(b) $\mathbf{q} = \begin{bmatrix} 0 & 1 & 0 & 0 & 1 \end{bmatrix}^T$ = geometry + symmetry.
(c) $\mathbf{q} = \begin{bmatrix} 1 & 0 & 0 & 1 & 0 \end{bmatrix}^T$ = dimension + statistics.

11. Referring to the data in Table 1.4, let us say that two viewers have compatible taste in movies when the angle between their rating vectors is small.

    (a) Compute the *cosine* of the angle between each pair of movie rating vectors shown in Table 1.4.
    (b) Determine which pairs of viewers have the most compatible taste in movies.
    (c) Determine which pairs of viewers have the least compatible taste in movies.

12. The cities of Macapa (Brazil) and Pontianak (Borneo) and the mountain Mt. Kenya (also called Kirinyaga) all lie on the equator (latitude 0). Their respective longitudes are 51° West, 109.3° East, and 37.3° East. Determine the great circle distances between these places. (*Hint:* The equator itself is the great circle that connects these places.)
13. The city of Philadelphia (USA) has longitude 75° West and latitude 40° North.

    (a) Determine the great circle distance from Philadelphia to the north pole. (*Hint:* Travel along the meridian.)
    (b) Determine the great circle distance from Philadelphia to the south pole. (*Hint:* Travel along the meridian.)
    (c) Determine the great circle distance from Philadelphia to Phnom Penh (Cambodia). (*Hint:* Phnom Penh is at longitude 105° East and latitude 11.5° North. The meridians of Philadelphia and Phnom Penh form a great circle.)

14. The cities of Philadelphia (USA) and Ankara (Turkey) both lie on the parallel at latitude 40° North. Their respective longitudes are 75° West and 32.9° East.

(a) Determine the distance between Philadelphia and Ankara traveling along the parallel at 40° North. (*Hint:* The parallel at latitude ϕ on the sphere of radius R is a circle of radius $R\cos(\phi)$.)
(b) Determine the great circle distance between Philadelphia and Ankara.

15. Determine whether or not each of the following sets of vectors is *mutually orthogonal*. (Note: A set of vectors is *mutually orthogonal* when each vector in the set is orthogonal to all other vectors in the set.)

(a) $\mathbf{a} = \begin{bmatrix} 1 \\ 4 \\ 2 \\ 0 \end{bmatrix}$, $\mathbf{b} = \begin{bmatrix} -2 \\ 0 \\ 1 \\ 7 \end{bmatrix}$, and $\mathbf{c} = \begin{bmatrix} 3 \\ 7 \\ 6 \\ 2 \end{bmatrix}$.

(b) $\mathbf{u} = \begin{bmatrix} 0 \\ 0 \\ 0 \\ 1 \end{bmatrix}$, $\mathbf{v} = \begin{bmatrix} 0 \\ 1 \\ 1 \\ 0 \end{bmatrix}$, and $\mathbf{w} = \begin{bmatrix} 1 \\ 1 \\ -1 \\ 0 \end{bmatrix}$.

(c) $\mathbf{p} = \begin{bmatrix} 2 \\ 1 \\ 2 \\ 1 \end{bmatrix}$, $\mathbf{q} = \begin{bmatrix} 3 \\ -6 \\ 1 \\ -2 \end{bmatrix}$, and $\mathbf{r} = \begin{bmatrix} 3 \\ -1 \\ -5 \\ 5 \end{bmatrix}$.

16. Verify by direct calculation that every point on the line generated by the vector $\begin{bmatrix} 2 & 11 & 10 \end{bmatrix}^T$ lies in both planes $x_1 - 2x_2 + 2x_3 = 0$ and $3x_1 + 4x_2 - 5x_3 = 0$.
17. Compute the area of the parallelogram generated by each given pair of vectors.

(a) $\mathbf{a} = \begin{bmatrix} 1 \\ 1 \end{bmatrix}$; $\mathbf{b} = \begin{bmatrix} 1 \\ -1 \end{bmatrix}$, in $\mathbb{R}^2$. (Draw a sketch.)

(b) $\mathbf{a} = \begin{bmatrix} 1 \\ 2 \end{bmatrix}$; $\mathbf{b} = \begin{bmatrix} 3 \\ 1 \end{bmatrix}$, in $\mathbb{R}^2$. (Draw a sketch.)

(c) $\mathbf{a} = \begin{bmatrix} -1 \\ 4 \\ 8 \end{bmatrix}$; $\mathbf{b} = \begin{bmatrix} 12 \\ 1 \\ 2 \end{bmatrix}$, in $\mathbb{R}^3$.

(d) $\mathbf{a} = \begin{bmatrix} 3 \\ -6 \\ 1 \\ -2 \end{bmatrix}$; $\mathbf{b} = \begin{bmatrix} 3 \\ -1 \\ -5 \\ 5 \end{bmatrix}$.

(e) $\mathbf{a} = \begin{bmatrix} 2 \\ 1 \\ 2 \\ 1 \end{bmatrix}$; $\mathbf{b} = \begin{bmatrix} 3 \\ 7 \\ 6 \\ 2 \end{bmatrix}$.

18. Find the projection $\mathrm{proj}_{\mathbf{a}}\mathbf{b}$ of the vector $\mathbf{b} = \begin{bmatrix} 2 \\ 3 \end{bmatrix}$ along the vector $\mathbf{a} = \begin{bmatrix} 1 \\ 4 \end{bmatrix}$. Sketch the vectors $\mathbf{a}$, $\mathbf{b}$, and $\mathrm{proj}_{\mathbf{a}}\mathbf{b}$.
19. Find the projection $\mathrm{proj}_{\mathbf{a}}\mathbf{b}$ of $\mathbf{b} = \begin{bmatrix} 12 \\ 1 \\ 2 \end{bmatrix}$ along $\mathbf{a} = \begin{bmatrix} -1 \\ 4 \\ 8 \end{bmatrix}$.
20. Let $\mathbf{a} = \begin{bmatrix} 1 \\ 1 \end{bmatrix}$ and $\mathbf{b} = \begin{bmatrix} 1 \\ -1 \end{bmatrix}$, and $\mathbf{v} = \begin{bmatrix} v_1 \\ v_2 \end{bmatrix}$.
    (a) Find the vectors $\mathrm{proj}_{\mathbf{a}}\mathbf{v}$ and $\mathrm{proj}_{\mathbf{b}}\mathbf{v}$.
    (b) Compute the vector $\mathrm{proj}_{\mathbf{b}}\left(\mathrm{proj}_{\mathbf{a}}\mathbf{v}\right)$.
    (c) Compute the vector $\mathrm{proj}_{\mathbf{a}}\left(\mathrm{proj}_{\mathbf{a}}\mathbf{v}\right)$.
21. Verify that, for all $\mathbf{a}$ and $\mathbf{b}$ in $\mathbb{R}^N$, we have

$$\mathrm{refl}_{\mathbf{a}}(\mathrm{refl}_{\mathbf{a}}\mathbf{b}) = \mathbf{b}\,.$$

    That is, $\mathbf{b}$ is the reflection across $\mathbf{a}$ of its own reflection across $\mathbf{a}$.
22. *(Proof Problem)* Provide proofs, using vector coordinates, of the following properties of the inner product.
    (a) *Homogeneity (1.22)* For all $\mathbf{a}$ and $\mathbf{b}$ in $\mathbb{R}^N$ and all scalars r,

$$(r\,\mathbf{a})^T\mathbf{b} = r \cdot (\mathbf{a}^T\mathbf{b}) = \mathbf{a}^T(r\mathbf{b})\,.$$

    (b) *Additivity (1.23)* For $\mathbf{a}$, $\mathbf{b}$, and $\mathbf{c}$ in $\mathbb{R}^N$,

$$(\mathbf{a}+\mathbf{b})^T\mathbf{c}\mathbf{a}^T\mathbf{c} + \mathbf{b}^T\mathbf{c}\,.$$

    Then equation (1.24) follows by symmetry.
23. *(Proof Problem)* Suppose the vector $\mathbf{u}$ is orthogonal to the vectors $\mathbf{v}$ and $\mathbf{w}$. Show that $\mathbf{u}$ is orthogonal to every vector of the form $r\mathbf{v} + s\mathbf{w}$, where r and s are real numbers.
24. *(Proof Problem)* **Pythagorean theorem:** Show that, if $\mathbf{x}$ and $\mathbf{y}$ are orthogonal vectors in $\mathbb{R}^N$, then

$$||\mathbf{x}+\mathbf{y}||^2 = ||\mathbf{x}-\mathbf{y}||^2 = ||\mathbf{x}||^2 + ||\mathbf{y}||^2\,.$$

    Sketch a parallelogram that illustrates this.
25. *(Proof Problem)* Prove the **polarization identity:** For all vectors $\mathbf{x}$ and $\mathbf{y}$ in $\mathbb{R}^N$,

$$||\mathbf{x}+\mathbf{y}||^2 - ||\mathbf{x}-\mathbf{y}||^2 = 4 \cdot \mathbf{x}^T\mathbf{y}\,.$$

26. *(Proof Problem)* (a) Prove the **parallelogram law:** For all vectors $\mathbf{x}$ and $\mathbf{y}$ in $\mathbb{R}^N$,

$$||\mathbf{x}+\mathbf{y}||^2 + ||\mathbf{x}-\mathbf{y}||^2 = 2\left(||\mathbf{x}||^2 + ||\mathbf{y}||^2\right). \tag{1.43}$$

   (b) Show that, when $\mathbf{x}$ and $\mathbf{y}$ are orthogonal, the parallelogram law (1.43) reduces to the *Pythagorean theorem*:

$$||\mathbf{x}+\mathbf{y}||^2 = ||\mathbf{x}||^2 + ||\mathbf{y}||^2 .$$

27. *(Proof Problem)* The **cross product** of two vectors $\mathbf{a}$ and $\mathbf{b}$ in $\mathbb{R}^3$ is defined by

$$\mathbf{a}\times\mathbf{b} = \left[\,(a_2b_3 - a_3b_2)\;(a_3b_1 - a_1b_3)\;(a_1b_2 - a_2b_1)\,\right]^T ,$$

   as in formula (1.36) from Example 1.28.

   (a) Verify that the vector $\mathbf{a}\times\mathbf{b}$ is orthogonal to both $\mathbf{a}$ and $\mathbf{b}$.
   (b) Verify that $\mathbf{b}\times\mathbf{a} = -\mathbf{a}\times\mathbf{b}$.
   (c) Verify that, for all constants r, $(r\cdot\mathbf{a})\times\mathbf{b} = r\cdot(\mathbf{a}\times\mathbf{b})$.
   (d) Verify that, for all vectors $\mathbf{a}$, $\mathbf{b}$, and $\mathbf{c}$ in $\mathbb{R}^3$,

$$\mathbf{a}\times(\mathbf{b}+\mathbf{c}) = \mathbf{a}\times\mathbf{b} + \mathbf{a}\times\mathbf{c}.$$

1.14 Projects

Project 1.1 (Using Correlation to Detect Structural Changes)
Table 1.6 shows aggregate data for disposable personal income and savings, in billions of $, for the years 1970–1993 in the United States.

In 1982, during Ronald Reagan's first term as President, the United States experienced its worst peacetime recession up to that point, including an official unemployment rate of 9.7%. In this project, we will explore how a change in the relationship between savings and disposable personal income might indicate the occurrence of a structural change like a recession.

1. Figure 1.16 shows a scatter plot for the data in Table 1.6 for the years 1970 to 1993. Discuss any differences you observe in the point clouds in the left and right halves of the plot.
2. Create separate scatter plots for the data from 1970–1981 and from 1982–1993. Do these two plots suggest different strengths in the correlation between savings and disposable personal income? Explain.

Table 1.6 Disposable personal income (x) and savings (y), for the United States, from 1970 to 1993 (in billions of $) (see Table 8.9 in [9])

Disposable personal income (x) **and savings** (y); **1970–1993** (in billions of $)

Year	1970	1971	1972	1973	1974	1975
x_i	727.1	790.2	855.3	965.0	1054.2	1159.2
y_i	61.0	68.6	63.6	89.6	97.6	104.4
Year	1976	1977	1978	1979	1980	1981
x_i	1273.0	1401.4	1580.1	1769.5	1973.3	2200.2
y_i	96.4	92.5	112.6	130.1	161.8	199.1
Year	1982	1983	1984	1985	1986	1987
x_i	2347.3	2522.4	2810.0	3002.0	3187.6	3363.1
y_i	205.5	167.0	235.7	206.2	196.5	168.4
Year	1988	1989	1990	1991	1992	1993
x_i	3640.8	3894.5	4166.8	4343.7	4613.7	4790.2
y_i	189.1	187.8	208.7	246.4	272.6	214.4

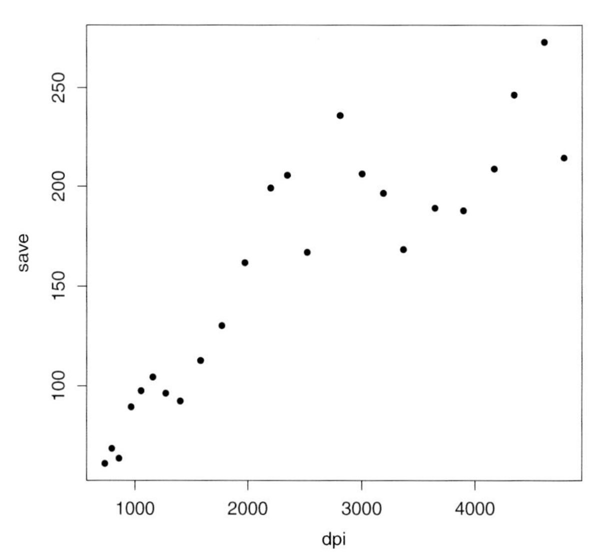

Fig. 1.16 The scatter plot displays savings (y-axis) and disposable personal income (x-axis) for the years 1970 to 1993 in the United States

3. Compute the Pearson correlation coefficient comparing savings to disposable personal income for the years 1970 to 1993.
4. Compute separate Pearson correlation coefficients comparing savings to disposable personal income for the years 1970 to 1981 and from 1982 to 1993.
5. Discuss whether the correlation coefficients you just computed support the observations you made about the scatter plots.
6. (optional) Find similar data for a period of years surrounding the 2008 financial collapse. Apply a similar analysis to see if a significant change in correlation occurred after 2008.

Project 1.2 (Categorized Climbs in the Tour de France)
In Example 1.17, we found a *Pearson correlation coefficient* close to 0 between the distance (in kilometers) and the average percent gradient for all categorized climbs from the 2018 Tour de France. We wondered if the correlation might be stronger within each individual category, since climbs in the same category are considered by the race organizers to present comparable overall difficulty for the riders. Table 1.7 shows the data for categories 1 through 4. For each separate category, carry out the following analysis:

1. Create a plot that shows the *distance* of each climb on the horizontal axis and the *average gradient* of each climb on the vertical axis.
2. From the plot you created, does there appear to be a weak correlation or a strong correlation between distance and gradient of climbs within the category? Explain your thinking.
3. Compute the Pearson correlation coefficient comparing the distance (in kilometers) and the average percent gradient for all climbs within the category.
4. Does the correlation coefficient just computed make sense to you based on the data plot you created? Explain your thinking.
5. Once you have analyzed all four categories separately, discuss your overall findings. Did the strength of the correlation change from category to category? If so, does that make sense to you? Were any of the correlation coefficients *negative* (less than 0)? If so, what does that mean? Does that make sense to you? Do you think some climbs may have been assigned the wrong category by the race organizers? Explain.

Table 1.7 Distance and gradient data for categorized climbs in the 2018 Tour de France

2018 Tour de France										
Category 1 climbs										
Dist. (km)	11.3	8.8	7.5	17.6	12.3	6.9	8.3	14.9	7.4	12
% gradient	7	8.9	8.5	5.8	6.3	8.1	7.1	6.7	8.3	6.5
Category 2 climbs										
Dist. (km)	8.6	5.4	3.0	9.1	3.4	5.7				
% gradient	5.8	7.1	10.2	5.3	8.2	6.5				
Category 3 climbs										
Dist. (km)	1.9	3	2.2	1.5	1.9	2.4	3.1			
% gradient	6.6	6.3	5.9	7	6.9	6.9	5.9			
Category 4 climbs										
Dist. (km)	1.0	0.8	1.6	1.7	2.5	1.5	2.3	2.0	1.5	
% gradient	3.9	7.8	5.6	7.1	3.5	3.9	4.3	4.3	5.6	
Dist. (km)	1.5	1.0	2.4	2.3	1.2	2.1	1.8	3.4		
% gradient	4.9	7.4	4.9	5.8	7.0	4.6	7.2	5.1		

Project 1.3 (Make Your Own Movie Rating Service)
This project is completely DIY. Generate a list of at least ten films and have at least ten friends rate them using the system $+1$ = like, -1 = do not like, 0 = did not see, as in Table 1.4, in Section 1.6. Then compute the cosines of the angles between everyone's rating vectors. Find out which friends have similar tastes and which friends can help each other to expand their horizons.

Chapter 2
Matrices

2.1 Matrices

Computational problems with vectors can be cumbersome as soon as more than just a few vectors are involved. For instance, we looked at the Pearson correlation coefficient for paired observations of two quantities. What if we have a set of observations of three, ten, or one hundred different quantities? How could we efficiently compute the correlation coefficients of all possible pairs of these quantities? Similarly, our basic method for information retrieval involves computing the cosine of the angle between a query vector and a set of document vectors. What if there are many documents and we wish to handle many queries simultaneously? For both of these situations, it is convenient to store all of the relevant vectors in a single array called a *matrix*.

Definition 2.1 A **matrix** is a rectangular array of numbers, arranged into rows and columns. An array with M rows and N columns is called an $M \times N$ matrix (read as "M-by-N"). For positive integers M and N, the set of all $M \times N$ matrices whose entries are real numbers is denoted by $\mathbb{M}_{M\times N}(\mathbb{R})$, or, more simply, by $\mathbb{M}_{M\times N}$. The plural of the word *matrix* is *matrices*.

It is customary to name a given matrix by an uppercase letter, such as A, B, or Q, and to enclose the entries of the matrix in square brackets [ ]. To label each individual entry in a matrix, the common practice is to use double subscripts indicating the row and column in which the entry appears, with the row label coming first. It is also common to name the matrix entries using the lowercase version of the matrix name. Thus, the label $a_{2,5}$ refers to the number in row #2 and column #5 of the matrix

---

Supplementary Information The online version contains supplementary material available at https://doi.org/10.1007/978-3-031-39562-8_2.

T. G. Feeman, *Applied Linear Algebra and Matrix Methods*,
Springer Undergraduate Texts in Mathematics and Technology,
https://doi.org/10.1007/978-3-031-39562-8_2

named A, while $a_{i,\,j}$ refers to the matrix entry in row i and column j. The notation $A = \left[a_{i,\,j}\right]$ means that A is a matrix whose individual entries are denoted by $a_{i,\,j}$. Programming languages and computational software often use the notation $A[i,\,j]$ to denote the entry in row i and column j of the matrix named A. This notation has the advantage of keeping the typing on one line, instead of using subscripts. We will often use this notation, especially when it adds clarity. The rows of a matrix are numbered from top to bottom, while the columns are numbered from left to right. Thus, for an $M \times N$ matrix, the $[1,\,1]$ entry is in the top left corner of the matrix and the $[M,\,N]$ entry is in the bottom right corner. A generic matrix A in $\mathbb{M}_{M\times N}$ can be depicted schematically as

$$A = \begin{bmatrix} a_{1,\,1} & a_{1,\,2} & \cdots & a_{1,\,N} \\ a_{2,\,1} & a_{2,\,2} & \cdots & a_{2,\,N} \\ \vdots & \vdots & \ddots & \vdots \\ a_{M,\,1} & a_{M,\,2} & \cdots & a_{M,\,N} \end{bmatrix}. \tag{2.1}$$

A set of vectors $\mathbf{a}_1$, $\mathbf{a}_2$, up to $\mathbf{a}_N$, all in $\mathbb{R}^M$, can be stored as the columns of an $M \times N$ matrix A. We depict this as

$$A = \left[\,\mathbf{a}_1 \middle| \mathbf{a}_2 \middle| \cdots \middle| \mathbf{a}_N\,\right].$$

In this situation, the entry in row i and column j of the matrix A corresponds to the ith coordinate of the (column) vector $\mathbf{a}_j$.

Remark Every column vector $\mathbf{a} \in \mathbb{R}^M$ is also an $M \times 1$ matrix. An entire $M \times N$ matrix can be converted into a single vector in $\mathbb{R}^{MN}$ by placing each column directly beneath the previous one to form a single long column with MN entries.

The arithmetic operations of addition and multiplication by a scalar are defined on the set $\mathbb{M}_{M\times N}(\mathbb{R})$, very much as with $\mathbb{R}^N$.

- *Addition.* We add two elements of $\mathbb{M}_{M\times N}(\mathbb{R})$ "entry by entry". Thus,

$$\begin{bmatrix} a_{1,\,1} & \cdots & a_{1,\,N} \\ a_{2,\,1} & \cdots & a_{2,\,N} \\ \vdots & \ddots & \vdots \\ a_{M,\,1} & \cdots & a_{M,\,N} \end{bmatrix} + \begin{bmatrix} b_{1,\,1} & \cdots & b_{1,\,N} \\ b_{2,\,1} & \cdots & b_{2,\,N} \\ \vdots & \ddots & \vdots \\ b_{M,\,1} & \cdots & b_{M,\,N} \end{bmatrix}$$

$$= \begin{bmatrix} a_{1,\,1} + b_{1,\,1} & \cdots & a_{1,\,N} + b_{1,\,N} \\ a_{2,\,1} + b_{2,\,1} & \cdots & a_{2,\,N} + b_{2,\,N} \\ \vdots & \ddots & \vdots \\ a_{M,\,1} + b_{M,\,1} & \cdots & a_{M,\,N} + b_{M,\,N} \end{bmatrix}.$$

That is, if $A = \left[a_{i,j}\right]$ and $B = \left[b_{i,j}\right]$, then $A + B = \left[(a_{i,j} + b_{i,j})\right]$. Written differently, $(A + B)[i, j] = A[i, j] + B[i, j]$. The sum is an element of $\mathbb{M}_{M \times N}(\mathbb{R})$.

- *Multiplication by a scalar.* Multiplication of a matrix $A \in \mathbb{M}_{M \times N}(\mathbb{R})$ by any real number r is computed by multiplying *every entry* of the matrix by r. Thus,

$$r \cdot \begin{bmatrix} a_{1,1} & \cdots & a_{1,N} \\ a_{2,1} & \cdots & a_{2,N} \\ \vdots & \ddots & \vdots \\ a_{M,1} & \cdots & a_{M,N} \end{bmatrix} = \begin{bmatrix} r \cdot a_{1,1} & \cdots & r \cdot a_{1,N} \\ r \cdot a_{2,1} & \cdots & r \cdot a_{2,N} \\ \vdots & \ddots & \vdots \\ r \cdot a_{M,1} & \cdots & r \cdot a_{M,N} \end{bmatrix}.$$

In other words, if $A = \left[a_{i,j}\right]$ is an $M \times N$ matrix and r is a real number, then $r \cdot A = \left[r \cdot a_{i,j}\right]$. That is, $(r \cdot A)[i, j] = r \cdot A[i, j]$. The result is again an $M \times N$ matrix.

Example 2.2 Let $A = \begin{bmatrix} 2 & 0 & -3 \\ -1 & 5 & 6 \end{bmatrix}$ and $B = \begin{bmatrix} -2 & 3 & 1 \\ 7 & 1 & 5 \end{bmatrix}$.

(i) $A + B = \begin{bmatrix} 2 + (-2) & 0 + 3 & -3 + 1 \\ -1 + 7 & 5 + 1 & 6 + 5 \end{bmatrix} = \begin{bmatrix} 0 & 3 & -2 \\ 6 & 6 & 11 \end{bmatrix}$.

(ii) $5 \cdot A = \begin{bmatrix} 5(2) & 5(0) & 5(-3) \\ 5(-1) & 5(5) & 5(6) \end{bmatrix} = \begin{bmatrix} 10 & 0 & -15 \\ -5 & 25 & 30 \end{bmatrix}$.

(iii) $-3 \cdot B = \begin{bmatrix} -3(-2) & -3(3) & -3(1) \\ -3(7) & -3(1) & -3(5) \end{bmatrix} = \begin{bmatrix} 6 & -9 & -3 \\ -21 & -3 & -15 \end{bmatrix}$.

(iv) $5 \cdot A - 3 \cdot B = \begin{bmatrix} 10 + 6 & 0 - 9 & -15 - 3 \\ -5 - 21 & 25 - 3 & 30 - 15 \end{bmatrix} = \begin{bmatrix} 16 & -9 & -18 \\ -26 & 22 & 15 \end{bmatrix}$.

Example 2.3 For $M \times N$ matrices

$$A = \left[\, \mathbf{a}_1 \middle| \mathbf{a}_2 \middle| \cdots \middle| \mathbf{a}_N \,\right] \text{ and } B = \left[\, \mathbf{b}_1 \middle| \mathbf{b}_2 \middle| \cdots \middle| \mathbf{b}_N \,\right],$$

and real numbers s and t,

$$s \cdot A + t \cdot B = \left[\, s \cdot \mathbf{a}_1 + t \cdot \mathbf{b}_1 \middle| s \cdot \mathbf{a}_2 + t \cdot \mathbf{b}_2 \middle| \cdots \middle| s \cdot \mathbf{a}_N + t \cdot \mathbf{b}_N \,\right].$$

Example 2.4 Suppose three Girl Scout troops are selling cookies. Each week, each troop tallies how many packages it sold of each type of cookie. We record this information in a matrix for each week. Suppose the sales matrices for three weeks are as follows:

Week #1: $W_1 =$

	Troop 1	Troop 2	Troop 3
trefoils	18	21	16
Thin mints	30	40	33
lemonades	24	23	19
Adventurefuls	28	30	33

;

Week #2: $W_2 =$

	Troop 1	Troop 2	Troop 3
Trefoils	25	25	18
Thin mints	35	37	41
Lemonades	31	30	22
Adventurefuls	31	19	31

;

Week #3: $W_3 =$

	Troop 1	Troop 2	Troop 3
Trefoils	21	18	15
Thin mints	27	30	32
Lemonades	23	21	18
Adventurefuls	20	15	27

.

In this context, the matrix $W_1 + W_2 + W_3$ shows the total number of packages sold by each troop of each type of cookie for the three weeks combined.

$W_1 + W_2 + W_3 =$

	Troop 1	Troop 2	Troop 3
Trefoils	64	64	49
Thin mints	92	107	106
Lemonades	78	74	59
Adventurefuls	79	64	91

.

Each package of cookies sells for \$5. Thus, the matrices $5 \cdot W_1$, $5 \cdot W_2$, and $5 \cdot W_3$ show the sales revenue for each troop for each type of cookie for the given week. Combining these, we get the total sales revenue for each troop for each type of cookie for all three weeks.

$$5 \cdot W_1 + 5 \cdot W_2 + 5 \cdot W_3 = 5 \cdot (W_1 + W_2 + W_3)$$

$=$

	Troop 1	Troop 2	Troop 3
Trefoils	320	320	245
Thin mints	460	535	530
Lemonades	390	370	295
Adventurefuls	395	320	455

.

2.1.1 Algebraic Properties of Matrix Arithmetic

The matrix operations of addition and scalar multiplication enjoy the same algebraic properties as vector arithmetic.

- The addition operation on $\mathbb{M}_{M \times N}(\mathbb{R})$ is both *associative* and *commutative*. That is, for arbitrary $M \times N$ matrices A, B, and C, we have

$$(A + B) + C = A + (B + C) ,$$

and

$$A + B = B + A .$$

- Scalar multiplication is *distributive* over addition. That is, for all scalars r and all $M \times N$ matrices A and B, $r \cdot (A + B) = r \cdot A + r \cdot B$, and for all scalars r and s and every matrix A, $(r + s) \cdot A = r \cdot A + s \cdot A$.
- Scalar multiplication is both *associative* and *commutative*. Thus, for all scalars r and s and every matrix A, we have

$$r (sA) = (rs)A = (sr)A = s(rA) .$$

- The matrix $\begin{bmatrix} 0 & \cdots & 0 \\ \vdots & \ddots & \vdots \\ 0 & \cdots & 0 \end{bmatrix}$ that has all of its entries equal to 0 is called the *zero matrix* and is denoted by $\mathbf{0}$. The zero matrix acts as an *additive identity*, meaning that, for every matrix A, we have $A + \mathbf{0} = \mathbf{0} + A = A$. Notice that each of the spaces $\mathbb{M}_{M\times N}(\mathbb{R})$ technically has its own zero matrix, since the shape of the array is different for different values of M and N.
- Every matrix has an *additive inverse*, since $A + (-1) \cdot A = \mathbf{0}$. Thus, $(-1) \cdot A$, which we also write as $-A$, is the additive inverse of A.
- The number 1 is a *scalar multiplicative identity*, since $1 \cdot A = A$ for every matrix A.

The fact that $\mathbb{R}^N$ and $\mathbb{M}_{M\times N}(\mathbb{R})$ both come equipped with the arithmetic of addition and multiplication by a scalar and that those operations share the other properties listed above means that both of these sets are examples of *vector spaces*, a term that has acquired broader meaning over the years to include any set of objects that can be equipped with this type of arithmetic structure. Other examples of vector spaces include the set of all continuous real-valued functions on the interval $[0, 1]$, the set of all absolutely summable sequences of real numbers, and the set of all solutions to a homogeneous linear differential equation.

2.2 Matrix Multiplication

To add matrices or to multiply a matrix by a scalar, we imitate the same arithmetic operations for vectors. In the same vein, we use the *inner product* of vectors to formulate a general method for multiplying two matrices together.

To see how, consider two examples.

Example 2.5 Look at Exercise 10 in Chapter 1, Section 1.13. The numerical portion of each column of Table 1.5 is a document vector for one of five documents using a library of five terms. Each part of the exercise asks us to compute the *cosine* of the angle between each of these document vectors and a given query vector. To do that, we must compute the inner product of each document vector with the given query vector. For convenience, label the document vectors as $\mathbf{B_1}$ through $\mathbf{B_5}$ and the three query vectors from the different parts of the problem as $\mathbf{q}_1$, $\mathbf{q}_2$, and $\mathbf{q}_3$. Using our definition 1.9 of inner product, we compute, for instance,

$$\mathbf{B_1}^T\mathbf{q}_1 = \begin{bmatrix} 0 & 1 & 1 & 0 & 0 \end{bmatrix} \cdot \begin{bmatrix} 0 \\ 1 \\ 1 \\ 0 \\ 0 \end{bmatrix} = 0 \cdot 0 + 1 \cdot 1 + 1 \cdot 1 + 0 \cdot 0 + 0 \cdot 0 = 2\,.$$

To record in one tableau the inner products of all five document vectors with the query vector $\mathbf{q}_1$, we form the matrix whose *rows* are $\mathbf{B_1}^T$, at the top, through $\mathbf{B_5}^T$, at the bottom. Then we compute the inner products and record the answers like so.

$$\begin{bmatrix} \mathbf{B_1}^T\mathbf{q}_1 \\ \mathbf{B_2}^T\mathbf{q}_1 \\ \mathbf{B_3}^T\mathbf{q}_1 \\ \mathbf{B_4}^T\mathbf{q}_1 \\ \mathbf{B_5}^T\mathbf{q}_1 \end{bmatrix} = \begin{bmatrix} 0 & 1 & 1 & 0 & 0 \\ 1 & 1 & 1 & 1 & 0 \\ 1 & 1 & 1 & 0 & 1 \\ 0 & 0 & 1 & 1 & 0 \\ 0 & 1 & 0 & 0 & 1 \end{bmatrix} \cdot \begin{bmatrix} 0 \\ 1 \\ 1 \\ 0 \\ 0 \end{bmatrix} = \begin{bmatrix} 2 \\ 2 \\ 2 \\ 1 \\ 1 \end{bmatrix}. \tag{2.2}$$

To compute all of the necessary inner products for all three parts of the problem and record these results in a single tableau, we also form a matrix whose *columns* are the three query vectors. Then repeat the process from (2.2) column by column. In this way, we will generate a 5×3 matrix where the entry in the ith row and jth column is the inner product of the ith row vector $\mathbf{B_i}^T$ and the jth column vector $\mathbf{q}_j$. It looks like this.

$$\begin{bmatrix} 0 & 1 & 1 & 0 & 0 \\ 1 & 1 & 1 & 1 & 0 \\ 1 & 1 & 1 & 0 & 1 \\ 0 & 0 & 1 & 1 & 0 \\ 0 & 1 & 0 & 0 & 1 \end{bmatrix} \cdot \begin{bmatrix} 0 & 0 & 1 \\ 1 & 1 & 0 \\ 1 & 0 & 0 \\ 0 & 0 & 1 \\ 0 & 1 & 0 \end{bmatrix} = \begin{bmatrix} 2 & 1 & 0 \\ 2 & 1 & 2 \\ 2 & 2 & 1 \\ 1 & 0 & 1 \\ 1 & 2 & 0 \end{bmatrix}. \tag{2.3}$$

For example, the [3, 2] entry (in the 3rd row and 2nd column) on the right-hand side of (2.3) is

$$\mathbf{B_3}^T\mathbf{q}_2 = \begin{bmatrix} 1 & 1 & 1 & 0 & 1 \end{bmatrix} \cdot \begin{bmatrix} 0 \\ 1 \\ 0 \\ 0 \\ 1 \end{bmatrix} = 2\,.$$

To generalize this process, suppose we have an $M \times K$ matrix A and a $K \times N$ matrix B. Then the matrix product, denoted AB, is the $M \times N$ matrix whose $[i,\ j]$ entry is equal to the inner product of the ith row of A with the jth column of B. Notice that the number K of columns of A *must be the same as* the number K of rows of B. Otherwise, we cannot compute the inner products of row vectors with column vectors.

Example 2.6 This notion of matrix multiplication, where we compute inner products of the rows of one matrix with the columns of the other, has a geometric interpretation. For example, take two nonzero, noncollinear vectors

$$\mathbf{a}_1 = \begin{bmatrix} a_{1,1} \\ a_{2,1} \\ \vdots \\ a_{M,1} \end{bmatrix} \text{ and } \mathbf{a}_2 = \begin{bmatrix} a_{1,2} \\ a_{2,2} \\ \vdots \\ a_{M,2} \end{bmatrix} \text{ in } \mathbb{R}^M\,.$$

Each of these vectors generates a line in $\mathbb{R}^M$, and the two lines together generate a plane. Every vector in this plane has the form $s \cdot \mathbf{a}_1 + t \cdot \mathbf{a}_2$, for some numbers s and t. See Figure 2.1. Expressed using coordinates,

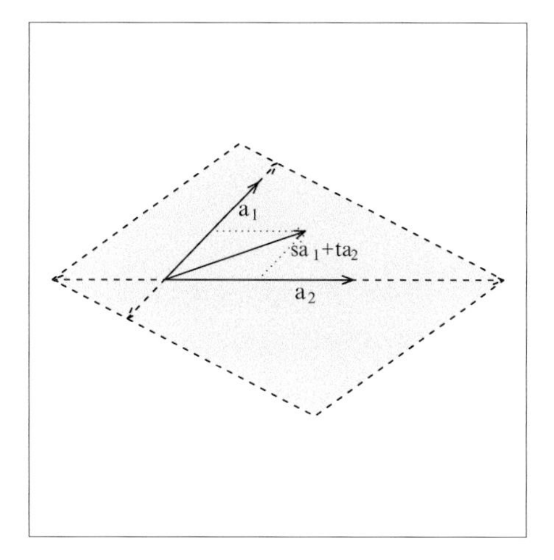

Fig. 2.1 Two vectors in $\mathbb{R}^M$ generate a plane

$$
\begin{aligned}
s\cdot\mathbf{a}_1+t\cdot\mathbf{a}_2 &= s\cdot\begin{bmatrix} a_{1,1}\\ a_{2,1}\\ \vdots\\ a_{M,1}\end{bmatrix}+t\cdot\begin{bmatrix} a_{1,2}\\ a_{2,2}\\ \vdots\\ a_{M,2}\end{bmatrix}\\
&= \begin{bmatrix} s\cdot a_{1,1}+t\cdot a_{1,2}\\ s\cdot a_{2,1}+t\cdot a_{2,2}\\ \vdots\\ s\cdot a_{M,1}+t\cdot a_{M,2}\end{bmatrix}\\
&= \begin{bmatrix} a_{1,1} & a_{1,2}\\ a_{2,1} & a_{2,2}\\ \vdots & \vdots\\ a_{M,1} & a_{M,2}\end{bmatrix}\begin{bmatrix} s\\ t\end{bmatrix}\\
&= \left[\mathbf{a}_1 \middle| \mathbf{a}_2\right]\begin{bmatrix} s\\ t\end{bmatrix}.
\end{aligned}
$$

In other words, every vector in the plane generated by $\mathbf{a}_1$ and $\mathbf{a}_2$ can be expressed as the product of the $M \times 2$ matrix A, whose columns are $\mathbf{a}_1$ and $\mathbf{a}_2$, with a 2×1 column vector $\begin{bmatrix} s\\ t\end{bmatrix}$, whose entries are the respective scalar multiples.

Here is the formal definition of matrix multiplication.

Definition 2.7 (Matrix Multiplication) Suppose A is an $M \times K$ matrix and B is a $K \times N$ matrix. In particular, the number of columns of A must be the same as the number of rows of B. Then the **matrix product** of A and B is the $M \times N$ matrix, denoted AB, whose $[i,\ j]$ entry, in row i and column j, is the inner product of the ith row vector of A and the jth column vector of B. That is, for $1 \le i \le M$ and $1 \le j \le N$,

$$
\begin{aligned}
(AB)\,[i,\ j] &= \begin{bmatrix} a_{i,1}\ a_{i,2}\ \cdots\ a_{i,K}\end{bmatrix}\begin{bmatrix} b_{1,j}\\ b_{2,j}\\ \vdots\\ b_{K,j}\end{bmatrix}\\
&= a_{i,1}b_{1,j}+a_{i,2}b_{2,j}+\cdots+a_{i,K}b_{K,j}\,. \qquad (2.4)
\end{aligned}
$$

Remark The expression in (2.4) can be abbreviated using "$\sum$" summation notation as

$$
(AB)\,[i,\ j] = \sum_{k=1}^{K} a_{i,k}b_{k,j}\,.
$$

Just as the inner product opened the door to interesting applications of vector geometry, this method of multiplying matrices will open up a vast array of powerful applications for analyzing and manipulating large data sets. In the rest of this chapter and in the next chapter, we begin to explore a variety of contexts in which matrix multiplication provides useful information and helps to answer deep questions.

Example 2.8 Suppose we have just two vectors, **a** and **b**, in $\mathbb{R}^N$, for some N. Laying the vector **a** on its side gives us the transpose vector $\mathbf{a}^T$. This is a $1 \times N$ matrix:

$$\mathbf{a}^T = \left[a_1 \; a_2 \; \cdots \; a_N \right].$$

The other vector **b** is an $N \times 1$ matrix, with N rows and just one column. We now identify the vector inner product $\mathbf{a}^T\mathbf{b}$ with the *matrix product* of a $1 \times N$ matrix times an $N \times 1$ matrix. Namely,

$$\mathbf{a}^T\mathbf{b} = \left[a_1 \; a_2 \; \cdots \; a_N \right] \begin{bmatrix} b_1 \\ b_2 \\ \vdots \\ b_N \end{bmatrix}$$

$$= [a_1 b_1 + a_2 b_2 + \cdots a_N b_N].$$

Until now, we interpreted the inner product of two vectors as a number. We now see that our notation $\mathbf{a}^T\mathbf{b}$ also corresponds to a 1×1 matrix. When it matters that we are dealing with a 1×1 matrix, the square bracket notation [ ] will be used. But we will also blur the distinction when it is convenient to do so. To see why this is okay, consider this computation:

$$\begin{bmatrix} u_1 \\ u_2 \\ \vdots \\ u_N \end{bmatrix} \cdot [r] = \begin{bmatrix} r \cdot u_1 \\ r \cdot u_2 \\ \vdots \\ r \cdot u_N \end{bmatrix} = r \cdot \begin{bmatrix} u_1 \\ u_2 \\ \vdots \\ u_N \end{bmatrix}. \tag{2.5}$$

Thus, we will equate $\mathbf{u}\left[\mathbf{a}^T\mathbf{b}\right] = \left(\mathbf{a}^T\mathbf{b}\right) \cdot \mathbf{u}$, where, on the left-hand side, $\left[\mathbf{a}^T\mathbf{b}\right]$ is a 1×1 matrix, and, on the right, $\left(\mathbf{a}^T\mathbf{b}\right)$ is a number.

Example 2.9 As a purely computational example, take

$$A = \begin{bmatrix} 1 & 2 & 3 \\ -2 & 5 & 3 \end{bmatrix} \text{ and } B = \begin{bmatrix} 7 & 4 \\ -2 & 3 \\ 1 & 5 \end{bmatrix}.$$

Notice that A has 3 columns and B has 3 rows. Thus, the matrix product AB is defined. In this case, AB is the 2×2 matrix computed like so.

$$AB = \begin{bmatrix} \begin{bmatrix} 1 & 2 & 3 \end{bmatrix} \begin{bmatrix} 7 \\ -2 \\ 1 \end{bmatrix} & \begin{bmatrix} 1 & 2 & 3 \end{bmatrix} \begin{bmatrix} 4 \\ 3 \\ 5 \end{bmatrix} \\ \begin{bmatrix} -2 & 5 & 3 \end{bmatrix} \begin{bmatrix} 7 \\ -2 \\ 1 \end{bmatrix} & \begin{bmatrix} -2 & 5 & 3 \end{bmatrix} \begin{bmatrix} 4 \\ 3 \\ 5 \end{bmatrix} \end{bmatrix}$$

$$= \begin{bmatrix} 7 - 4 + 3 & 4 + 6 + 15 \\ -14 - 10 + 3 & -8 + 15 + 15 \end{bmatrix}$$

$$= \begin{bmatrix} 6 & 25 \\ -21 & 22 \end{bmatrix}.$$

Example 2.10 We saw in Example 2.6 that the product of an $M \times 2$ matrix times a column vector in $\mathbb{R}^2$ is a linear combination of the column vectors of the $M \times 2$ matrix. More generally, suppose $A = \left[\mathbf{a}_1 \middle| \cdots \middle| \mathbf{a}_K \right]$ is an $M \times K$ matrix whose columns are vectors in $\mathbb{R}^M$, and $\mathbf{s}$ is a column vector in $\mathbb{R}^K$. Then

$$A\mathbf{s} = \left[\mathbf{a}_1 \middle| \cdots \middle| \mathbf{a}_K \right] \begin{bmatrix} s_1 \\ \vdots \\ s_K \end{bmatrix} = s_1 \cdot \mathbf{a}_1 + \cdots + s_K \cdot \mathbf{a}_K .$$

This vector is a linear combination of the column vectors of A. By considering all possible vectors $\mathbf{s}$ in $\mathbb{R}^K$, we obtain every possible such combination of the column vectors of A. That is, we obtain the *span* of the collection of column vectors of A. We call this the **column space** of the matrix A, denoted $Col(A)$.

Definition 2.11 The **column space** of a matrix A, denoted $Col(A)$, is the set of all linear combinations of the column vectors of A. That is, $Col(A)$ is the span of the column vectors of A.

In the most general setting, where A is as above and $S = \left[\mathbf{s}_1 \middle| \cdots \middle| \mathbf{s}_N \right]$ is a $K \times N$ matrix whose columns are vectors in $\mathbb{R}^K$, then AS is an $M \times N$ matrix whose jth column is equal to $A\mathbf{s}_j$. Thus, every column of AS is a linear combination of the columns of A with the scalars coming from the corresponding column of S.

Example 2.12 To illustrate the geometry just described, let

$$A = \left[\mathbf{a}_1 \middle| \mathbf{a}_2 \middle| \mathbf{a}_3\right] = \begin{bmatrix} 1 & -1 & 1 \\ 2 & 4 & 2 \\ 3 & 6 & 0 \\ 5 & -3 & 2 \end{bmatrix},$$

and

$$S = \left[\mathbf{s}_1 \middle| \mathbf{s}_2\right] = \begin{bmatrix} 1 & 4 \\ 2 & 1 \\ -1 & -1 \end{bmatrix}.$$

Then

$$AS = \left[A\mathbf{s}_1 \middle| A\mathbf{s}_2\right] = \begin{bmatrix} -2 & 2 \\ 8 & 10 \\ 15 & 18 \\ -12 & 15 \end{bmatrix}.$$

The first column of AS is the same as $1 \cdot \mathbf{a}_1 + 2 \cdot \mathbf{a}_2 - 1 \cdot \mathbf{a}_3$, while the second column of AS is equal to $4 \cdot \mathbf{a}_1 + 1 \cdot \mathbf{a}_2 - 1 \cdot \mathbf{a}_3$.

Example 2.13 Suppose we have an $M \times K$ matrix A, a $K \times N$ matrix B, and a vector $\mathbf{x}$ in $\mathbb{R}^N$. Certainly, we would like to have $A(B\mathbf{x}) = (AB)\mathbf{x}$, which is a form of the associative property for matrix multiplication. Let us check this for a small example.

Take $A = \begin{bmatrix} 1 & 2 & 3 \\ -2 & 5 & 3 \end{bmatrix}$ and $B = \begin{bmatrix} 7 & 4 \\ -2 & 3 \\ 1 & 5 \end{bmatrix}$. Then

$$B \begin{bmatrix} x_1 \\ x_2 \end{bmatrix} = \begin{bmatrix} 7 & 4 \\ -2 & 3 \\ 1 & 5 \end{bmatrix} \begin{bmatrix} x_1 \\ x_2 \end{bmatrix} = \begin{bmatrix} 7x_1 + 4x_2 \\ -2x_1 + 3x_2 \\ x_1 + 5x_2 \end{bmatrix}.$$

This leads us to

$$A\left(B\begin{bmatrix} x_1 \\ x_2 \end{bmatrix}\right) = \begin{bmatrix} 1 & 2 & 3 \\ -2 & 5 & 3 \end{bmatrix} \begin{bmatrix} 7x_1 + 4x_2 \\ -2x_1 + 3x_2 \\ x_1 + 5x_2 \end{bmatrix}$$

$$= \begin{bmatrix} 1 \cdot (7x_1 + 4x_2) + 2 \cdot (-2x_1 + 3x_2) + 3 \cdot (x_1 + 5x_2) \\ -2 \cdot (7x_1 + 4x_2) + 5 \cdot (-2x_1 + 3x_2) + 3 \cdot (x_1 + 5x_2) \end{bmatrix}$$

$$= \begin{bmatrix} (7-4+3)\cdot x_1 + (4+6+15)\cdot x_2 \\ (-14-10+3)\cdot x_1 + (-8+15+15)\cdot x_2 \end{bmatrix}$$

$$= \begin{bmatrix} 6\cdot x_1 + 25\cdot x_2 \\ -21\cdot x_1 + 22\cdot x_2 \end{bmatrix}$$

$$= \begin{bmatrix} 6 & 25 \\ -21 & 22 \end{bmatrix} \begin{bmatrix} x_1 \\ x_2 \end{bmatrix}.$$

Thus, in order to have $A(B\mathbf{x}) = (AB)\mathbf{x}$, we need

$$AB = \begin{bmatrix} 1 & 2 & 3 \\ -2 & 5 & 3 \end{bmatrix} \begin{bmatrix} 7 & 4 \\ -2 & 3 \\ 1 & 5 \end{bmatrix} = \begin{bmatrix} 6 & 25 \\ -21 & 22 \end{bmatrix}.$$

Sure enough, this is the result of Example 2.9.

In general, with $A = [a_{i,\,k}]$ and $B = [b_{k,\,j}]$, we get

$$B\mathbf{x} = \begin{bmatrix} b_{1,\,1}x_1 + b_{1,\,2}x_2 + \cdots + b_{1,\,N}x_N \\ b_{2,\,1}x_1 + b_{2,\,2}x_2 + \cdots + b_{2,\,N}x_N \\ \vdots \\ b_{K,\,1}x_1 + b_{K,\,2}x_2 + \cdots + b_{K,\,N}x_N \end{bmatrix}.$$

Computing the inner product of this with the ith row of A gives us the expression

$$\begin{aligned} & a_{i,\,1}(b_{1,\,1}x_1 + b_{1,\,2}x_2 + \cdots + b_{1,\,N}x_N) \\ & + a_{i,\,2}(b_{2,\,1}x_1 + b_{2,\,2}x_2 + \cdots + b_{2,\,N}x_N) \\ & + \cdots + a_{i,\,K}(b_{K,\,1}x_1 + b_{K,\,2}x_2 + \cdots + b_{K,\,N}x_N) \end{aligned}$$

$$\begin{aligned} & = (a_{i,\,1}b_{1,1} + a_{i,\,2}b_{2,1} + \cdots + a_{i,\,K}b_{K,1})\cdot x_1 \\ & + (a_{i,\,1}b_{1,2} + a_{i,\,2}b_{2,2} + \cdots + a_{i,\,K}b_{K,2})\cdot x_2 \\ & + \cdots + (a_{i,\,1}b_{1,N} + a_{i,\,2}b_{2,N} + \cdots + a_{i,\,K}b_{K,N})\cdot x_N\,. \end{aligned}$$

The coefficient of x_j in this last expression is exactly the inner product of the ith row of A with the jth column of B, which is the $[i,\ j]$ entry of the matrix product AB. It follows that the ith entry of $A(B\mathbf{x})$ is the same as the ith entry of $(AB)\mathbf{x}$. Hence, $A(B\mathbf{x}) = (AB)\mathbf{x}$, as we hoped.

2.2.1 Algebraic Properties of Matrix Multiplication

When we apply the computation of Example 2.13 column by column to matrices A and B and $C = \left[\, \mathbf{x}_1 \middle| \cdots \middle| \mathbf{x}_J \,\right]$, all of suitable dimensions, we get

$$\begin{aligned} A(BC) &= A(B\left[\, \mathbf{x}_1 \middle| \cdots \middle| \mathbf{x}_J \,\right]) \\ &= A(\left[\, B\mathbf{x}_1 \middle| \cdots \middle| B\mathbf{x}_J \,\right] \\ &= \left[\, A(B\mathbf{x}_1) \middle| \cdots \middle| A(B\mathbf{x}_J) \,\right] \\ &= \left[\, (AB)\mathbf{x}_1 \middle| \cdots \middle| (AB)\mathbf{x}_J \,\right] \\ &= (AB)C\,. \end{aligned}$$

Thus, $A(BC) = (AB)C$ for all matrices of suitable dimensions. This is the *associative property* for matrix multiplication.

The *distributive property* for matrix multiplication and addition asserts that $A(B + C) = AB + AC$ and $(A + B)C = AC + BC$, whenever the matrices have suitable dimensions. The reader is encouraged to check these claims. Notice that, in both cases, the two matrices being added must have the same size array.

Unlike numerical multiplication, matrix multiplication is not commutative. That is, the products AB and BA need not be the same! Indeed, these products might not both make sense, or they might have different dimensions. For example, if A is 2×3 and B is 3×4, then AB is 2×4 but BA is not defined. If A is 2×3 and B is 3×2, then AB is 2×2, but BA is 3×3. Even if A and B are square matrices of the same size, the products AB and BA can be different. This is to be expected because, to compute AB, we form inner products of rows of A with columns of B. The product BA reverses that, using the rows of B and the columns of A. So we should not be surprised if the results are different.

Example 2.14 The following examples illustrate some of the ways in which matrix multiplication differs from numerical multiplication.

(i) Let $A = \begin{bmatrix} 1 & -1 & 1 \\ 1 & 1 & 1 \\ 4 & 2 & 1 \end{bmatrix}$ and $B = \begin{bmatrix} 3 & -1 \\ 2 & 4 \\ 0 & 5 \end{bmatrix}$. Then

$$AB = \begin{bmatrix} 1 & -1 & 1 \\ 1 & 1 & 1 \\ 4 & 2 & 1 \end{bmatrix} \begin{bmatrix} 3 & -1 \\ 2 & 4 \\ 0 & 5 \end{bmatrix} = \begin{bmatrix} 1 & 0 \\ 5 & 8 \\ 16 & 9 \end{bmatrix}.$$

The product BA is not defined since B has 2 columns but A has 3 rows.

(ii) With $C = \begin{bmatrix} 3 & 5 & 1 \\ 1 & 2 & 1 \\ 2 & 6 & 7 \end{bmatrix}$ and $D = \begin{bmatrix} 1 & -1 & 2 \\ 2 & 1 & 11 \\ 4 & -3 & 10 \end{bmatrix}$, we get

$$CD = \begin{bmatrix} 17 & -1 & 71 \\ 9 & -2 & 34 \\ 42 & -17 & 140 \end{bmatrix}, \text{ but } DC = \begin{bmatrix} 6 & 15 & 14 \\ 29 & 78 & 80 \\ 29 & 74 & 71 \end{bmatrix}.$$

Thus, the products CD and DC are both defined and even have the same dimensions, but $CD \neq DC$.

(iii) Let $T = \begin{bmatrix} 1 & 2 & 1 \\ -2 & 0 & -6 \\ 5 & -3 & 18 \end{bmatrix}$, $U_1 = \begin{bmatrix} 4 & -1 \\ 3 & -5 \\ 2 & 0 \end{bmatrix}$, and $U_2 = \begin{bmatrix} 1 & 5 \\ 4 & -7 \\ 3 & -2 \end{bmatrix}$. We can calculate that

$$TU_1 = TU_2 = \begin{bmatrix} 12 & -11 \\ -20 & 2 \\ 47 & 10 \end{bmatrix}.$$

That means that, if we let $U_3 = U_1 - U_2 = \begin{bmatrix} 3 & -6 \\ -1 & 2 \\ -1 & 2 \end{bmatrix}$, then

$$TU_3 = T\,(U_1 - U_2) = TU_1 - TU_2 = \begin{bmatrix} 0 & 0 \\ 0 & 0 \\ 0 & 0 \end{bmatrix}.$$

Therefore, $TU_1 = TU_2$ but $U_1 \neq U_2$. This shows that we may not simply "cancel" a matrix on both sides of an equation. Also, $TU_3 = \mathbf{0}$ even though $T \neq \mathbf{0}$ and $U_3 \neq \mathbf{0}$. In other words, the $\mathbf{0}$ matrix can be factored into nonzero factors. Of course, the number 0 cannot be factored like this using regular real or complex number multiplication. In the language of abstract algebra, we say that matrix multiplication admits **zero divisors**.

2.3 The Identity Matrix, $\mathcal{I}$

The number 1 has the special property that $1 \cdot x = x$ for every number x. Suppose now that $\mathbf{x} = \begin{bmatrix} x_1 & x_2 & \cdots & x_N \end{bmatrix}^T$ is a vector in $\mathbb{R}^N$. Is there a *matrix*, call it $\mathcal{I}$, for which $\mathcal{I}\mathbf{x} = \mathbf{x}$? For this equation to make sense, $\mathcal{I}$ has to be a square matrix of size $N \times N$. Next, if $\mathcal{I}$ has entries $[\mathcal{I}_{i,\,j}]$, then, for each i from 1 to N, we require

$$\mathcal{I}_{i,\,1}x_1 + \mathcal{I}_{i,\,2}x_2 + \cdots \mathcal{I}_{i,\,N}x_N = x_i\,.$$

We want this to work for *every vector* $\mathbf{x}$ in $\mathbb{R}^N$. In particular, suppose $\mathbf{x}$ has coordinate 1 in the jth entry and all other coordinates equal to 0. Then, for each i between 1 and N,

$$\mathcal{I}_{i,1}x_1 + \mathcal{I}_{i,2}x_2 + \cdots \mathcal{I}_{i,N}x_N = \mathcal{I}_{i,j}x_j = \mathcal{I}_{i,j} \,.$$

But we require that this is equal to x_i. Thus, we find that $\mathcal{I}_{i,j} = 0$ when $i \neq j$ and $\mathcal{I}_{j,j} = 1$. This holds for every value of i. In short, $\mathcal{I}$ is an $N \times N$ matrix that has all 1s on the *main diagonal*, where the row number and column number are the same, and all 0s everywhere else. We call this the **identity matrix**. Technically, there is *a different identity matrix* $\mathcal{I}$ for each N. They all have the same format, with 1s in the diagonal entries and 0s everywhere else.

Definition 2.15 For each natural number N, the $N \times N$ **identity matrix** is defined by

$$\mathcal{I}_N = \begin{bmatrix} 1 & 0 & \cdots & 0 \\ 0 & 1 & \cdots & 0 \\ \vdots & \vdots & \ddots & \vdots \\ 0 & 0 & \cdots & 1 \end{bmatrix} \tag{2.6}$$

That is, for all i and j from 1 to N, the $[i, j]$ entry of the identity matrix $\mathcal{I}_N$ is given by

$$\mathcal{I}_N[i, j] = \begin{cases} 1 & \text{if } i = j; \text{ and} \\ 0 & \text{if } i \neq j. \end{cases} \tag{2.7}$$

We usually write simply $\mathcal{I}$ when the value of N is clear from the context.

The defining property of $\mathcal{I}_N$ is that $\mathcal{I}_N\mathbf{x} = \mathbf{x}$, for every vector $\mathbf{x}$ in $\mathbb{R}^N$. It follows that $\mathcal{I}_N X = X$ for every matrix X that has N rows, regardless of the number of columns. It is also true that $X\mathcal{I}_N = X$, for every $M \times N$ matrix X, as the reader should verify. In the case where X is a square $N \times N$ matrix, then the $N \times N$ identity matrix $\mathcal{I}_N$ satisfies both $\mathcal{I}_N X = X$ and $X\mathcal{I}_N = X$.

The individual column vectors of $\mathcal{I}_N$ are denoted by $\mathbf{e}_1, \ldots, \mathbf{e}_N$. Thus, $\mathbf{e}_j$ has a 1 in the jth coordinate and 0s in all other coordinates. These vectors, whose exact expression depends on N, are often called the **standard basis vectors** for $\mathbb{R}^N$. Notice that, for any vector $\mathbf{a}$ in $\mathbb{R}^N$ with coordinates $[a_j]$, we have

$$\mathbf{a} = \mathcal{I}_N\mathbf{a} = a_1\mathbf{e}_1 + \cdots a_N\mathbf{e}_N \,.$$

2.4 Matrix Inverses

Two numbers a and b are called *multiplicative inverses* if their product $a \cdot b = 1$. In that same vein, we consider the possibility that two matrices A and B have the product $AB = \mathcal{I}$, the identity matrix. We have already seen that one product AB might make sense while the other product BA is either not defined or gives a different output. To avoid these complications, we restrict our attention to square matrices and make the following definition.

Definition 2.16 An $N \times N$ matrix A is said to be **invertible** if there is an $N \times N$ matrix B for which

$$AB = BA = \mathcal{I}_N \,. \tag{2.8}$$

Such a matrix B is called the **inverse** of A and is denoted by A^{-1}.

Notice that, if A is invertible with inverse B, then B is also invertible with inverse A. Thus, $\left(A^{-1}\right)^{-1} = A$ whenever A is invertible.

To justify referring to A^{-1} as *the* inverse of A, we should verify that a matrix cannot have two different inverses. To see this, suppose temporarily that two matrices B and C both serve as inverses for a matrix A. Then,

$$C = C\mathcal{I} = C(AB) = (CA)B = \mathcal{I}B = B \,.$$

In other words, $C = B$, so the inverse really is unique, if it exists at all.

In Chapter 4, we explore how to decide whether or not a given matrix is invertible and how to find its inverse if it is.

Example 2.17 For a 2×2 matrix, the existence (or not) of an inverse can be determined fairly easily. Given a generic 2×2 matrix $\begin{bmatrix} a & b \\ c & d \end{bmatrix}$, we wish to find a matrix $\begin{bmatrix} w & x \\ y & z \end{bmatrix}$ such that

$$\begin{bmatrix} a & b \\ c & d \end{bmatrix} \begin{bmatrix} w & x \\ y & z \end{bmatrix} = \begin{bmatrix} 1 & 0 \\ 0 & 1 \end{bmatrix} .$$

This implies the four equations

$$\begin{cases} aw + by = 1 \\ cw + dy = 0 \end{cases} \qquad \begin{cases} ax + bz = 0 \\ cx + dz = 1 \end{cases} .$$

The equation $cw + dy = 0$ is satisfied by taking $w = d$ and $y = -c$. In that case, we have $aw + by = ad - bc$. We want to have $aw + by = 1$. As long as $ad - bc \neq 0$, we can achieve this by setting $w = d/(ad - bc)$ and $y = -c/(ad - bc)$ instead.

Similarly, with $x = -b/(ad - bc)$ and $z = a/(ad - bc)$, we get $ax + bz = 0$ and $cx + dz = 1$, as desired.

Thus, the matrix

$$\begin{bmatrix} w & x \\ y & z \end{bmatrix} = \begin{bmatrix} d/(ad-bc) & -b/(ad-bc) \\ -c/(ad-bc) & a/(ad-bc) \end{bmatrix} = \frac{1}{ad-bc} \cdot \begin{bmatrix} d & -b \\ -c & a \end{bmatrix}$$

does what we want. The reader should verify that this also gives

$$\begin{bmatrix} w & x \\ y & z \end{bmatrix} \begin{bmatrix} a & b \\ c & d \end{bmatrix} = \mathcal{I}_2 .$$

Referring back to Example 1.27, the quantity $|ad - bc|$ represents the area of the parallelogram formed by the vectors $\begin{bmatrix} a \\ c \end{bmatrix}$ and $\begin{bmatrix} b \\ d \end{bmatrix}$. When $ad - bc = 0$, these vectors are collinear (or $\mathbf{0}$) and do not generate a proper parallelogram. Collinearity means that there is some number t where $\begin{bmatrix} a \\ c \end{bmatrix} = t \cdot \begin{bmatrix} b \\ d \end{bmatrix}$. This gives us

$$\begin{bmatrix} a & b \\ c & d \end{bmatrix} \cdot \begin{bmatrix} 1 \\ -t \end{bmatrix} = \begin{bmatrix} 0 \\ 0 \end{bmatrix} .$$

In Example 2.21 below, we will see that this means that $\begin{bmatrix} a & b \\ c & d \end{bmatrix}$ has no inverse.

To sum up, we have the following rule.

Formula 2.18 The 2×2 matrix $\begin{bmatrix} a & b \\ c & d \end{bmatrix}$ is invertible if, and only if, $ad - bc \neq 0$. In that case, the inverse is given by

$$\begin{bmatrix} a & b \\ c & d \end{bmatrix}^{-1} = \frac{1}{ad-bc} \cdot \begin{bmatrix} d & -b \\ -c & a \end{bmatrix} . \tag{2.9}$$

So, to produce the inverse matrix, we switch the two diagonal entries (a and d), change the signs of the other entries (b and c), and divide the whole thing by $ad-bc$.

Matrix inverses interact nicely with scalar multiplication and matrix products. First, suppose A is an invertible matrix and r is a nonzero number. Then the matrix $r \cdot A$ is also invertible, and

$$(r \cdot A)^{-1} = (1/r) \cdot A^{-1} . \tag{2.10}$$

We leave it as an exercise for the reader to confirm this fact.

Next, suppose S and T are invertible $N \times N$ matrices. Then $T^{-1}S^{-1}$ is defined. We compute

$$(ST)(T^{-1}S^{-1}) = S(TT^{-1})S^{-1} = S\mathcal{I}S^{-1} = SS^{-1} = \mathcal{I}. \tag{2.11}$$

Similarly, $(T^{-1}S^{-1})(ST) = \mathcal{I}$. Thus, ST is also invertible and $(ST)^{-1} = T^{-1}S^{-1}$. We record this finding in the following formula:

Formula 2.19 Suppose the $N \times N$ matrices S and T are both invertible. Then the product ST is also invertible and

$$(ST)^{-1} = T^{-1}S^{-1}. \tag{2.12}$$

In words, the inverse of a product is equal to the product of the inverses in the other order.

Example 2.20 For any square matrix A, the product $A^2 = AA$ is defined. Similarly, we can compute the power A^k, for every counting number $k = 1, 2, 3, \ldots$. In case A is invertible, then, by repeated application of the formula (2.12), we see that

$$\left(A^k\right)^{-1} = \left(A^{k-1}A\right)^{-1} = A^{-1}\left(A^{k-1}\right)^{-1} = \cdots = \left(A^{-1}\right)^k. \tag{2.13}$$

We usually write A^{-k} rather than $\left(A^k\right)^{-1}$ or $\left(A^{-1}\right)^k$.

Example 2.21 (Cancellation in Matrix Equations) In Example 2.14 (iii), we saw matrices T, U_1, and U_2, where $TU_1 = TU_2$ but $U_1 \neq U_2$. This shows that, in general, we are not able to cancel a common matrix factor from both sides of an equation. However, if the matrix T is *invertible* and $TU_1 = TU_2$, then we get

$$TU_1 = TU_2 \Rightarrow T^{-1}TU_1 = T^{-1}TU_2 \Rightarrow \mathcal{I}U_1 = \mathcal{I}U_2 \Rightarrow U_1 = U_2.$$

In other words, an *invertible* matrix factor can be cancelled from both sides. *Caution:* The invertible matrix factor must be on the *same side* of both expressions in the equation, because matrix multiplication is not commutative. Thus, if $TU = VT$, then $U = T^{-1}VT$, not simply V.

We have seen that one can have two nonzero matrices whose product is the all-0s matrix. That cannot happen if one of the factors is invertible. Indeed, suppose A is invertible and that $AX = \mathbf{0}$. Multiplying both sides by A^{-1} on the left yields $X = A^{-1}\mathbf{0} = \mathbf{0}$.

For products of a matrix and a vector, we can interpret these cancellation results as follows:

(i) Suppose S is an invertible matrix and that $S\mathbf{x} = S\mathbf{y}$, for vectors $\mathbf{x}$ and $\mathbf{y}$. Then $\mathbf{x} = \mathbf{y}$, by cancellation of S. Equivalently, if $\mathbf{x} \neq \mathbf{y}$, then $S\mathbf{x} \neq S\mathbf{y}$.

(ii) Suppose the matrix S is invertible and that $S\mathbf{x} = \mathbf{0}$, for some vector $\mathbf{x}$. Then $\mathbf{x} = \mathbf{0}$, too. In other words, if S is invertible, then the *only vector* $\mathbf{x}$ for which $S\mathbf{x} = \mathbf{0}$ is the zero vector $\mathbf{x} = \mathbf{0}$.

2.5 Transpose of a Matrix

As discussed at the beginning of Section 2.2, if we wish to compute all inner products of the vectors in one set with the vectors in another set, we make the first set of vectors into the *rows* of a matrix and the other set of vectors into the columns of a second matrix. Then we multiply these two matrices together. Suppose, however, that we have already stored each set of vectors as the columns of a matrix. That is, suppose we already have the matrices

$$A = \left[\mathbf{a}_1 \middle| \cdots \middle| \mathbf{a}_M\right] \text{ and } B = \left[\mathbf{b}_1 \middle| \cdots \middle| \mathbf{b}_N\right],$$

where A is $K \times M$ and B is $K \times N$. (So all column vectors of both A and B are in $\mathbb{R}^K$; A has M columns and B has N columns.) To compute the desired inner products, we flip the matrix A onto its side. We do this by transposing each column vector into a row vector. The resulting matrix is called the **transpose** of the matrix A, denoted by A^T. That is, A^T is the $M \times K$ matrix

$$A^T = \left[\mathbf{a}_1 \middle| \cdots \middle| \mathbf{a}_M\right]^T = \begin{bmatrix} \mathbf{a}_1^T \\ \hline \vdots \\ \hline \mathbf{a}_M^T \end{bmatrix}. \tag{2.14}$$

Entry by entry,

$$\text{if } A = \left[\begin{bmatrix} a_{1,1} \\ a_{2,1} \\ \vdots \\ a_{K,1} \end{bmatrix} \begin{bmatrix} a_{1,2} \\ a_{2,2} \\ \vdots \\ a_{K,2} \end{bmatrix} \begin{matrix} \cdots \\ \cdots \\ \cdots \\ \cdots \end{matrix} \begin{bmatrix} a_{1,M} \\ a_{2,M} \\ \vdots \\ a_{K,M} \end{bmatrix} \right], \text{ then}$$

$$A^T = \begin{bmatrix} \left[a_{1,1}\ a_{2,1} \ \cdots \ a_{K,1}\right] \\ \left[a_{1,2}\ a_{2,2} \ \cdots \ a_{K,2}\right] \\ \vdots\ \vdots\ \vdots\ \vdots \\ \left[a_{1,M}\ a_{2,M} \ \cdots \ a_{K,M}\right] \end{bmatrix}. \tag{2.15}$$

With this setup, the $\left(A^T B\right)[i,\ j] = \mathbf{a}_i{}^T \mathbf{b}_j$, for $1 \le i \le M$ and $1 \le j \le N$. Note that the matrix A^T is $M \times K$, so that $A^T B$ has dimensions $M \times N$.

Definition 2.22 For a given $M \times N$ matrix A, the **transpose** of A, denoted by A^T, is the $N \times M$ matrix whose entry in row i and column j is the same as the entry of A in row j and column i. In other words, column j of A^T is the same as row j of A, and row i of A^T is the same as column i of A. In double index notation, $A^T[i, j] = A[j, i]$, for all suitable values of i and j.

Example 2.23 The transpose of $A = \begin{bmatrix} 4 & 0 \\ -3 & 2 \\ 1 & 5 \end{bmatrix}$ is $A^T = \begin{bmatrix} 4 & -3 & 1 \\ 0 & 2 & 5 \end{bmatrix}$.

The transpose of $B = \begin{bmatrix} 1 & 2 & 3 \\ 8 & 7 & 6 \end{bmatrix}$ is $B^T = \begin{bmatrix} 1 & 8 \\ 2 & 7 \\ 3 & 6 \end{bmatrix}$.

Notice that $\left(A^T\right)^T = A$, for every matrix A. Also, for real numbers s and t and $M \times N$ matrices A and B, we get

$$(s \cdot A + t \cdot B)^T = s \cdot A^T + t \cdot B^T .$$

To see how transposition interacts with matrix multiplication, start with an example.

Example 2.24 Take $A = \begin{bmatrix} 1 & 2 & 3 \\ -2 & 5 & 3 \end{bmatrix}$ and $B = \begin{bmatrix} 7 & 4 \\ -2 & 3 \\ 1 & 5 \end{bmatrix}$, as in Example 2.13.

Again we compute

$$\begin{aligned} AB &= \begin{bmatrix} 1 & 2 & 3 \\ -2 & 5 & 3 \end{bmatrix} \begin{bmatrix} 7 & 4 \\ -2 & 3 \\ 1 & 5 \end{bmatrix} \\ &= \begin{bmatrix} (7 - 4 + 3) & (4 + 6 + 15) \\ (-14 - 10 + 3) & (-8 + 15 + 15) \end{bmatrix} \\ &= \begin{bmatrix} 6 & 25 \\ -21 & 22 \end{bmatrix} . \end{aligned}$$

Thus, the transpose of the product (AB) is $(AB)^T = \begin{bmatrix} 6 & -21 \\ 25 & 22 \end{bmatrix}$. Now look at the product $B^T A^T$ of the two transposes, but in the other order. Notice that this product will also be a 2×2 matrix, like $(AB)^T$. In fact,

$$B^T A^T = \begin{bmatrix} 7 & -2 & 1 \\ 4 & 3 & 5 \end{bmatrix} \begin{bmatrix} 1 & -2 \\ 2 & 5 \\ 3 & 3 \end{bmatrix}$$

$$= \begin{bmatrix} 7-4+3 & -14-10+3 \\ 4+6+15 & -8+15+15 \end{bmatrix}$$

$$= \begin{bmatrix} 6 & -21 \\ 25 & 22 \end{bmatrix}.$$

This is the same as $(AB)^T$.

In general, suppose A is an $M \times K$ matrix and B is a $K \times N$ matrix. Then both $(AB)^T$ and $B^T A^T$ are $N \times M$ matrices. Moreover, the $[i,\ j]$ entry of $(AB)^T$ is equal to the inner product of the jth row of A with the ith column of B. This is the same as the inner product of the ith row of B^T with the jth column of A^T. In turn, that equals the $[i,\ j]$ entry of the product $B^T A^T$. Thus, $(AB)^T = B^T A^T$. We record this finding as the following formula:

Formula 2.25 Let A be an $M \times K$ matrix and let B be a $K \times N$ matrix. Then

$$(AB)^T = B^T A^T. \tag{2.16}$$

In words, the transpose of a product is equal to the product of the transposes in the other order.

Example 2.26 For the matrices $A = \begin{bmatrix} 4 & 0 \\ -3 & 2 \\ 1 & 5 \end{bmatrix}$ and $B = \begin{bmatrix} 1 & 2 & 3 \\ 8 & 7 & 6 \end{bmatrix}$ from Example 2.23, we compute

$$AB = \begin{bmatrix} 4 & 8 & 12 \\ 13 & 8 & 3 \\ 41 & 37 & 33 \end{bmatrix}, \text{ while } B^T A^T = \begin{bmatrix} 4 & 13 & 41 \\ 8 & 8 & 37 \\ 12 & 3 & 33 \end{bmatrix}.$$

Thus, $(AB)^T = B^T A^T$, as expected.

Example 2.27 For any square matrix A, the product $A^2 = AA$ is defined. Similarly, we can compute the power A^k, for every counting number $k = 1, 2, 3, \ldots$. By repeated application of the formula (2.16), we see that

$$\left(A^k\right)^T = \left(A^{k-1} A\right)^T = A^T \left(A^{k-1}\right)^T = \cdots = \left(A^T\right)^k. \tag{2.17}$$

Example 2.28 Suppose the square matrix A is invertible. Then $AA^{-1} = A^{-1}A = \mathcal{I}$. Taking transposes yields $(A^{-1})^T A^T = A^T (A^{-1})^T = \mathcal{I}$, since $\mathcal{I}$ is its own transpose. It follows from this that the transpose matrix A^T is also invertible and that $(A^T)^{-1} = (A^{-1})^T$. In words, the inverse of the transpose is the transpose of the inverse.

Example 2.29 For two vectors $\mathbf{a} = \begin{bmatrix} a_1 \\ a_2 \\ \vdots \\ a_N \end{bmatrix}$ and $\mathbf{b} = \begin{bmatrix} b_1 \\ b_2 \\ \vdots \\ b_N \end{bmatrix}$ in $\mathbb{R}^N$, the $N \times N$ matrix

$$\mathbf{a}\mathbf{b}^T = \begin{bmatrix} a_1 \\ a_2 \\ \vdots \\ a_N \end{bmatrix} \begin{bmatrix} b_1 \; b_2 \cdots b_N \end{bmatrix} = \begin{bmatrix} a_1 b_1 & a_1 b_2 & \cdots & a_1 b_N \\ a_2 b_1 & a_2 b_2 & \cdots & a_2 b_N \\ \vdots & \vdots & \ddots & \vdots \\ a_N b_1 & a_N b_2 & \cdots & a_N b_N \end{bmatrix} \tag{2.18}$$

is called the **outer product** of the two vectors.

Notice that each row of $\mathbf{a}\mathbf{b}^T$ is a scalar multiple of $\mathbf{b}^T$ and each column is a scalar multiple of $\mathbf{a}$. Also, while the inner products $\mathbf{a}^T\mathbf{b}$ and $\mathbf{b}^T\mathbf{a}$ are the same, the outer products $\mathbf{a}\mathbf{b}^T$ and $\mathbf{b}\mathbf{a}^T$ are generally different. In fact, these outer products are each other's transposes:

$$\left(\mathbf{a}\mathbf{b}^T\right)^T = \left(\mathbf{b}^T\right)^T \mathbf{a}^T = \mathbf{b}\mathbf{a}^T .$$

Outer products play an important part in matrix geometry. When we study the singular value decomposition of a matrix, we will see that every matrix can be expressed as a sum of outer products, each of which represents a different one-dimensional "layer" of the matrix. Highlighting the most significant among these layers is a key feature of data compression.

Example 2.30 For every $M \times N$ matrix A, the products $A^T A$ and AA^T are both defined. The first of these products is $N \times N$ and the second is $M \times M$. Each of these matrices is equal to its own transpose, by virtue of Formula 2.25. Indeed,

$$\left(A^T A\right)^T = A^T \left(A^T\right)^T = A^T A \,, \text{ and } \left(AA^T\right)^T = \left(A^T\right)^T A^T = AA^T .$$

Moreover, for any vector $\mathbf{x}$ in $\mathbb{R}^N$, we calculate

$$\mathbf{x}^T \left(A^T A\mathbf{x}\right) = \left(\mathbf{x}^T A^T\right)(A\mathbf{x}) = (A\mathbf{x})^T (A\mathbf{x}) = ||A\mathbf{x}||^2 \geq 0 . \tag{2.19}$$

Thus, the inner product $\mathbf{x}^T \left(A^T A\mathbf{x}\right)$ is never negative. It follows that the angle between $\mathbf{x}$ and $A^T A\mathbf{x}$ is between 0 and $\pi/2$.

Definition 2.31 (a) A square matrix A is said to be **symmetric** provided that $A^T = A$. Thus, for a symmetric matrix, the ith row is the transpose of the ith column; $a_{i,j} = a_{j,i}$, for all suitable i and j.

(b) A symmetric $M \times M$ matrix B is said to be **positive semidefinite** provided that $\mathbf{x}^T(B\mathbf{x}) \geq 0$, for every vector $\mathbf{x}$ in $\mathbb{R}^M$.

Example 2.30 shows that the matrices $A^T A$ and AA^T are symmetric and positive semidefinite, for every matrix A of any size array. We explore such matrices in greater depth in Chapter 9.

2.6 Exercises

1. Consider the matrices

$$A = \begin{bmatrix} 3 & -2 \\ 6 & 5 \end{bmatrix}, \quad B = \begin{bmatrix} 1 & 3 \\ 7 & 5 \end{bmatrix}, \quad C = \begin{bmatrix} 3 & -1 & 1 \\ 1 & 4 & 2 \end{bmatrix},$$

$$R = \begin{bmatrix} 1 & 5 & 2 \\ -1 & 1 & 2 \\ 4 & 2 & 3 \end{bmatrix}, \quad S = \begin{bmatrix} 6 & -1 & 4 \\ 5 & 0 & 2 \\ 1 & 1 & 1 \end{bmatrix}.$$

Compute the following matrices.

(a) $5A - 2B$ (b) $3R + S$ (c) $7C$
(d) AB (e) $(A + B)C$ (f) $C(R - S)$
(g) $C^T A$ (h) RC^T (i) $SC^T B$

2. Let $A = \begin{bmatrix} 1 & 0 & -3 \\ 3 & 2 & 4 \\ 2 & -3 & 5 \end{bmatrix}$ and $B = \begin{bmatrix} 7 & -4 & 3 \\ 1 & 5 & -2 \\ 0 & 3 & 9 \end{bmatrix}$.

(a) Compute the matrices AB and BA.

(b) Verify that $A^{-1} = \dfrac{1}{61} \begin{bmatrix} 22 & 9 & 6 \\ -7 & 11 & -13 \\ -13 & 3 & 2 \end{bmatrix}$.

(c) Verify that $B^{-1} = \dfrac{1}{402} \begin{bmatrix} 51 & 45 & -7 \\ -9 & 63 & 17 \\ 3 & -21 & 39 \end{bmatrix}$.

(d) Compute the matrix inverses $(AB)^{-1}$ and $(BA)^{-1}$.

3. Let $A = \begin{bmatrix} 1 & 0 & 3 \\ 2 & -5 & 4 \end{bmatrix}$ and $B = \begin{bmatrix} 3 & 0 \\ -1 & 4 \\ 6 & 5 \end{bmatrix}$.

(a) Compute the matrices AB and BA.
(b) Write down the transposes A^T and B^T.

(c) Verify that $B^T A^T = (AB)^T$.
(d) Verify that $A^T B^T = (BA)^T$.

4. Let $A = \begin{bmatrix} 1 & 0 & 3 \\ 2 & -5 & 4 \end{bmatrix}$.

   (a) Compute AA^T.

   (b) Compute $\mathbf{x}^T(AA^T)\mathbf{x}$ for a generic vector $\mathbf{x} = \begin{bmatrix} x_1 \\ x_2 \end{bmatrix}$ in $\mathbb{R}^2$.

   (c) Conclude that $10x_1^2 + 28x_1x_2 + 45x_2^2 \geq 0$ for all values of x_1 and x_2. (See (2.19).)

5. Let $B = \begin{bmatrix} 3 & 0 \\ -1 & 4 \\ 6 & 5 \end{bmatrix}$.

   (a) Compute $B^T B$.

   (b) Compute $\mathbf{x}^T(B^T B)\mathbf{x}$ for a generic vector $\mathbf{x} = \begin{bmatrix} x_1 \\ x_2 \end{bmatrix}$ in $\mathbb{R}^2$.

   (c) Conclude that $46x_1^2 + 52x_1x_2 + 41x_2^2 \geq 0$ for all values of x_1 and x_2. (See (2.19).)

6. A street vendor charges \$4.50 for a burrito, \$4.00 for a taco, and \$1.50 for a soft drink. On Monday through Friday of a given week, labeled as days 1 through 5, suppose the vendor sells a_j burritos, b_j tacos, and c_j soft drinks on the jth day. Now set

$$\mathbf{p} = \begin{bmatrix} 4.50 \\ 4 \\ 1.50 \end{bmatrix} \text{ and } V = \begin{bmatrix} a_1 & a_2 & a_3 & a_4 & a_5 \\ b_1 & b_2 & b_3 & b_4 & b_5 \\ c_1 & c_2 & c_3 & c_4 & c_5 \end{bmatrix}.$$

   Explain the meaning of each entry of the 1×5 matrix $\mathbf{p}^T V$.

7. In Example 2.4, we looked at the hypothetical cookie sales for three Girl Scout troops. The matrices W_1, W_2 and W_3 showed how many packages of each type of cookie each troop sold in three different weeks.

   (a) Compute the product $(W_1 + W_2 + W_3)\mathbf{1}_3$, where $\mathbf{1}_3$ is the "all-1s" vector in $\mathbb{R}^3$. What does each entry in this product represent?
   (b) Compute the product $\mathbf{1}_4^T(W_1 + W_2 + W_3)$, where $\mathbf{1}_4$ is the "all-1s" vector in $\mathbb{R}^4$. What does each entry in this product represent?
   (c) Compute $\mathbf{1}_4^T(W_1 + W_2 + W_3)\mathbf{1}_3$. This results in a 1×1 matrix, i.e., a number. What does this number represent?
   (d) The cookies are sold for \$5 per package. Multiply each of the results of parts (a), (b), and (c) by 5. What do the new results represent?

8. Let $A = \begin{bmatrix} 0 & 1 \\ -1 & 0 \end{bmatrix}$.

   (a) Compute A^2, A^3, and A^4.
   (b) Using your work from part (a), determine the values of A^{4n}, A^{4n+1}, A^{4n+2}, and A^{4n+3}, for every positive integer n.

9. Let $A = \begin{bmatrix} 1 & 1 \\ 0 & 1 \end{bmatrix}$.

   (a) Compute A^2, A^3, and A^4.
   (b) *(Proof Problem)* Using your work from part (a), form a conjecture about the value of A^k for every positive integer k. Use induction to prove your conjecture.

10. Suppose A is an $N \times M$ matrix where every column is a unit vector. Suppose B is an $N \times K$ matrix where every column is a unit vector. What does each individual entry of the product $A^T B$ represent geometrically? Explain.
11. This exercise shows that *cancellation* does not always work for matrix multiplication.

$$\text{Let } A = \begin{bmatrix} 1 & 2 \\ 2 & 4 \end{bmatrix}; \; B = \begin{bmatrix} 3 & 2 \\ -1 & 7 \end{bmatrix}; \; C = \begin{bmatrix} 5 & -4 \\ -2 & 10 \end{bmatrix}.$$

   (a) Compute AB and AC.
   (b) Compute $B^T A$ and $C^T A$.
   (c) Thus, show that $A(B - C) = (B^T - C^T)A = \begin{bmatrix} 0 & 0 \\ 0 & 0 \end{bmatrix}$. This shows that the all-0s matrix can be *factored* into nonzero factors.

12. This exercise shows that *cancellation* does not always work for matrix multiplication.

$$\text{Let } U = \begin{bmatrix} 4 & 1 & -2 & 7 \\ 3 & 1 & -1 & 5 \end{bmatrix}; \; V = \begin{bmatrix} 1 & 5 \\ 3 & -1 \\ -2 & 4 \\ 2 & 3 \end{bmatrix}; \; W = \begin{bmatrix} 3 & 4 \\ 2 & 1 \\ -2 & 3 \\ 1 & 3 \end{bmatrix}.$$

   (a) Compute UV and UW.
   (b) Thus, show that $U(V - W) = \begin{bmatrix} 0 & 0 \\ 0 & 0 \end{bmatrix}$. This shows that the all-0s matrix can be *factored* into nonzero factors.

13. *(Proof Problem)* Prove that, if the matrix A is invertible and $AB = AC$ for some matrices B and C, then, in fact, $B = C$. Thus, cancellation does work if the common matrix factor is invertible.
14. *(Proof Problem)* Suppose the $N \times N$ matrix A is invertible. Prove the following:

    (a) The matrix $A^T A$ is also invertible.

    (b) $\left(A^T A\right)^{-1} A^T = A^{-1}$.

15. *(Proof Problem)* A matrix A, with entries $a_{i,j}$, is **upper triangular** if $a_{i,j} = 0$ whenever $i > j$. Show that a product of two upper triangular $N \times N$ matrices is also an upper triangular $N \times N$ matrix.
16. *(Proof Problem)* Let A and B be matrices of the form

$$A = \left[\begin{array}{c|c} A_1 & A_2 \\ \hline 0 & A_3 \end{array}\right] \text{ and } B = \left[\begin{array}{c|c} B_1 & B_2 \\ \hline 0 & B_3 \end{array}\right],$$

where A_1 and B_1 are square $K \times K$ matrices, A_3 and B_3 are square $M \times M$ matrices, A_2 and B_2 are $K \times M$ matrices, and 0 represents an $M \times K$ block of 0s. Such matrices are called **block upper triangular** matrices. Show that the product AB is also block upper triangular.

Chapter 3
Matrix Contexts

3.1 Digital Images

Every digital image can be represented as a matrix. The image consists of a grid of pixels where each pixel is assigned a color, where the spectrum of colors is translated into a set of numbers. For black and white images, the colors are different shades of gray on a scale from *black*, usually assigned the value 0, to *white*, assigned the value 1. Each number between 0 and 1 corresponds to a darker or lighter gray. (This scale can easily be reversed by subtracting the assigned value from 1.) The array of numerical gray values gives us a matrix associated to the given image. In the case of color images, each pixel is assigned a set of three numbers, labeled RGB, corresponding to the contributions of *red*, *green*, and *blue* to the color in that pixel. In digital image processing, each of the three RGB layers can be manipulated separately. The results are recombined to give an overall image.

For example, the image in Figure 3.1, with 100 pixels in a 10×10 grid, corresponds to the matrix of numerical gray values given by

Supplementary Information The online version contains supplementary material available at https://doi.org/10.1007/978-3-031-39562-8_3.

T. G. Feeman, *Applied Linear Algebra and Matrix Methods*,
Springer Undergraduate Texts in Mathematics and Technology,
https://doi.org/10.1007/978-3-031-39562-8_3

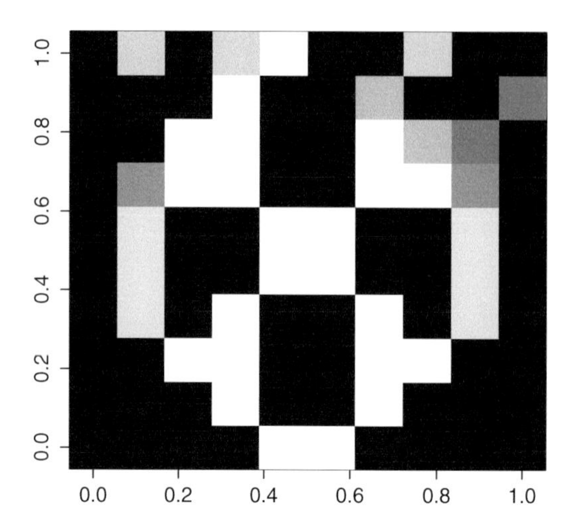

Fig. 3.1 The matrix in (3.1) produces this grayscale image

$$A = \begin{bmatrix} 0 & 0.7 & 0 & 0.7 & 1 & 0 & 0 & 0.7 & 0 & 0 \\ 0 & 0 & 0 & 1 & 0 & 0 & 0.6 & 0 & 0 & 0.3 \\ 0 & 0 & 1 & 1 & 0 & 0 & 1 & 0.6 & 0.3 & 0 \\ 0 & 0.4 & 1 & 1 & 0 & 0 & 1 & 1 & 0.4 & 0 \\ 0 & 0.8 & 0 & 0 & 1 & 1 & 0 & 0 & 0.8 & 0 \\ 0 & 0.8 & 0 & 0 & 1 & 1 & 0 & 0 & 0.8 & 0 \\ 0 & 0.8 & 0 & 1 & 0 & 0 & 1 & 0 & 0.8 & 0 \\ 0 & 0 & 1 & 1 & 0 & 0 & 1 & 1 & 0 & 0 \\ 0 & 0 & 0 & 1 & 0 & 0 & 1 & 0 & 0 & 0 \\ 0 & 0 & 0 & 0 & 1 & 1 & 0 & 0 & 0 & 0 \end{bmatrix}. \tag{3.1}$$

Many national flags consist of three stripes, arranged either horizontally or vertically. Both types of flags can be represented as outer products. For example, let $\mathbf{a} = \begin{bmatrix} 0.2 \\ 0.8 \\ 0.5 \end{bmatrix}$ and $\mathbf{1} = \begin{bmatrix} 1 \\ 1 \\ 1 \end{bmatrix}$. Then

$$\mathbf{a}\mathbf{1}^T = \begin{bmatrix} 0.2 & 0.2 & 0.2 \\ 0.8 & 0.8 & 0.8 \\ 0.5 & 0.5 & 0.5 \end{bmatrix} \text{ and } \mathbf{1}\mathbf{a}^T = \begin{bmatrix} 0.2 & 0.8 & 0.5 \\ 0.2 & 0.8 & 0.5 \\ 0.2 & 0.8 & 0.5 \end{bmatrix}.$$

Figure 3.2 shows these flags.

Adding the numerical matrices for two images corresponds to overlaying the images, one on top of the other. When we multiply an image matrix by a constant c, with $0 \le c < 1$, we darken the colors in the image. Figure 3.3 shows three intensities of patterns that overlay horizontal and vertical stripes. In fact, every digital image can be decomposed into layers represented as outer products. In Chapter 10, we

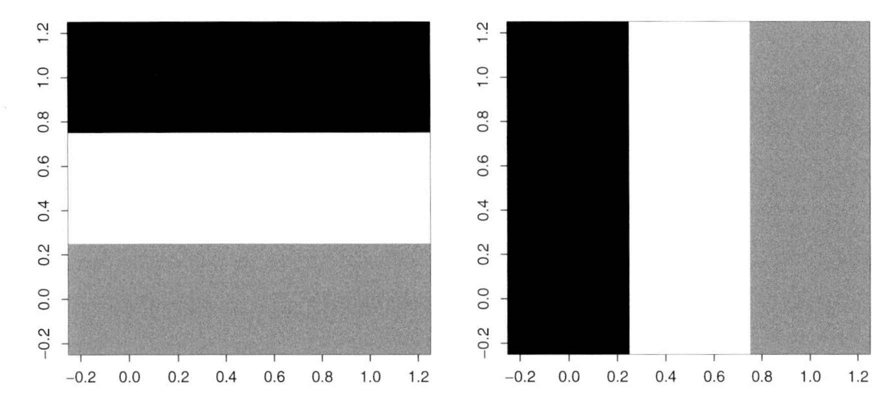

Fig. 3.2 A tricolor flag can be represented as an outer product $\mathbf{a}\mathbf{1}^T$, for horizontal stripes, or $\mathbf{1}\mathbf{a}^T$, for vertical stripes

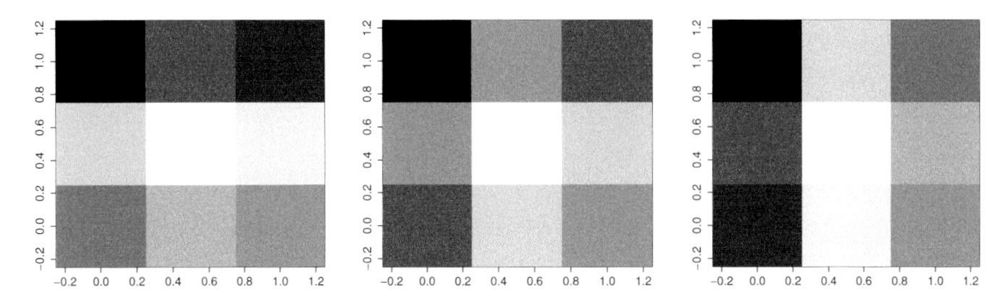

Fig. 3.3 Three checker patterns: $(0.75) * \mathbf{a}\mathbf{1}^T + (0.25) * \mathbf{1}\mathbf{a}^T$, $(0.5) * \mathbf{a}\mathbf{1}^T + (0.5) * \mathbf{1}\mathbf{a}^T$, and $(0.25) * \mathbf{a}\mathbf{1}^T + (0.75) * \mathbf{1}\mathbf{a}^T$

explore how the *singular value decomposition* can help us to identify the layers that contain the most important information in the image. Using these key layers, we can retain a faithful version of the image while reducing the amount of data to be stored.

3.2 Information Retrieval Revisited

As in Section 1.6 above, suppose we have a library of documents and a dictionary of terms. For each document, we form a vector showing which terms in the dictionary appear in the document. We interpret a search query as a sort of document and form a vector for it, too. To evaluate a search, we compute the inner products of all of the vectors for the different documents with the vector for the query. We can organize this work more efficiently using matrices. Table 1.3 shows a term–document matrix where each column corresponds to the vector of terms for one of the documents in the library. Call this matrix L, for *library*, say. As a first step, convert each column vector of L into a unit vector by dividing by its length; call the resulting matrix L_c. Now flip L_c on its side to form the transpose matrix L_c^T. Thus, L_c^T is an 8×18 matrix each row of which looks like $\mathbf{a}_j{}^T/||\mathbf{a}_j||$ where $\mathbf{a}_j$ is the vector of terms for the jth document. A query vector $\mathbf{q}$ will be a column vector in $\mathbb{R}^{18}$. Set $\widehat{\mathbf{q}} = \mathbf{q}/||\mathbf{q}||$, a unit vector that represents the same query. The product $L_c^T \widehat{\mathbf{q}}$ now gives us in a single column all of the cosine values $\cos(\theta_j)$ that we computed in Algorithm 1.18 above. As before, the search recommends those documents whose cosine values exceed some preselected threshold.

To evaluate several queries at once, we first normalize each query vector and then store these unit vectors as columns of a matrix Q, say. The matrix product $L_c^T Q$ shows, column by column, the results for the different queries.

Example 3.1 (Comparing Library Documents) We can compare the documents in our library with each other by computing the matrix $L_c^T L_c$. For $i \neq j$, the $[i,\ j]$ entry of $L_c^T L_c$ is the cosine of the angle between the normalized vectors of terms for documents i and j. When two documents are extremely similar, we can save computation time and storage by removing one of them from the search analysis and automatically recommending it whenever the retained document is recommended.

Example 3.2 (Term–Term Comparisons) We might wish to know which pairs of terms in our dictionary tend to occur together in the same documents. We can quantify this by comparing *rows* of a term–document matrix like the one in Table 1.3. Begin by normalizing each row vector of the library matrix, L, calling the resulting matrix L_r, say. Then the $[i,\ j]$ entry of $L_r L_r^T$ equals the cosine of the angle between the row vectors corresponding to terms i and j. If this value is above a certain threshold, then the terms tend to appear in the same documents. In the example from Table 1.3, there are 153 pairs of different terms to be compared. For 93 of these pairs, the cosine of the angle between the respective rows is at least 0.5. For instance, the terms *algebra* (#1) and *permutation* (#13) are close to all of the terms except *analysis* (#2), *Riemann* (#14), and *sequence* (#15).

Example 3.3 (Movie Ratings) As we discussed in Section 1.6, the term–document approach also applies to analyzing movie ratings of different viewers. Table 1.4 shows the matrix of ratings of five films by six viewers. Call this matrix R and let R_c and R_r, respectively, denote the matrices obtained by normalizing the columns or rows of R. The $[i,\ j]$ entry of $R_c^T R_c$ is the cosine of the angle between the rating vectors of viewers i and j. If this is close to 1, then these viewers have similar tastes in movies. The $[k,\ \ell]$ entry of $R_r R_r^T$ is the cosine of the angle between the vectors of viewer ratings for movies k and ℓ. If this is close to 1, then most viewers give the two movies the same rating. This helps a lot if you run a streaming service, since, if a new viewer likes movie k, you can feel confident about recommending movie ℓ to them as well.

3.3 Markov Processes: A First Look

Imagine a fictional metropolitan area consisting of the city of Shamborough and its suburbs. Suppose that, over the course of any given year, $1/5$ of the current city dwellers will relocate to the suburbs while $4/5$ of them remain in the city.

Similarly, suppose $52/55$ (approximately 94.5%) of those living in the suburbs stay in the suburbs, with $3/55$ of suburbanites moving into the city. For simplicity, we assume that the total population remains constant at 1 million people year in and year out. We would like to understand the long-term relative populations of the city and suburbs, perhaps in order to make demographic projections that will affect the allocation of resources throughout the region.

Looking just from one year to the next, suppose the current populations of the city and of the suburbs, respectively, are C_0 and S_0, measured in millions of people. Thus, $C_0 + S_0 = 1$. These values are also the *proportions* of the total population living in each part. According to our information about population movements, the corresponding populations next year will be

$$\begin{aligned} C_1 &= (4/5) \cdot C_0 + (3/55) \cdot S_0 \text{ and} \\ S_1 &= (1/5) \cdot C_0 + (52/55) \cdot S_0 \,. \end{aligned}$$

This is the same as the matrix/vector product

$$\begin{bmatrix} C_1 \\ S_1 \end{bmatrix} = \begin{bmatrix} 4/5 & 3/55 \\ 1/5 & 52/55 \end{bmatrix} \begin{bmatrix} C_0 \\ S_0 \end{bmatrix} = T \begin{bmatrix} C_0 \\ S_0 \end{bmatrix}. \tag{3.2}$$

Notice that $C_1 + S_1 = C_0 + S_0 = 1$, representing 1 million people.

The equation (3.2) shows that we can analyze the population shifts using a matrix T that we call the **transition matrix** for the system. This particular system has two *states*: living in the city or living in the suburbs. The entry in the ith row and jth column of the transition matrix records the *probability* that a randomly selected person who is *currently in* state j will *move to* state i during the year ahead. (In the language of probability theory, these are called *conditional probabilities*.) Because each person currently in state j must either stay there or move, the entries in each column of the transition matrix add up to 1. This also guarantees that, for any vector $\mathbf{x}$, the coordinates of the vector $T\mathbf{x}$ have the same sum as those of $\mathbf{x}$.

Equation (3.2) shows that if the population distribution for the current year is $\mathbf{p}_0$, then next year's population distribution will be $\mathbf{p}_1 = T\mathbf{p}_0$. We can extrapolate from this to see that the population distribution after two years will be $\mathbf{p}_2 = T\mathbf{p}_1 = T^2\mathbf{p}_0$. More generally, the population vector k years from now will be given by $\mathbf{p}_k = T\mathbf{p}_{k-1} = \cdots = T^k\mathbf{p}_0$. To compute this, we could multiply the matrix T by itself k times to get T^k. In practice, it is more efficient to start with $\mathbf{p}_0$ and compute the successive population vectors $\mathbf{p}_1$, $\mathbf{p}_2$, and so on.

For instance, suppose 400,000 people live in the city this year and 600,000 live in the suburbs. The total fixed population is 1 million people. Using just the population proportions, we get the population distribution vector $\mathbf{p}_0 = \begin{bmatrix} 0.4 \\ 0.6 \end{bmatrix}$. Next year's population distribution vector is computed to be

$$\mathbf{p}_1 = T\mathbf{p}_0 = \begin{bmatrix} 4/5 & 3/55 \\ 1/5 & 52/55 \end{bmatrix} \begin{bmatrix} 0.4 \\ 0.6 \end{bmatrix} \approx \begin{bmatrix} 0.3527273 \\ 0.6472727 \end{bmatrix}.$$

After that, we get

$$\mathbf{p}_2 = T\mathbf{p}_1 \approx \begin{bmatrix} 0.3174876 \\ 0.6825214 \end{bmatrix} \text{ and } \mathbf{p}_3 = T\mathbf{p}_2 \approx \begin{bmatrix} 0.291218 \\ 0.708782 \end{bmatrix}.$$

After 10 years, we have $\mathbf{p}_{10} = T\mathbf{p}_9 \approx \begin{bmatrix} 0.2241271 \\ 0.7758729 \end{bmatrix}$. Rounded off to the nearest whole person, this corresponds to 224,127 people in the city and 775,873 people in the suburbs.

In this example, as the city population declines, its future contributions to the population of the suburbs also decline. The reverse trend occurs in the suburbs, where, as its population grows, the absolute number of those leaving each year also grows. These trends together suggest that the *rate* at which the city population declines from year to year will slow down as fewer people leave the city *and* more come in from the suburbs. At the same time, the growth rate of the suburban population might also slow down as some sort of balance is achieved between the two parts of the metropolitan area. Of course, the interactions are entangled, so our intuition may be imperfect. Nonetheless, we have arrived at an interesting question: **Is there a scenario where the populations of the city and the suburbs are stable and stay the same from year to year?** To say that the populations stay the same does not mean that no one is moving. It means only that the numbers moving between subregions exactly balance out.

It seems reasonable to imagine that the population distribution will evolve towards stability over time. Since the population distribution k years from now is $\mathbf{p}_k = T\mathbf{p}_{k-1} = \cdots = T^k\mathbf{p}_0$, we can continue the string of computations we already started. Then we stop when things do not change much. For instance, with $\mathbf{p}_0 = \begin{bmatrix} 0.4 \\ 0.6 \end{bmatrix}$, we get

$$\mathbf{p}_{40} = \begin{bmatrix} 0.214287 \\ 0.785713 \end{bmatrix} \text{ and } \mathbf{p}_{50} = \begin{bmatrix} 0.214286 \\ 0.785714 \end{bmatrix}.$$

These proportions have been rounded to six digits, corresponding to the nearest whole person in a total population of 1 million. The change from year 40 to year 50 is just one person moving from the city to the suburbs. It seems we have reached an equilibrium, at least for all practical purposes. One downside of this iterative approach is that we have no way of knowing ahead of time how large k has to be before $T^k\mathbf{p}_0$ settles down.

A stable population distribution for Shamborough would correspond to a vector $\mathbf{p}_*$ for which $T\mathbf{p}_* = \mathbf{p}_*$. Setting $\mathbf{p}_* = \begin{bmatrix} C_* \\ S_* \end{bmatrix}$, we want

$$\begin{bmatrix} 4/5 & 3/55 \\ 1/5 & 52/55 \end{bmatrix} \begin{bmatrix} C_* \\ S_* \end{bmatrix} = \begin{bmatrix} C_* \\ S_* \end{bmatrix} \text{ and } C_* + S_* = 1.$$

This is the same as

$$\begin{cases} (4/5)C_* + (3/55)S_* &= C_* \\ (1/5)C_* + (52/55)S_* &= S_* \end{cases} \text{ and } C_* + S_* = 1\,.$$

Combining like terms in the first two equations, we get the relationship $(1/5)C_* = (3/55)S_*$, or $C_* = (3/11)S_*$. Since we also require $C_* + S_* = 1$, the solution is $C_* = 3/14$ and $S_* = 11/14$. For a total population of 1 million, these proportions correspond to an equilibrium of 214,286 people in the city and 785,714 people in the suburbs. This is the same as $\mathbf{p}_{50}$ computed above.

Now let us refine our scenario by splitting the suburbs of Shamborough into two subregions that we call the suburbs and the exurbs. This means a randomly selected resident of the metropolitan area may potentially be in one of three *states*: living in the city, living in the suburbs, or living in the exurbs. Movements between subregions become more intricate as a city dweller may now move to the suburbs or to the exurbs. There is also movement between the two parts of the previously combined suburban region. For example, suppose the transition matrix is now given by

$$T = \begin{bmatrix} 0.8 & 0.1 & 0 \\ 0.15 & 0.8 & 0.15 \\ 0.05 & 0.1 & 0.85 \end{bmatrix}.$$

As before, the entry in the ith row and jth column gives the probability that someone who is currently in state j will move to state i during the year ahead. For instance, 10% of this year's suburban dwellers will move to the city next year and 10% will move to the exurbs. People living in the exurbs never move directly into the city the next year. Notice that the entries in each column of T sum to 1, as before.

The matrix T allows us to compute the change in the population distribution over time. Suppose $\mathbf{p}_k = \begin{bmatrix} C_k & S_k & E_k \end{bmatrix}^T$ is the population distribution after k years. Then, after $k+1$ years, we compute

$$\mathbf{p}_{k+1} = \begin{bmatrix} C_{k+1} \\ S_{k+1} \\ E_{k+1} \end{bmatrix} = \begin{bmatrix} 0.8 & 0.1 & 0 \\ 0.15 & 0.8 & 0.15 \\ 0.05 & 0.1 & 0.85 \end{bmatrix} \begin{bmatrix} C_k \\ S_k \\ E_k \end{bmatrix} = T\mathbf{p}_k\,. \tag{3.3}$$

Notice that $C_{k+1} + S_{k+1} + E_{k+1} = C_k + S_k + E_k$.

To explore the long-term trend, start with $\mathbf{p}_0 = \begin{bmatrix} 0.4 \; 0.4 \; 0.2 \end{bmatrix}^T$. Next year's population distribution is then

$$\mathbf{p}_1 = T\mathbf{p}_0 = \begin{bmatrix} 0.8 & 0.1 & 0 \\ 0.15 & 0.8 & 0.15 \\ 0.05 & 0.1 & 0.85 \end{bmatrix} \begin{bmatrix} 0.4 \\ 0.4 \\ 0.2 \end{bmatrix} = \begin{bmatrix} 0.36 \\ 0.41 \\ 0.23 \end{bmatrix}.$$

After that, we get

$$\mathbf{p}_2 = T\mathbf{p}_1 = \begin{bmatrix} 0.329 \\ 0.4165 \\ 0.2545 \end{bmatrix} \text{ and } \mathbf{p}_3 = T\mathbf{p}_2 = \begin{bmatrix} 0.304850 \\ 0.420725 \\ 0.274425 \end{bmatrix}.$$

Further on, we find

$$\begin{aligned} \mathbf{p}_{10} = T\mathbf{p}_9 &\approx \begin{bmatrix} 0.232438\ 0.428187\ 0.339375 \end{bmatrix}^T, \\ \mathbf{p}_{50} &\approx \begin{bmatrix} 0.214288\ 0.428571\ 0.357141 \end{bmatrix}^T, \text{ and} \\ \mathbf{p}_{60} &\approx \begin{bmatrix} 0.214286\ 0.428571\ 0.357143 \end{bmatrix}^T. \end{aligned}$$

Years 50 and 60 differ by a shift of just two people from the city to the exurbs.

A stable population distribution $\mathbf{p}_*$ must satisfy $T\mathbf{p}_* = \mathbf{p}_*$. Even with just three states instead of two, this is harder to find. Later on, we will explore how to determine if such a vector exists and how to compute it if it does. For now, observe that the vector

$$\mathbf{p}_* = \begin{bmatrix} 3/14 \\ 6/14 \\ 5/14 \end{bmatrix} \approx \begin{bmatrix} 0.214286 \\ 0.428571 \\ 0.357143 \end{bmatrix} = \mathbf{p}_{60}$$

has the properties we are looking for: $T\mathbf{p}_* = \mathbf{p}_*$ and the coordinates of $\mathbf{p}_*$ add up to 1. The vector $\mathbf{p}_{60}$ we computed above is virtually the same as this.

The population dynamics of the fictional Shamborough metropolitan area provide an example of a **Markov process** or **Markov chain**. These terms are used to describe the dynamics of any system that has some specific finite number of possible states and where step-by-step movement from one state to another takes place according to prescribed rules and in such a way that the transition to a new state depends only on the current state of the system and not on the entire transition history. In that sense, the dynamics of a Markov process are memoryless.

In a Markov process, the transitions between states are described using probabilities, or, more precisely, conditional probabilities: If we are currently in state j, there is a fixed probability that we will move next to state i. The value of this probability can be stored as the $[i,\ j]$ entry of a matrix T, called the **transition matrix**. If the system has M states in all, then the transition matrix is an $M \times M$ matrix. The entries in each column of the transition matrix add up to 1. This ensures that, for every vector $\mathbf{x}$, the coordinates of $\mathbf{x}$ and of $T\mathbf{x}$ have the same sum. We will look at Markov processes again in Chapter 8. There, we explore conditions under which there is guaranteed to be an equilibrium, known as the **stationary distribution** of the process. When the number of states is large, the problem of calculating the stationary distribution can be daunting. We will discuss the efficacy of using high powers of the transition matrix to estimate it, as we did in our small examples here. For now, we mention one more example.

Example 3.4 The World Wide Web consists of billions of web pages. A person, or a *bot*, surfing the web at random could potentially find themselves on any page. They choose the next page to visit by randomly selecting one of the links available on the current page. (We will not allow clicking on $\leftarrow$ to automatically return to the previously visited page.) This is the framework of a Markov process. In Chapter 8, we will look at the problem of estimating the stationary distribution of a large team of random web surfers as they move from page to page on the World Wide Web by clicking on hyperlinks.

3.4 Graphs and Networks

A graph is a collection of vertices and edges. Each **vertex**, also called a **node**, can be thought of as a point. Each **edge** can be thought of as a line segment or arc that connects two vertices. We only consider graphs that are *undirected*, meaning that we can travel in either direction along any edge. Also, we only consider graphs that have a finite number of vertices. Communication networks, such as networks of cell towers or airports, and social networks where friendships are mutual are two real-world examples that can be represented by graphs. In the first example, each cell tower represents a vertex and two vertices are joined by an edge if there is an open communication channel between the two towers. In a social network, each subscriber corresponds to a vertex and two vertices are connected by an edge when the two subscribers are friends.

Many real-world graphs and networks have intrinsic structure. Nodes that are particularly well connected serve as communication hubs. A cluster of vertices that has many internal links but relatively few connections to nodes outside the cluster operates like its own subnetwork. Still other vertices might play a crucial role in connecting different clusters. It might be difficult to discern any of this structure just by looking at the graph, especially if the number of vertices is in the hundreds or even millions. It is not even clear how we might draw a useful diagram of a graph that has a huge number of vertices. Without some guidance, our drawing could easily be a chaotic tangle of vertices and edges with no visible structure.

Happily, we can represent any given graph by a matrix. Suppose we have a graph with N vertices. Label these as $\nu_1, \ldots, \nu_N$. (We label vertices with the Greek letter ν [*nu*], perhaps to suggest *node*.) The **adjacency matrix** of the graph is the $N \times N$ matrix $A = \left[a_{i,\,j}\right]$ whose $[i,\ j]$ entry, $a_{i,\,j}$, is equal to 1, if there is an edge connecting ν_i and ν_j, and 0, if there is no such edge in the graph. The graph is undirected, so an edge that connects ν_i and ν_j also connects ν_j to ν_i. That means that $a_{i,\,j} = a_{j,\,i}$, for all i and j. In other words, the adjacency matrix A is *symmetric*: $A^T = A$. We assume that no edge connects a vertex to itself, meaning that the diagonal entries of the adjacency matrix are all 0s: $a_{i,\,i} = 0$ for all i.

Definition 3.5 The **adjacency matrix** of a graph with N vertices, labeled v_1, v_2, $\ldots$, v_N, is the $N \times N$ matrix $A = \left[a_{i,\,j}\right]$, where

$$a_{i,\,j} = \begin{cases} 1 & \text{, if an edge connects } v_i \text{ and } v_j \text{, or} \\ 0 & \text{, if } v_i \text{ and } v_j \text{ are not connected by an edge.} \end{cases} \tag{3.4}$$

By convention, $a_{i,\,i} = 0$, for all i. For an undirected graph, $a_{i,\,j} = a_{j,\,i}$, for all i and j, and $A^T = A$.

If we add up all the entries in the ith row of A, we get the number of edges that have vertex v_i as one end point. This is called the **degree** of the vertex v_i, denoted by $\deg(v_i)$. That is, for every $i = 1, \ldots, N$, the degree of v_i is

$$\deg(v_i) = a_{i,\,1} + a_{i,\,2} + \cdots + a_{i,\,N}\,. \tag{3.5}$$

We obtain the degrees of all vertices at once by computing the matrix/vector product $A\mathbf{1}$, where $\mathbf{1}$ is the vector of all 1s. Since each edge connects two vertices, the sum of all the degrees is equal to twice the number of edges. We compute this like so:

$$\text{number of edges} = \frac{1}{2}\mathbf{1}^T A\mathbf{1}\,. \tag{3.6}$$

Example 3.6 Consider the graph with $N = 5$ vertices and adjacency matrix $A = \begin{bmatrix} 0\,1\,1\,0\,1 \\ 1\,0\,0\,0\,1 \\ 1\,0\,0\,1\,0 \\ 0\,0\,1\,0\,0 \\ 1\,1\,0\,0\,0 \end{bmatrix}$. We compute

$$A\mathbf{1} = \begin{bmatrix} 0\,1\,1\,0\,1 \\ 1\,0\,0\,0\,1 \\ 1\,0\,0\,1\,0 \\ 0\,0\,1\,0\,0 \\ 1\,1\,0\,0\,0 \end{bmatrix} \begin{bmatrix} 1 \\ 1 \\ 1 \\ 1 \\ 1 \end{bmatrix} = \begin{bmatrix} 3 \\ 2 \\ 2 \\ 1 \\ 2 \end{bmatrix}, \text{ and}$$

$$\mathbf{1}^T A\mathbf{1} = \begin{bmatrix} 1\,1\,1\,1\,1 \end{bmatrix} \begin{bmatrix} 3 \\ 2 \\ 2 \\ 1 \\ 2 \end{bmatrix} = 10\,.$$

Thus, this graph has $\frac{1}{2}\mathbf{1}^T A\mathbf{1} = (1/2) \cdot 10 = 5$ edges. Three of these edges have the first vertex v_1 as an endpoint. The fourth vertex v_4 has degree 1, which means that v_4 is connected to only one other vertex (v_3, in this case). It is not immediately obvious how to draw even this small graph in a nice way.

Two vertices connected by an edge are called **neighbors**. The degree of each vertex is the same as its number of neighbors. To explore deeper connections in a graph, we start by asking whether two given vertices have any *common neighbors*. If vertex v_k is a common neighbor of both v_i and v_j, then there is an edge joining v_i to v_k and another edge connecting v_k to v_j. For the adjacency matrix, this means that $a_{i,k}a_{k,j} = 1$. If v_k is not a common neighbor of v_i and v_j, then $a_{i,k}a_{k,j} = 0$. Thus, the total number of common neighbors of v_i and v_j is equal to the sum

$$a_{i,1}a_{1,j} + a_{i,2}a_{2,j} + \cdots + a_{i,N}a_{N,j} = \sum_{k=1}^{N} a_{i,k}a_{k,j} . \tag{3.7}$$

The sum in (3.7) is exactly equal to the $[i,\ j]$ entry of the matrix A^2.

Every common neighbor of vertices v_i and v_j provides a path along two edges connecting v_i to v_j. In this way, the $[i,\ j]$ entry of A^2 counts the total number of paths along two edges connecting v_i and v_j.

Technically, the case where $i = j$ must be considered separately. For any vertex v_i, having a common neighbor with itself is the same as having a neighbor at all. If v_k is a neighbor of v_i, then we can travel from v_i to v_k and back to v_i along the same edge. That makes a path from v_i to itself along two edges. Computationally, if $i = j$, then, for each k, $a_{i,k}a_{k,j} = a_{i,k}^2 = a_{i,k}$. Hence, the $[i,\ i]$ entry of A^2 is the same as the sum of the entries in the ith row of A, which is the degree of v_i. That is,

$$A^2_{[i,\,i]} = \deg(v_i)\,, \text{ for every } i\,. \tag{3.8}$$

Remark Since $A^2_{[i,\,i]}$ is equal to the number of edges that have one end at vertex v_i, and every edge has two endpoints, it follows that the sum of the diagonal entries of A^2 counts every edge of the graph twice. The sum of the diagonal entries of any matrix is called the *trace* of the matrix. Thus,

$$\begin{aligned} \text{The number of edges } &= \frac{1}{2} \cdot \left(A^2_{[1,\,1]} + \cdots + A^2_{[N,\,N]} \right) \\ &= \frac{1}{2} \cdot \operatorname{trace}(A^2) \end{aligned} \tag{3.9}$$

Remark Two vertices can be neighbors even if they do not appear to be close in a particular drawing of the graph. Indeed, it would be nice to find a way to draw a graph so that neighboring vertices are actually close to each other.

Example 3.7 Return to the graph with 5 vertices in Example 3.6. The square of the adjacency matrix is

$$A^2 = \begin{bmatrix} 0&1&1&0&1 \\ 1&0&0&0&1 \\ 1&0&0&1&0 \\ 0&0&1&0&0 \\ 1&1&0&0&0 \end{bmatrix} \begin{bmatrix} 0&1&1&0&1 \\ 1&0&0&0&1 \\ 1&0&0&1&0 \\ 0&0&1&0&0 \\ 1&1&0&0&0 \end{bmatrix} = \begin{bmatrix} 3&1&0&1&1 \\ 1&2&1&0&1 \\ 0&1&2&0&1 \\ 1&0&0&1&0 \\ 1&1&1&0&2 \end{bmatrix} .$$

The diagonal entries of A^2 are the degrees of the different vertices. The other entries count paths of length 2 between the corresponding vertices. For example, there is one path of length 2 from v_1 to v_4, passing through v_3 in this case. There are no paths of length 2 from v_2 to v_4. These vertices are not neighbors, either, so any communication between these vertices must pass through multiple intermediaries. Using Formula (3.9), we see that the total number of edges in the graph is $(1/2) \cdot \text{trace}(A^2) = (1/2) \cdot (3 + 2 + 2 + 1 + 2) = 5$.

Higher powers of the adjacency matrix provide additional information about the connections in a graph. For instance, since $A^3 = A(A^2)$, the $[i, j]$ entry of A^3 is given by

$$A^3_{[i,j]} = a_{i,1} \cdot A^2_{[1,j]} + a_{i,2} \cdot A^2_{[2,j]} + \cdots + a_{i,N} \cdot A^2_{[N,j]} .$$

Moreover, for each number ℓ between 1 and N, we have

$$A^2_{[\ell,j]} = a_{\ell,1}a_{1,j} + a_{\ell,2}a_{2,j} + \cdots + a_{\ell,N}a_{N,j} .$$

Sorting out these sums, we see that

$$A^3_{[i,j]} \text{ is the sum of all terms } a_{i,\ell} \cdot a_{\ell,k} \cdot a_{k,j} , \tag{3.10}$$

for ℓ and k between 1 and N.

Each of the products $a_{i,\ell} \cdot a_{\ell,k} \cdot a_{k,j}$ is equal to either 0 or 1. The value 1 occurs when vertex v_i is connected to vertex v_j via the intermediary vertices v_ℓ and v_k. This forms a *walk along three edges* to get from v_i to v_j. Adding up over all ℓ and k, we see that **the matrix entry $A^3_{[i,j]}$ is equal to the total number of ways to walk from v_i to v_j along three edges.**

Remark A walk from one vertex to another along three edges might not be the shortest possible walk between the two vertices. For example, if v_i is directly connected to v_j by an edge in the graph, then we can take $\ell = j$ and $k = i$ and walk from v_i to v_j, back to v_i, and back to v_j. That still counts as three edges. Also, if v_i is directly connected to v_j and v_k is any other vertex connected to v_j, then the walk from v_i to v_j to v_k and back to v_j traverses three edges.

A walk along three edges that starts and ends at the same vertex makes a *triangle* in the graph. Each triangle at v_i can be traversed in two opposite directions. Therefore, each diagonal entry $A^3_{[i,i]}$ is twice the number of triangles based at v_i. Each triangle has three vertices, of course, so we can *count the total number of triangles in a graph* by adding up the diagonal entries of A^3 and dividing by 6.

$$\text{The number of triangles in a graph } = (1/6) \cdot \text{trace}(A^3) \tag{3.11}$$

Example 3.8 Looking again at the graph from Examples 3.6 and 3.7, the cube of its adjacency matrix is

$$A^3 = A\,A^2 = \begin{bmatrix} 0\,1\,1\,0\,1 \\ 1\,0\,0\,0\,1 \\ 1\,0\,0\,1\,0 \\ 0\,0\,1\,0\,0 \\ 1\,1\,0\,0\,0 \end{bmatrix} \begin{bmatrix} 3\,1\,0\,1\,1 \\ 1\,2\,1\,0\,1 \\ 0\,1\,2\,0\,1 \\ 1\,0\,0\,1\,0 \\ 1\,1\,1\,0\,2 \end{bmatrix} = \begin{bmatrix} 2\,4\,4\,0\,4 \\ 4\,2\,1\,1\,3 \\ 4\,1\,0\,2\,1 \\ 0\,1\,2\,0\,1 \\ 4\,3\,1\,1\,2 \end{bmatrix}.$$

The diagonal entries of A^3 indicate that vertices v_1, v_2, and v_5 each lie on one triangle. Indeed, this is the only triangle in this small graph. The other entries count walks along 3 edges between the corresponding vertices. For example, there are three such walks from v_2 to v_5 but only one from v_2 to v_4.

Similar computations show that, for each natural number n, the total number of walks that start at the vertex v_i, end at the vertex v_j, and cover n edges is equal to the $[i, j]$ entry of the power A^n of the adjacency matrix. It is important to note that any walk that covers two or more edges may not be the shortest possible route between the two vertices.

A walk along four edges that starts and ends at the same vertex *and* that does not repeat any edge makes a "square" in the graph. More properly, this is called a **4-cycle**. Every 4-cycle that starts and ends at vertex v_i will be included in the matrix entry $A^4_{[i,i]}$. That entry also includes repetitive paths, such as walks that go back and forth twice between v_i and any neighboring vertex and walks that go out along two edges and then return along the same edges. To count just the 4-cycles, observe that a walk of the form $v_i \to v_j \to v_k \to v_\ell \to v_i$ that does not repeat any vertex occurs when two vertices, in this case v_i and v_k, have more than one common neighbor. The matrix entry $A^2_{[i,k]}$ counts common neighbors of v_i and v_k. Every pair of common neighbors generates one 4-cycle. The number of pairs chosen from n objects is $\binom{n}{2} = \frac{n(n-1)}{2}$. So, the number of 4-cycles with v_i and v_k at opposite corners is equal to $(1/2)A^2_{[i,k]}\left(A^2_{[i,k]} - 1\right)$. Summing up over all values of k not equal to i will give us a count of the number of 4-cycles that have v_i at one corner.

$$\text{The number of 4-cycles at } v_i = (1/2)\sum_{k \neq i} A^2_{[i,k]}\left(A^2_{[i,k]} - 1\right). \tag{3.12}$$

For convenience, we record our findings in one place.

Formula 3.9 Let $\mathcal{G}$ be a graph with N vertices, labeled $v_1, v_2, \ldots, v_N$. Let A be the adjacency matrix of $\mathcal{G}$.

- For each i between 1 and N, the *degree* of vertex v_i is equal to the sum of the entries in the ith row of A.
- The total number of edges in $\mathcal{G}$ is equal to one-half of the sum of the degrees of all the vertices. Equivalently,

$$\text{the number of edges} = \frac{1}{2}\mathbf{1}^T A \mathbf{1}. \tag{3.13}$$

- For all i and j, the $[i, j]$ entry of A^2 is equal to the number of paths of length 2 between vertices v_i and v_j. This is the same as the number of common neighbors of these vertices in the graph.
- For each i, the entry $A^2_{[i,i]}$ is equal to the degree of vertex v_i.
- The total number of edges in the graph is equal to one-half the sum of the diagonal entries of A^2. That is,

$$\text{the number of edges} = \frac{1}{2} \cdot \text{trace}(A^2). \tag{3.14}$$

- For each i and j, the matrix entry $A^3_{[i,j]}$ is equal to the total number of paths from v_i to v_j along three edges. In particular, $A^3_{[i,i]}$ is equal to twice the number of triangles that include vertex v_i.
- The total number of triangles in the graph is equal to one-sixth of the sum of the diagonal entries of A^3. That is,

$$\text{the number of triangles in } \mathcal{G} = (1/6) \cdot \text{trace}(A^3). \tag{3.15}$$

- For each natural number n and each pair of vertices v_i and v_j, the total number of walks that start at v_i, end at v_j, and cover n edges is equal to the $[i, j]$ entry of the power A^n of the adjacency matrix.
- For vertices v_i and v_k, with $i \neq k$, the number of 4-cycles with v_i and v_k at opposite corners is equal to

$$(1/2) \cdot A^2_{[i,k]} \cdot \left(A^2_{[i,k]} - 1\right). \tag{3.16}$$

- For each i between 1 and N, the number of 4-cycles having v_i at one corner is equal to

$$(1/2) \sum_{k \neq i} A^2_{[i,k]} \left(A^2_{[i,k]} - 1\right). \tag{3.17}$$

  Note that the sum is over all values of k not equal to i.

Example 3.10 Given a dictionary of terms and a library of documents, we can form a term–document matrix, like the one in Table 1.3. As in Example 3.1, we compare two documents to each other by computing the cosine of the angle between their respective vectors of terms. This gives rise to a graph where each document is a vertex and there is an edge when the corresponding cosine value is above a selected threshold. Finding connections within this graph can inform the results of a search query. For instance, if a certain document is deemed relevant to a search, then documents connected to it in the graph might also be deemed relevant. Similarly, in Example 3.2, we compare terms in the dictionary by computing the cosine of the angle between their respective document vectors. A larger cosine identifies terms that are more likely to occur together in the same documents. We can use this information to make a graph whose vertices are the terms in the dictionary, with an edge joining two vertices when the corresponding cosine is large. Connections within this graph can help to suggest additional search terms to strengthen a query. In Chapter 9, Section 9.4, we explore how to find clusters of terms that tend to appear in similar documents.

Example 3.11 The Hungarian mathematician Paul Erdös (1913–1996) was famously collaborative in his research, generously sharing many brilliant insights and conjectures. His approximately 1500 research articles involved hundreds of coauthors. We can form a collaboration graph of mathematicians by considering every author of a mathematical research article as a vertex, with an edge between two authors if they have coauthored together. Each author then has an *Erdös number* defined as the length of the shortest path that connects to Paul Erdös. For those who coauthored with Erdös himself, this number is 1; coauthors of those coauthors have Erdös number 2, and so on. An author who has no path connecting to Erdös has number ∞, though this is rare. It is fair to say that every mathematician knows their Erdös number. (I am one of the more than 12,000 people with an Erdös number of 2.)

A similar concept in the film industry connects actors who have appeared in the same movie together. In this way, every actor can compute their *Bacon number* as the shortest path linking them to the prolific and versatile actor Kevin Bacon. Interestingly, the mathematician Daniel Kleitman has an Erdös number of 1 and a Bacon number of 2, having appeared along with Minnie Driver in the film *Good Will Hunting*. The actress and mathematics writer Danica McKellar has an Erdös number of 4 and a Bacon number of 2.

Example 3.12 Suppose we have a collection of objects, each identified by the values of two parameters or two coordinates. Think of each object as a vertex and join two vertices with an edge if they have the same value for at least one coordinate. In this case, a vertex v_k is a common neighbor of v_i and v_j if it has one coordinate in common with each of them. For example, $v_k = (\alpha, \gamma)$ is a common neighbor of $v_i = (\alpha, \beta)$ and $v_j = (\delta, \gamma)$. Three vertices will form a triangle if, and only if, they all have the same value in one coordinate. In the same vein, the path

$$v_i = (\alpha, \beta) \mapsto v_k = (\alpha, \gamma) \mapsto v_j = (\delta, \gamma) \mapsto v_\ell = (\delta, \beta) \mapsto v_i$$

forms a 4-cycle in this graph.

One graph like this is formed by taking a group of people and connecting two people if they share the initial letter of either their given name or their family name.

Counting paths of various lengths in a graph helps to distinguish certain vertices, or pairs of vertices, that are especially well connected. Think of every walk between two vertices as a communication link. A message sent along a link that does not pass through too many intermediaries is more likely to get through clearly without much deterioration. So, walks along two, three, or four edges are advantageous. Also, having several such paths to choose from gives us an advantage over a would-be interceptor, who might have a harder time spying on or blocking multiple pathways. This suggests assigning, to each pair of vertices, a connection quality score that counts relatively short paths between them. Given that walks along three and four edges can involve repeated edges, we down-weight such paths in computing the score.

A vertex that is part of many triangles and 4-cycles represents an important communication hub in the graph. Some vertices may only be able to link to each other via such a hub. Thus, removing a hub could damage the overall connectivity of the graph. This suggests assigning a hub score to each vertex by adding up its number of neighbors, the number of triangles it belongs to, and the number of 4-cycles it is part of.

$$\text{Hub score of } v_i = A^2_{[i,\,i]} + (1/2)\cdot A^3_{[i,\,i]} + (1/2)\cdot \sum_{k\neq i} A^2_{[i,\,k]} \cdot \left(A^2_{[i,\,k]} - 1\right). \tag{3.18}$$

In Section 3.9, we look at some other possible formulas for connection quality scores and hub scores.

3.5 Simple Linear Regression

Suppose there are two measurements, X and Y, that can be observed for each member of some population. Let us say we have collected a set of M data points in the xy-plane, $\{(x_i,\ y_i)\ :\ i = 1, \ldots,\ M\}$, where each data point represents the observed values of X and Y on a particular member of the population. A plot of the data might suggest an approximately linear relationship between the variables X and Y, with $\alpha + \beta X \approx Y$ for some choice of numbers α and β. To test this idea, we use our sample data to find a line $\alpha + \beta x = y$, called the *regression line*, that best fits the data. We can use such a line to make predictions about other data. For example, if we randomly observed the value x_*, say, then we would expect the corresponding value y_* to be close to $\alpha + \beta x_*$.

It is very unlikely that any given line $\alpha + \beta x = y$ will fit our data exactly. That would require $\alpha + \beta x_i = y_i$, for all i from 1 to M. That amounts to imposing M constraints on just two unknowns, α and β. In any case, real-world data measurements may be rounded off or otherwise approximated. That makes a perfect fit for our data unrealistic. But maybe we can find a *good* fit. For that, we can choose tentative values for α and β and, for each data point, compare the predicted value $\widehat{y_i} = \alpha + \beta x_i$ to the measured value y_i. Specifically, given α and β, compute the **sum of square errors** (SSE) defined by

$$\begin{aligned} SSE &= SSE(\alpha, \beta) \\ &= (\widehat{y}_1 - y_1)^2 + (\widehat{y}_2 - y_2)^2 + \cdots + (\widehat{y}_M - y_M)^2 \\ &= (\alpha + \beta x_1 - y_1)^2 + (\alpha + \beta x_2 - y_2)^2 + \cdots + (\alpha + \beta x_M - y_M)^2 . \end{aligned} \tag{3.19}$$

We seek the pair of values for α and β that minimize the sum of the square errors. This is called a **least squares problem**. Solving this problem gives us a regression line that fits our data the best compared to all other lines in the plane.

One approach to minimizing the expression in (3.19) uses multivariable calculus. First, compute the partial derivatives with respect to α and β. Then determine where these are both equal to 0. This entails solving a system of two equations in the two unknowns α and β.

A different approach uses an orthogonal projection to help us. For this, form the vectors $\mathbf{x}$ and $\mathbf{y}$ in $\mathbb{R}^M$ with coordinates x_i and y_i, respectively. Also, let $\mathbf{1}$ denote the vector in $\mathbb{R}^M$ with coordinates all 1s, and take A to be the $M \times 2$ matrix whose columns are the vectors $\mathbf{1}$ and $\mathbf{x}$. That is,

$$A = \begin{bmatrix} 1 & x_1 \\ \vdots & \vdots \\ 1 & x_M \end{bmatrix} . \tag{3.20}$$

In this context, for each α and β, we have

$$A \begin{bmatrix} \alpha \\ \beta \end{bmatrix} = \alpha \cdot \mathbf{1} + \beta \cdot \mathbf{x} .$$

This is a vector in the plane generated by $\mathbf{1}$ and $\mathbf{x}$, the column vectors of the matrix A, as shown in Figure 3.4.

The sum of square errors in (3.19) is now equivalent to

$$SSE = SSE(\alpha, \beta) = \left\| A \begin{bmatrix} \alpha \\ \beta \end{bmatrix} - \mathbf{y} \right\|^2 . \tag{3.21}$$

It follows that minimizing the SSE amounts to finding the values α and β for which the vector $A \begin{bmatrix} \alpha \\ \beta \end{bmatrix}$ is *closer to* $\mathbf{y}$ than any other vector in the plane generated by $\mathbf{1}$ and $\mathbf{x}$. Our knowledge of geometry and right triangles suggests that, of all vectors $\mathbf{v}$ lying in this plane, the one closest to $\mathbf{y}$ should have the property that the difference $\mathbf{v} - \mathbf{y}$ is orthogonal to the plane. In other words, $\mathbf{v}$ and $\mathbf{v} - \mathbf{y}$ should be the adjacent sides of a right triangle that has hypotenuse $\mathbf{y}$. The vector $\mathbf{v}$ will be the orthogonal projection of $\mathbf{y}$ into the plane. See Figure 3.4.

In order for the vector $\mathbf{v} - \mathbf{y}$ to be orthogonal to the given plane, it must be orthogonal to both $\mathbf{1}$ and $\mathbf{x}$. That is, $\mathbf{1}^T(\mathbf{v} - \mathbf{y}) = 0$ and $\mathbf{x}^T(\mathbf{v} - \mathbf{y}) = 0$. Taken

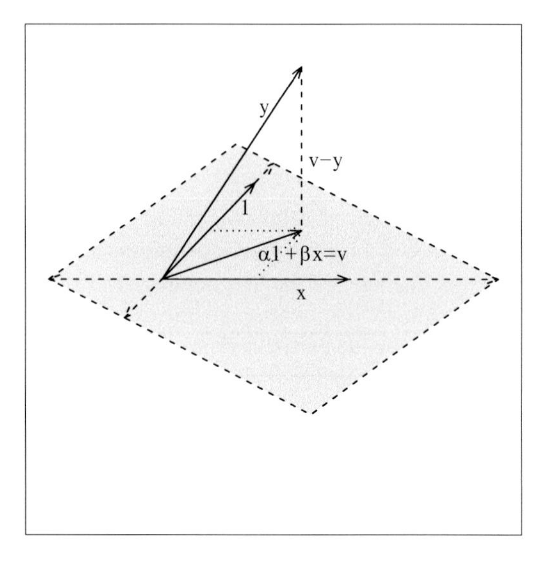

Fig. 3.4 The vector $\mathbf{v} = \alpha\mathbf{1} + \beta\mathbf{x}$ is the closest vector to $\mathbf{y}$ in the plane generated by $\mathbf{1}$ and $\mathbf{x}$. The vector $\mathbf{v} - \mathbf{y}$ is orthogonal to this plane

together, these two equations yield $A^T(\mathbf{v} - \mathbf{y}) = \mathbf{0}$. Also, $\mathbf{v}$ must have the form $\mathbf{v} = A\begin{bmatrix}\alpha \\ \beta\end{bmatrix}$ for some values of α and β.

Thus, to minimize the SSE (3.19) and (3.21), the vector $\begin{bmatrix}\alpha \\ \beta\end{bmatrix}$ must satisfy

$$A^T\left(A\begin{bmatrix}\alpha \\ \beta\end{bmatrix} - \mathbf{y}\right) = \mathbf{0}.$$

We rewrite this as

$$A^T A\begin{bmatrix}\alpha \\ \beta\end{bmatrix} = A^T\mathbf{y}. \tag{3.22}$$

This formula (3.22) is called the **normal equation** for the given least squares problem. The 2×2 matrix $A^T A$ has an inverse, assuming $\mathbf{x}$ is not a constant vector. (See Example 3.16 and the remark following that.) Hence, the normal equation (3.22) has the unique solution

$$\begin{bmatrix}\alpha \\ \beta\end{bmatrix} = \left(A^T A\right)^{-1} A^T\mathbf{y}. \tag{3.23}$$

Example 3.13 In Example 1.15, we examined the relationship between the rate at which a cricket chirps and the surrounding air temperature. For a small sample of observations, we found a Pearson correlation coefficient of 0.958. Since that number is near 1, we concluded that there was a strong correlation between these two quantities. When we plot the data, it appears that the relationship between chirp rate and temperature may be close to linear. To examine this more closely, consider the five observations listed in Table 1.1 above.

We seek a regression line $\alpha + \beta x = y$ that best fits our data, where x is the temperature and y is the chirp rate. To this end, let

$$A = \begin{bmatrix} 1 & 88.6 \\ 1 & 93.3 \\ 1 & 75.2 \\ 1 & 80.6 \\ 1 & 82.6 \end{bmatrix},$$

as in (3.20). In accordance with the normal equation (3.22), we seek a vector $\begin{bmatrix} \alpha \\ \beta \end{bmatrix}$ such that

$$A^T A \begin{bmatrix} \alpha \\ \beta \end{bmatrix} = A^T \begin{bmatrix} 20.0 \\ 19.8 \\ 15.5 \\ 17.1 \\ 17.2 \end{bmatrix}.$$

In this example, $A^T A = \begin{bmatrix} 5.0 & 420.3 \\ 420.3 & 35529.01 \end{bmatrix}$. Thus,

$$\begin{bmatrix} \alpha \\ \beta \end{bmatrix} = (A^T A)^{-1} A^T \begin{bmatrix} 20.0 \\ 19.8 \\ 15.5 \\ 17.1 \\ 17.2 \end{bmatrix} = \begin{bmatrix} -4.1515 \\ 0.2626 \end{bmatrix}.$$

Figure 3.5 shows the data with the regression line $y = -4.1515 + 0.2626x$ that we just computed.

When we solve the normal equation (3.22), we are finding a vector $\begin{bmatrix} \alpha \\ \beta \end{bmatrix}$ for which the value $\left\| A \begin{bmatrix} \alpha \\ \beta \end{bmatrix} - \mathbf{y} \right\|^2$ is as small as possible. We are also claiming that the line $\alpha + \beta x = y$ is a good model for our data. We should be reluctant to make this claim without comparing the error in our regression to the overall variability in our measurements of Y just on their own, without taking X into account. In other words, if the errors that arise from estimating Y as $\alpha + \beta X$ are not much less than the intrinsic variability in Y itself, then our regression is not accomplishing much.

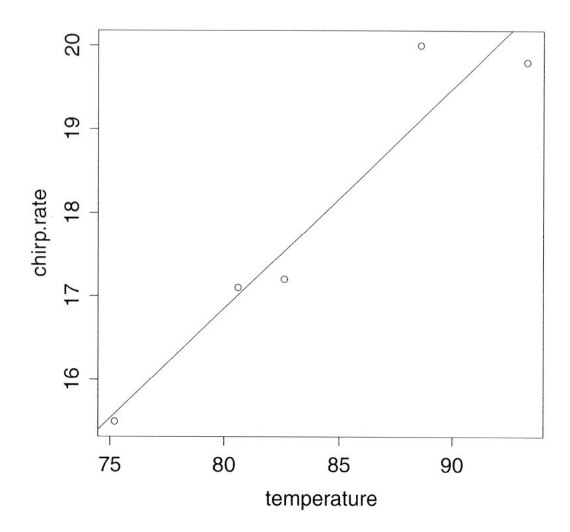

Fig. 3.5 Five observations of chirps per second of a cricket and air temperature along with the regression line based on these data

To compute the variability in the sampled observations of Y, we first compute the average, or mean, of the sample values. The mean of the set of observations $\{y_i \mid i = 1, \ldots, M\}$ is defined by $\bar{y} = (y_1 + \cdots + y_M)/M$. The variability in the sample is then the sum of the squares of the differences between each observed value of Y and the average value $\bar{y}$. This is called the **total sum of squares**, or TSS. That is,

$$TSS = (y_1 - \bar{y})^2 + \cdots (y_M - \bar{y})^2 = ||\mathbf{y} - \bar{y}\mathbf{1}||^2. \tag{3.24}$$

Statisticians use a value called R^2 to compare TSS to the error SSE that results from the least squares solution. Specifically,

$$\begin{aligned} R^2 &= 1 - \frac{SSE}{TSS} \\ &= 1 - \frac{\left\| A \begin{bmatrix} \alpha \\ \beta \end{bmatrix} - \mathbf{y} \right\|^2}{||\mathbf{y} - \bar{y}\mathbf{1}||^2} \\ &= 1 - \frac{(\alpha + \beta x_1 - y_1)^2 + \cdots + (\alpha + \beta x_M - y_M)^2}{(y_1 - \bar{y})^2 + \cdots (y_M - \bar{y})^2}. \end{aligned} \tag{3.25}$$

If our regression estimates are perfect, then $R^2 = 1$. If the regression errors are large, then R^2 will be close to 0.

Example 3.14 Returning to the relationship between air temperature and the rate at which crickets chirp, we get $R^2 = 0.9184$. This is near 1, which suggests that the regression line we found is an accurate estimator for the chirp rate of a cricket given the air temperature, at least based on our small sample of observed values.

Example 3.15 In Example 1.16, we looked at data from 1980 on child mortality and female literacy rates, as well as per capita gross national product, for 57 developing countries. The data are given in Table 1.2. Figure 1.8 shows the corresponding scatter plots. Equation (1.28) shows the different correlation coefficients we computed.

To compute the least squares regression line for child mortality as a function of the female literacy rate, first form the matrix A with all 1s in the first column and the FLR data in the second column. The normal equation is

$$A^T A \begin{bmatrix} \alpha \\ \beta \end{bmatrix} = A^T(CM) = \begin{bmatrix} 8681 \\ 335001 \end{bmatrix}. \tag{3.26}$$

The solution is

$$\begin{bmatrix} \alpha \\ \beta \end{bmatrix} = (A^T A)^{-1} \begin{bmatrix} 8681 \\ 335001 \end{bmatrix} = \begin{bmatrix} 261.561 \\ -2.276 \end{bmatrix}. \tag{3.27}$$

In particular, the slope of -2.276 indicates that we would expect a rise of 4% in the female literacy rate to be associated with a reduction in child mortality by about 9 children per 1000 births. For this example, we have $SSE \approx 101434.5$ and $TSS \approx 287377.9$. Thus, $R^2 = 1-(SSE/TSS) \approx 0.647$. That value of R^2 is not especially near 1, indicating that the female literacy rate by itself is not enough to explain the variability in the child mortality rate (see Figure 3.6).

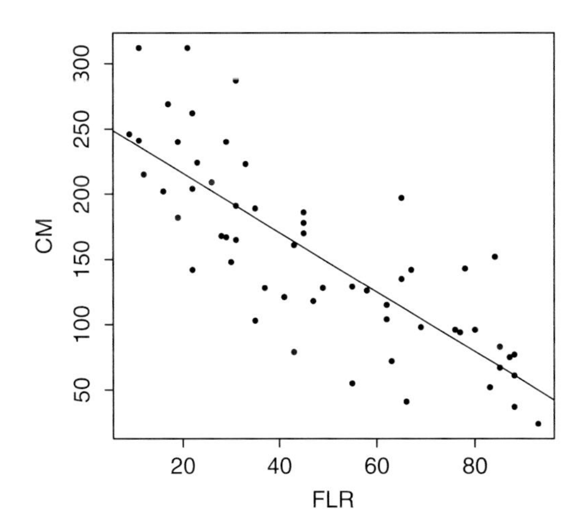

Fig. 3.6 The figure shows the least squares regression line and a scatter plot for data on child mortality and female literacy rates from 57 developing countries

Remark The vector $\bar{y}\mathbf{1}$ is the projection of the vector $\mathbf{y}$ along the all-1s vector $\mathbf{1}$. (See (1.41) in Section 1.12.) Thus, the vector $\mathbf{y} - \bar{y}\mathbf{1}$ is orthogonal to $\mathbf{1}$. The vector

$A\begin{bmatrix}\alpha\\ \beta\end{bmatrix}$ represents the projection of $\mathbf{y}$ into the plane generated by $\mathbf{1}$ and $\mathbf{x}$. Thus, $A\begin{bmatrix}\alpha\\ \beta\end{bmatrix} - \bar{y}\mathbf{1}$ lies in this plane, while $A\begin{bmatrix}\alpha\\ \beta\end{bmatrix} - \mathbf{y}$ is orthogonal to it. In short, we have a right triangle with adjacent sides $A\begin{bmatrix}\alpha\\ \beta\end{bmatrix} - \mathbf{y}$ and $A\begin{bmatrix}\alpha\\ \beta\end{bmatrix} - \bar{y}\mathbf{1}$ and hypotenuse $\mathbf{y} - \bar{y}\mathbf{1}$. It follows that

$$||\mathbf{y} - \bar{y}\mathbf{1}||^2 = ||A\begin{bmatrix}\alpha\\ \beta\end{bmatrix} - \mathbf{y}||^2 + ||A\begin{bmatrix}\alpha\\ \beta\end{bmatrix} - \bar{y}\mathbf{1}||^2 .$$

The expression $||A\begin{bmatrix}\alpha\\ \beta\end{bmatrix} - \bar{y}\mathbf{1}||^2$ is called the **sum of squares due to regression**, or SSR, for short. Thus, we have

$$TSS = SSE + SSR .$$

In particular, this shows that $SSE/TSS \leq 1$.

Example 3.16 The solution to the normal equation (3.22) can be expressed in terms of the observed values x_i and y_i. Specifically,

$$A^T A = \begin{bmatrix} 1 & \cdots & 1 \\ x_1 & \cdots & x_M \end{bmatrix} \begin{bmatrix} 1 & x_1 \\ \vdots & \vdots \\ 1 & x_M \end{bmatrix} = \begin{bmatrix} M & \sum x_i \\ \sum x_i & \sum {x_i}^2 \end{bmatrix} .$$

From Formula 2.18 and (2.9),

$$\left(A^T A\right)^{-1} = \frac{1}{M\left(\sum {x_i}^2\right) - \left(\sum x_i\right)^2} \begin{bmatrix} \sum {x_i}^2 & -\sum x_i \\ -\sum x_i & M \end{bmatrix} . \tag{3.28}$$

Also,

$$A^T \mathbf{y} = \begin{bmatrix} \sum y_i \\ \sum x_i y_i \end{bmatrix} .$$

Thus, the optimal values of α and β are given by

$$\begin{bmatrix}\alpha\\ \beta\end{bmatrix} = \left(A^T A\right)^{-1} A^T \mathbf{y} \tag{3.29}$$

$$= \frac{1}{M\left(\sum {x_i}^2\right) - \left(\sum x_i\right)^2} \begin{bmatrix} \left(\sum {x_i}^2\right)\left(\sum y_i\right) - \left(\sum x_i\right)\left(\sum x_i y_i\right) \\ M \cdot \left(\sum x_i y_i\right) - \left(\sum x_i\right)\left(\sum y_i\right) \end{bmatrix} .$$

It is possible to show that the optimal value of α satisfies $\alpha = \bar{y} - \beta\bar{x}$, where $\bar{y} = \sum y_i/M$ and $\bar{x} = \sum x_i/M$ are the average values of the sample observations.

Remark Note that the inverse matrix $\left(A^T A\right)^{-1}$, in (3.28), is defined unless $M\left(\sum {x_i}^2\right) = \left(\sum x_i\right)^2$. The left-hand side here is $||\mathbf{1}_M||^2 \cdot ||\mathbf{x}||^2$, while the right-hand side is $\left(\mathbf{1}_M^T\mathbf{x}\right)^2$. Equality of these expressions amounts to equality in the Cauchy–Schwarz inequality. As we remarked following the proof of Theorem 1.10, this will happen if, and only if, $\mathbf{x} = 0$ or $\mathbf{x}$ is a constant multiple of $\mathbf{1}$. Thus, as long as $\mathbf{x}$ is not a constant vector, then the inverse matrix in (3.28) is defined.

3.6 *k*-Means

A college admissions counselor is considering the SAT exam scores of a set of 12 high school graduates. To help advise the students about course placement, the counselor considers the verbal and math component scores separately, rather than only the combined score. Here are the exam scores of the 12 students.

Student #	1	2	3	4	5	6
Verbal	614	796	693	676	617	729
Math	604	796	604	735	512	515
Student #	7	8	9	10	11	12
Verbal	613	796	570	579	542	569
Math	669	786	786	777	584	556

The average verbal score for this group is 649.5, and the average math score is 660.33. Next, the counselor wants to create smaller subsets of students with similar scores. How to do that may not be clear from the scatter plot alone. See Figure 3.7.

The concept of k**-means** provides an algorithm for generating clusters of similar elements within a larger set. Here is how it works for the SAT scores. Suppose we wish to make $k = 2$ smaller subsets of students. Start by selecting two students to compare the others to. To be specific, suppose we select student #1, with scores [614, 604], and student #12, with scores [569, 556]. Now assign a student to *cluster A* if their score vector is closer to the score vector of student #1 and *cluster B* if their score vector is closer to that of student #12. That is, we take each student's score vector and compute the norms of the differences between it and each of the starting vectors. In this case, cluster A consists of students #1, 2, 3, 4, 6, 7, 8, 9, and 10, who are closer to student #1, and cluster B consists of students #5, 11, and 12, who are closer to student #12. For the next round, we compute the mean verbal and math scores for all members of each cluster. This gives [674.0, 696.89] for cluster A and [576.0, 550.67] for cluster B. These vectors are called the **cluster means**. Note that

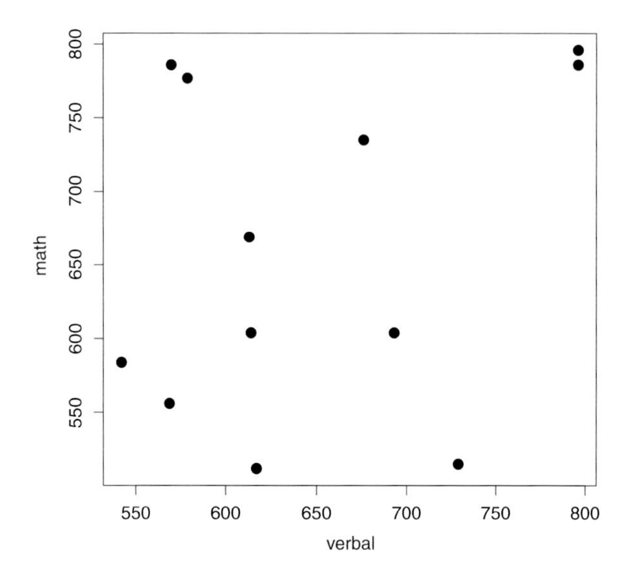

Fig. 3.7 Verbal and math SAT scores for a randomly selected set of 12 students

the cluster means are not the scores of any one student. Now, reassign each student to either cluster A or cluster B according to which of these new cluster means is closer to the student's own score vector. In this example, the new cluster A consists of students #2, 3, 4, 7, 8, 9, and 10. The new cluster B contains students #1, 5, 6, 11, and 12. We notice that students #1 and #6 switched clusters. Repeat the process using the cluster means for the new cluster configuration. Continue in this way until the clusters do not change.

We can separate the original data set into more than 2 clusters in a similar way. To make k clusters, we start by selecting k vectors from our set as the initial cluster means. If we want, we may select these at random from the full data set. Next, compute the square distance from every vector in the data set to each of the k cluster means. Assign each vector in the data set to one of k clusters according to which cluster mean was closest. That is the first step of the process. For the second step, compute the new cluster means by averaging the elements of each cluster. Next, compute the square distance from every vector in the data set to each of the k new cluster means. Form new clusters by assigning each vector in the data set to the cluster with the closest cluster mean. As the clusters re-form, the cluster means change, and data points can change affiliations accordingly. Eventually, the process will stop, though we will not try to prove that.

Even with only $k = 2$ clusters and a small set of vectors to sort out, it is burdensome to do the computations by hand. Fortunately, many statistical computing platforms have an algorithm for computing k-means built in. Figure 3.8 was generated in R by applying the $kmeans(A, k)$ command to the matrix A of verbal and math SAT scores shown above. With $k = 2$, the algorithm produced one set of seven students (#1, 3, 5, 6, 7, 11, and 12) with math scores at or below the average for the entire group and a second set of five students (#2, 4, 8, 9, and 10) whose math scores are above the average for the whole group. The cluster means of the two groups are $\left[\, 625.29\ 577.71 \,\right]^T$ and $\left[\, 683.4\ 776.0 \,\right]^T$, respectively.

For $k = 4$, the four clusters represent students with scores that are (i) above average in verbal but below average in math (students #3 and 6), (ii) above average in both (#2 and 8), (iii) below average in verbal but above average in math (#4, 7, 9, and 10), and (iv) below average in both (#1, 5, 11, and 12). The cluster means are $\left[\, 711.0\ 559.5 \,\right]^T$, $\left[\, 796.0\ 791.0 \,\right]^T$, $\left[\, 609.5\ 741.75 \,\right]^T$, and $\left[\, 585.5\ 564.0 \,\right]^T$.

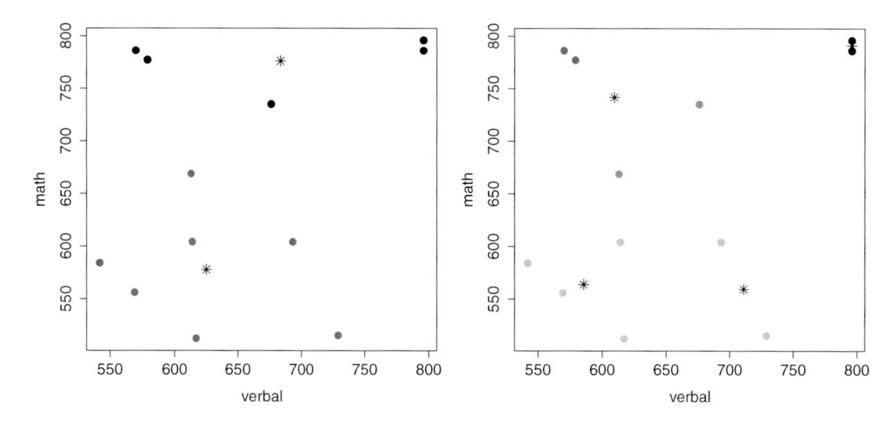

Fig. 3.8 The students are grouped into $k = 2$ clusters (left image) and $k = 4$ clusters (right image). The stars show the locations of the cluster means

Essentially, the k-means algorithm minimizes the sum of square distances between the points in the data set and their assigned cluster means. However, this may be what is called a *local minimum* in that there may be a different initial choice of means that would produce a different set of clusters with a smaller sum of square distances. For example, in the case of $k = 2$ clusters of the SAT data, moving student #7 to the other cluster gives an alternate stable solution. Picking an appropriate value of k is up to the user.

Example 3.17 (Information Retrieval) The method of k-means is useful in information retrieval, where we can cluster either documents that tend to have similar term contents, or terms that tend to appear in similar documents. One difficulty might be determining a suitable value of k to use. To illustrate, refer to the dictionary of terms and library of documents shown in Table 1.3, in Section 1.6. We will form clusters of terms by applying the k-means algorithm to the row vectors of the term–document matrix. It is not feasible to plot the clusters, as the vectors are in $\mathbb{R}^8$.

In the first round of the algorithm, the square of the distance between any two row vectors is the number of places where one vector has a 1 and the other has a 0. So, right from the start, we will tend to cluster terms whose vectors have the fewest disagreements.

Using a computer, with $k = 2$, we get clusters of four terms and 14 terms. The four terms *(analysis, derivative, Riemann, sequence)* all appear in the analysis-oriented texts $B3$ and $B7$. The cluster means are the row vectors of

$$\begin{bmatrix} 1/2 & 1/4 & 1 & 1/4 & 0 & 1/4 & 1 & 0 \\ 3/7 & 5/7 & 3/14 & 9/14 & 4/7 & 6/7 & 0 & 1 \end{bmatrix}.$$

Taking $k = 3$, the cluster of four analysis terms shows up again, but the cluster of 14 terms is split into the two clusters *(algebra, group, linear, matrix, permutation, symmetric, vector)* and *(applied, eigenvalue, factorization, inner product, nullspace, orthogonal, transpose)*. The cluster means now are the row vectors of

$$\begin{bmatrix} 1/2 & 1/4 & 1 & 1/4 & 0 & 1/4 & 1 & 0 \\ 6/7 & 5/7 & 2/7 & 1 & 1 & 6/7 & 0 & 1 \\ 0 & 5/7 & 1/7 & 2/7 & 1/7 & 6/7 & 0 & 1 \end{bmatrix}.$$

From this, we see that the two clusters of seven terms are distinguished mainly by which terms appear in documents $B1$, $B4$, and $B5$. Those are texts in abstract algebra as opposed to linear algebra.

3.7 Projection and Reflection Revisited

From Definition 1.29 and Formula (1.37), the projection of a vector $\mathbf{b}$ along a nonzero vector $\mathbf{a}$, both in the same space $\mathbb{R}^N$, is $\text{proj}_{\mathbf{a}}\mathbf{b} = \dfrac{\mathbf{a}^T\mathbf{b}}{||\mathbf{a}||^2} \cdot \mathbf{a}$. As discussed in Example 2.8 and Equation (2.5), we have

$$(\mathbf{a}^T\mathbf{b}) \cdot \mathbf{a} = \mathbf{a}\left[\mathbf{a}^T\mathbf{b}\right] = \left(\mathbf{a}\mathbf{a}^T\right)\mathbf{b}.$$

(The last step uses the associativity of matrix multiplication.) This shows that the action of projection along $\mathbf{a}$ is represented by the matrix

$$P_{\mathbf{a}} = \frac{1}{||\mathbf{a}||^2} \cdot (\mathbf{a}\mathbf{a}^T). \tag{3.30}$$

This is an example of an *outer product*, defined in Example 2.29. The projection of $\mathbf{b}$ along $\mathbf{a}$ is given by the matrix/vector product $P_{\mathbf{a}}\mathbf{b}$. Writing out the full $N \times N$ matrix for $P_{\mathbf{a}}$, we get

$$P_{\mathbf{a}} = \frac{1}{||\mathbf{a}||^2} \cdot \mathbf{a}\mathbf{a}^T = \frac{1}{||\mathbf{a}||^2} \cdot \begin{bmatrix} a_1 \\ a_2 \\ \vdots \\ a_M \end{bmatrix} \begin{bmatrix} a_1 & a_2 & \cdots & a_M \end{bmatrix}$$

$$= \frac{1}{||\mathbf{a}||^2} \cdot \begin{bmatrix} {a_1}^2 & a_1a_2 & \cdots & a_1a_M \\ a_2a_1 & {a_2}^2 & \cdots & a_2a_M \\ \vdots & \vdots & \ddots & \vdots \\ a_Ma_1 & a_Ma_2 & \cdots & {a_M}^2 \end{bmatrix}. \tag{3.31}$$

The entry in row i and column j of $P_{\mathbf{a}}$ is $(a_ia_j)/||\mathbf{a}||^2$. Notice that every row of $P_{\mathbf{a}}$ is a numerical multiple of $\mathbf{a}^T$ and every column is a numerical multiple of $\mathbf{a}$.

Visually, the projection $P_{\mathbf{a}}\mathbf{b}$ represents the shadow that $\mathbf{b}$ casts along the line determined by $\mathbf{a}$ in the presence of a distant light source whose rays all shine perpendicular to $\mathbf{a}$. Since the shadow cast by $\mathbf{b}$ is already its own shadow, we see that $P_{\mathbf{a}}\,(P_{\mathbf{a}}\mathbf{b}) = P_{\mathbf{a}}\mathbf{b}$. In matrix terms, we have $P_{\mathbf{a}} \cdot P_{\mathbf{a}} = P_{\mathbf{a}}^2 = P_{\mathbf{a}}$. In other words, $P_{\mathbf{a}}$ is equal to its own square! Indeed, the outer product formulation of $P_{\mathbf{a}}$, in (3.30), gives us

$$\begin{aligned} P_{\mathbf{a}}^2 &= \frac{1}{||\mathbf{a}||^4}(\mathbf{a}\mathbf{a}^T)(\mathbf{a}\mathbf{a}^T) \\ &= \frac{1}{||\mathbf{a}||^4}\mathbf{a}(\mathbf{a}^T\mathbf{a})\mathbf{a}^T \text{ (by associativity)} \\ &= \frac{\mathbf{a}^T\mathbf{a}}{||\mathbf{a}||^4}(\mathbf{a}\mathbf{a}^T) \text{ (since } \mathbf{a}^T\mathbf{a} \text{ is a scalar)} \\ &= \frac{1}{||\mathbf{a}||^2}(\mathbf{a}\mathbf{a}^T) \text{ (since } \mathbf{a}^T\mathbf{a} = ||\mathbf{a}||^2) \\ &= P_{\mathbf{a}} \text{ as claimed.} \end{aligned}$$

Definition 3.18 A square matrix A is said to be **idempotent** if $A^2 = A$.

Remark Suppose the matrix $A \in \mathbb{M}_{N\times N}(\mathbb{R})$ is idempotent, so that $A^2 = A$. Then, for all natural numbers k, $A^k = A$. Indeed, for $k \geq 2$, we have

$$A^k = A^{k-2} \cdot A^2 = A^{k-2} \cdot A = A^{k-1}.$$

Repeating this argument for A^{k-1}, if necessary, leads to the result that $A^k = A$ for all natural numbers k. The word *idempotent* literally means *same powers*.

Remark If the $N \times N$ matrix A is idempotent, then so is the matrix $\mathcal{I}_N - A$. We leave the proof of this as an exercise.

Example 3.19 In Section 1.12, we computed the projection of an arbitrary vector $\mathbf{x}$ along the "all-1s" vector $\mathbf{1}$ as $\text{proj}_{\mathbf{1}}\mathbf{x} = \bar{x}\mathbf{1}$, where $\bar{x}$ denotes the *mean*, or average, of the coordinates of $\mathbf{x}$. Since $||\mathbf{1}||^2 = N$, formula (3.30) shows that the matrix for projection along $\mathbf{1}$ is

$$P_{\mathbf{1}} = \frac{1}{N} \cdot \mathbf{1} \cdot \mathbf{1}^T = \frac{1}{N} \cdot \begin{bmatrix} 1 & 1 & \cdots & 1 \\ \vdots & \vdots & \ddots & \vdots \\ 1 & 1 & \cdots & 1 \end{bmatrix} .$$

In other words, the matrix $P_{\mathbf{1}}$ is the $N \times N$ matrix where every entry is the number $1/N$.

For any vector $\mathbf{x}$ in $\mathbb{R}^N$, the mean-centered vector $\mathbf{x} - \bar{x}\mathbf{1}$ can be expressed as $\mathbf{x} - \bar{x}\mathbf{1} = (\mathcal{I} - P_{\mathbf{1}})\mathbf{x}$. Notice that the matrix $\mathcal{I} - P_{\mathbf{1}}$ has entries $(\mathcal{I} - P_{\mathbf{1}})[i,\ j] = \begin{cases} (N-1)/N & \text{if } i = j\,, \\ -1/N & \text{if } i \neq j. \end{cases}$ Every row and every column of this matrix sums to 0.

Example 3.20 The reflection of a vector $\mathbf{b}$ across a nonzero vector $\mathbf{a}$ is

$$\text{refl}_{\mathbf{a}}\mathbf{b} = 2 \cdot \text{proj}_{\mathbf{a}}\mathbf{b} - \mathbf{b}\,,$$

as in Definition 1.30. It follows that the action of reflection across $\mathbf{a}$ has the matrix representation

$$R_{\mathbf{a}} = 2 \cdot P_{\mathbf{a}} - \mathcal{I}\,. \tag{3.32}$$

That is, $R_{\mathbf{a}}\mathbf{b} = \text{refl}_{\mathbf{a}}\mathbf{b}$, for all $\mathbf{b}$. Since every vector is the reflection of its own reflection, it must be the case that ${R_{\mathbf{a}}}^2 = \mathcal{I}$. Indeed,

$${R_{\mathbf{a}}}^2 = (2 \cdot P_{\mathbf{a}} - \mathcal{I}) \cdot (2 \cdot P_{\mathbf{a}} - \mathcal{I}) = 4 \cdot {P_{\mathbf{a}}}^2 - 4 \cdot P_{\mathbf{a}} + \mathcal{I} = \mathcal{I}\,,$$

since ${P_{\mathbf{a}}}^2 = P_{\mathbf{a}}$. This is exciting because it means that the matrix $R_{\mathbf{a}}$ acts like a square root of the identity matrix *for any nonzero* $\mathbf{a}$! With numbers, the number 1 has only two square roots: 1 and -1. But the identity matrix has infinitely many square roots.

3.8 Geometry of 2 × 2 Matrices

Every $M \times N$ matrix A represents a *function* with domain $\mathbb{R}^N$ and target space $\mathbb{R}^M$. Specifically, multiplication by A implements a mapping

$$\mathbf{x} \mapsto A\mathbf{x}\,, \tag{3.33}$$

in which an arbitrary input vector $\mathbf{x}$ in $\mathbb{R}^N$ is mapped to the output vector $A\mathbf{x}$ in $\mathbb{R}^M$.

A function defined in this way, as multiplication by a matrix, is called a **linear transformation.** Linear transformations of $\mathbb{R}^2$ and $\mathbb{R}^3$ are especially important in computer graphics. In this section, we consider the case where A is a 2×2 matrix. We will look at the general case in Chapter 5. Rotations of $\mathbb{R}^3$ are discussed in Chapter 7.

So, let $A = \begin{bmatrix} a & b \\ c & d \end{bmatrix}$. Then $A \begin{bmatrix} 1 \\ 0 \end{bmatrix} = \begin{bmatrix} a \\ c \end{bmatrix}$ and $A \begin{bmatrix} 0 \\ 1 \end{bmatrix} = \begin{bmatrix} b \\ d \end{bmatrix}$. For an arbitrary vector $\mathbf{x} = \begin{bmatrix} x_1 \\ x_2 \end{bmatrix}$ in $\mathbb{R}^2$, it follows that

$$\begin{aligned} A\mathbf{x} &= \begin{bmatrix} a & b \\ c & d \end{bmatrix} \begin{bmatrix} x_1 \\ x_2 \end{bmatrix} \\ &= \begin{bmatrix} ax_1 + bx_2 \\ cx_1 + dx_2 \end{bmatrix} \\ &= x_1 \cdot \begin{bmatrix} a \\ c \end{bmatrix} + x_2 \cdot \begin{bmatrix} b \\ d \end{bmatrix} \\ &= x_1 \cdot A \begin{bmatrix} 1 \\ 0 \end{bmatrix} + x_2 \cdot A \begin{bmatrix} 0 \\ 1 \end{bmatrix}. \end{aligned} \tag{3.34}$$

When we consider only vectors $\mathbf{x}$ with coordinates between 0 and 1, we see that the transformation $\mathbf{x} \mapsto A\mathbf{x}$ has the visual effect of mapping the unit square, with corners at $(0, 0)$, $(1, 0)$, $(0, 1)$, and $(1, 1)$, to the parallelogram with corners at $(0, 0)$, (a, c), (b, d), and $(a + b, c + d)$. (If $\begin{bmatrix} a \\ c \end{bmatrix}$ and $\begin{bmatrix} b \\ d \end{bmatrix}$ lie on the same line, then this image is a line segment.)

When viewed as implementing a linear transformation, the matrix product BA is the matrix for the composition of the transformations implemented by B and A individually. To illustrate, let $A = \begin{bmatrix} 3 & 5 \\ 1 & 7 \end{bmatrix}$ and $B = \begin{bmatrix} 1 & 2 \\ 6 & 5 \end{bmatrix}$. First applying A to a vector $\mathbf{x}$ and then applying B to the vector $A\mathbf{x}$, we get

$$\begin{aligned} B\left(A\begin{bmatrix} x_1 \\ x_2 \end{bmatrix}\right) &= B\left(\begin{bmatrix} 3x_1 + 5x_2 \\ x_1 + 7x_2 \end{bmatrix}\right) \\ &= \begin{bmatrix} 1 \cdot (3x_1 + 5x_2) + 2 \cdot (x_1 + 7x_2) \\ 6 \cdot (3x_1 + 5x_2) + 5 \cdot (x_1 + 7x_2) \end{bmatrix} \\ &= \begin{bmatrix} (1 \cdot 3 + 2 \cdot 1)x_1 + (1 \cdot 5 + 2 \cdot 7)x_2 \\ (6 \cdot 3 + 5 \cdot 1)x_1 + (6 \cdot 5 + 5 \cdot 7)x_2 \end{bmatrix} \\ &= (BA)\begin{bmatrix} x_1 \\ x_2 \end{bmatrix}. \end{aligned}$$

Thus, the composite mapping $\mathbf{x} \mapsto A\mathbf{x} \mapsto B(A\mathbf{x})$ is the same as the mapping $\mathbf{x} \mapsto (BA)\mathbf{x}$. Notice that BA and AB generally correspond to completely different transformations. We can use this interpretation of matrix multiplication to create more complicated transformations from simpler ones. Here is a list of some specific transformations that can be used as basic building blocks. We can visualize the effect of each transformation by considering its effect on the unit square.

1. $\begin{bmatrix} a & 0 \\ 0 & 1 \end{bmatrix} \begin{bmatrix} x \\ y \end{bmatrix} = \begin{bmatrix} ax \\ y \end{bmatrix}$. For $a > 0$, this scales the x direction by the factor a, transforming the unit square into a rectangle of dimensions a by 1.
2. $\begin{bmatrix} 1 & 0 \\ 0 & d \end{bmatrix} \begin{bmatrix} x \\ y \end{bmatrix} = \begin{bmatrix} x \\ dy \end{bmatrix}$. For $d > 0$, this scales the y direction by the factor d, transforming the unit square into a rectangle of dimensions 1 by d.
3. $\begin{bmatrix} 1 & 0 \\ 0 & -1 \end{bmatrix} \begin{bmatrix} x \\ y \end{bmatrix} = \begin{bmatrix} x \\ -y \end{bmatrix}$. This is the matrix for reflection across the x-axis. It leaves the x-axis alone, but flips the y-axis upside down.
4. $\begin{bmatrix} -1 & 0 \\ 0 & 1 \end{bmatrix} \begin{bmatrix} x \\ y \end{bmatrix} = \begin{bmatrix} -x \\ y \end{bmatrix}$. This is the matrix for reflection across the y-axis. It leaves the y-axis alone, but flips the x-axis over right to left.
5. $\begin{bmatrix} 0 & 1 \\ 1 & 0 \end{bmatrix} \begin{bmatrix} x \\ y \end{bmatrix} = \begin{bmatrix} y \\ x \end{bmatrix}$. This is the matrix for reflection across the line $y = x$. It maps the positive x-axis to the positive y-axis and *vice versa*.
6. $\begin{bmatrix} 1 & b \\ 0 & 1 \end{bmatrix} \begin{bmatrix} x \\ y \end{bmatrix} = \begin{bmatrix} x + by \\ y \end{bmatrix}$. For $b > 0$, this leaves the x-axis alone, but slides the point $(0, 1)$ over to the point $(b, 1)$. This is called a *shear* in the x-direction. The unit square becomes a parallelogram with corners at $(0, 0)$, $(1, 0)$, $(b, 1)$, and $(b + 1, 1)$.
7. $\begin{bmatrix} 1 & 0 \\ c & 1 \end{bmatrix} \begin{bmatrix} x \\ y \end{bmatrix} = \begin{bmatrix} x \\ cx + y \end{bmatrix}$. For $c > 0$, this is a *shear* in the y-direction that leaves the y-axis alone, but slides the point $(1, 0)$ up to the point $(1, c)$. The unit square becomes a parallelogram with corners at $(0, 0)$, $(1, c)$, $(0, 1)$, and $(1, c + 1)$.
8. $\begin{bmatrix} \cos(\theta) & -\sin(\theta) \\ \sin(\theta) & \cos(\theta) \end{bmatrix} \begin{bmatrix} x \\ y \end{bmatrix} = \begin{bmatrix} \cos(\theta)x - \sin(\theta)y \\ \sin(\theta)x + \cos(\theta)y \end{bmatrix}$. For $\theta > 0$, this implements a *counterclockwise rotation* by the angle θ about the origin.
9. $\begin{bmatrix} \cos(\theta) & \sin(\theta) \\ -\sin(\theta) & \cos(\theta) \end{bmatrix} \begin{bmatrix} x \\ y \end{bmatrix} = \begin{bmatrix} \cos(\theta)x + \sin(\theta)y \\ -\sin(\theta)x + \cos(\theta)y \end{bmatrix}$. For $\theta > 0$, this implements a *clockwise rotation* by the angle θ about the origin. This matrix is the inverse of the previous one.
10. $\begin{bmatrix} 1 & 0 \\ 0 & 0 \end{bmatrix} \begin{bmatrix} x \\ y \end{bmatrix} = \begin{bmatrix} x \\ 0 \end{bmatrix}$. This is the matrix for projection along the x-axis. The unit square is flattened onto the unit interval of the x-axis.

11. $\begin{bmatrix} 0 & 0 \\ 0 & 1 \end{bmatrix} \begin{bmatrix} x \\ y \end{bmatrix} = \begin{bmatrix} 0 \\ y \end{bmatrix}$. This is the matrix for projection along the y-axis. The unit square is flattened onto the unit interval of the y-axis.
12. $\begin{bmatrix} 1 & 0 \\ 0 & 1 \end{bmatrix} \begin{bmatrix} x \\ y \end{bmatrix} = \begin{bmatrix} x \\ y \end{bmatrix}$. This is $\mathcal{I}_2$, the identity matrix. This transformation leaves everything unchanged.

Example 3.21 The basic transformations listed above can be combined using matrix multiplication to form composite actions.

(i) For instance, $\begin{bmatrix} a & 0 \\ 0 & 1 \end{bmatrix} \cdot \begin{bmatrix} 1 & 0 \\ 0 & d \end{bmatrix} = \begin{bmatrix} a & 0 \\ 0 & d \end{bmatrix}$. For a and d both positive, this scales the x direction by the factor a and the y direction by the factor d, thus transforming the unit square into a rectangle of dimensions a by d.

(ii) Suppose $\mathbf{a} = \begin{bmatrix} r\cos(\theta) \\ r\sin(\theta) \end{bmatrix}$, for $r > 0$ and $0 \leq \theta < 2\pi$, and consider projection along $\mathbf{a}$. Notice that $||\mathbf{a}|| = r$. Clockwise rotation by θ maps $\mathbf{a}$ to the positive x-axis. Thus, if we take any vector $\mathbf{u}$ in $\mathbb{R}^2$, rotate it clockwise by θ, project the result along the x-axis, and then rotate that projection counterclockwise by θ, the end result will be the same as the projection of $\mathbf{u}$ along $\mathbf{a}$. That is,

$$\begin{aligned} &\begin{bmatrix} \cos(\theta) & -\sin(\theta) \\ \sin(\theta) & \cos(\theta) \end{bmatrix} \cdot \begin{bmatrix} 1 & 0 \\ 0 & 0 \end{bmatrix} \cdot \begin{bmatrix} \cos(\theta) & \sin(\theta) \\ -\sin(\theta) & \cos(\theta) \end{bmatrix} \\ &= \begin{bmatrix} \cos^2(\theta) & \cos(\theta)\sin(\theta) \\ \cos(\theta)\sin(\theta) & \sin^2(\theta) \end{bmatrix} \\ &= \frac{1}{r^2}\mathbf{a}\mathbf{a}^T = P_{\mathbf{a}} . \end{aligned}$$

(iii) The scaling matrix $\begin{bmatrix} a & 0 \\ 0 & b \end{bmatrix}$, with $a,\ b > 0$, maps the unit circle $x^2 + y^2 = 1$ to the ellipse $x^2/a^2 + y^2/b^2 = 1$. If we then rotate this ellipse counterclockwise by the angle θ, we get a tilted ellipse. We can map the unit circle onto this tilted ellipse all at once using the matrix product

$$\begin{bmatrix} \cos(\theta) & -\sin(\theta) \\ \sin(\theta) & \cos(\theta) \end{bmatrix} \cdot \begin{bmatrix} a & 0 \\ 0 & b \end{bmatrix} = \begin{bmatrix} a\cos(\theta) & -b\sin(\theta) \\ a\sin(\theta) & b\cos(\theta) \end{bmatrix} .$$

(iv) Figure 3.9 shows a letter **G** on the left. On the right, the **G** has been sheared, using the matrix $\begin{bmatrix} 1 & 1 \\ 0 & 1 \end{bmatrix}$, rotated counterclockwise by $\pi/4$ radians ($= 45°$) using $\begin{bmatrix} 1/\sqrt{2} & -1/\sqrt{2} \\ 1/\sqrt{2} & 1/\sqrt{2} \end{bmatrix}$, and reflected across the y-axis using $\begin{bmatrix} -1 & 0 \\ 0 & 1 \end{bmatrix}$.

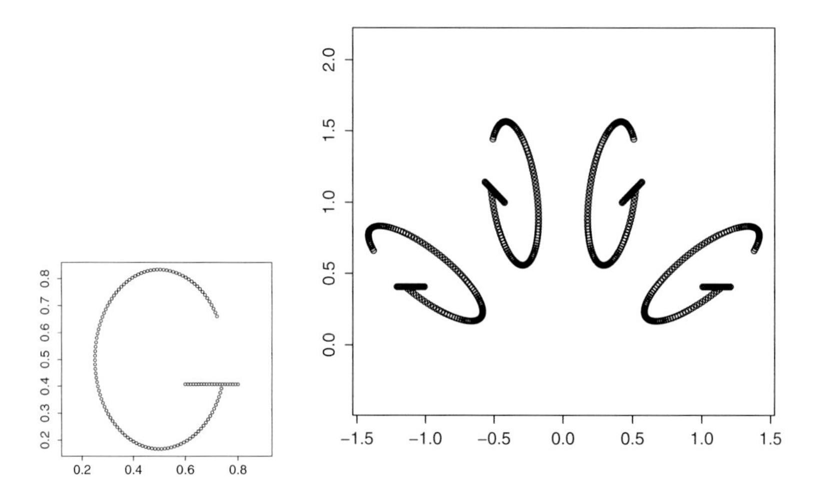

Fig. 3.9 The letter **G** (left image) has been subjected, in counterclockwise order, to shearing, rotation, reflection, and further rotation (right image)

3.9 The Matrix Exponential

Formula 3.9, in Section 3.4, includes the fact that, for a graph with adjacency matrix A, the number of walks along r edges that start at the vertex v_i and end at the vertex v_j is equal to the $[i, j]$ entry of the matrix power A^r. Each such walk represents a path along which information can be sent between the two vertices. To get a measure of how closely connected these two vertices are to each other, we might simply add up all walks of all lengths between them. However, we know that a walk along even as few as two edges might not be the shortest possible path between the vertices. From the point of view of secure communications, information is transmitted more efficiently over a shorter walk. A longer path can involve repetition, traversing and retraversing some of the same edges in opposite directions. A message sent over a longer path is more likely to get corrupted or intercepted along the way. In a social network, a path of three edges indicates a friend of a friend of a friend. A path that links to a friend of a friend of a friend of a friend of a ...—well, after enough steps, that could literally be almost anyone!

Taking these concerns into account, we can measure the quality of the connection between two vertices by assigning a weight to each walk, where a longer walk has a weight closer to 0. That way, when we add up all of the weighted walks between the two vertices, longer walks still matter, but shorter walks matter more.

From a mathematical point of view, it is interesting to assign the weight $1/k!$ to each walk that uses k edges. (The symbol ! denotes the *factorial* of the number; thus, $1! = 1$, $2! = 1 \cdot 2 = 2$, $3! = 1 \cdot 2 \cdot 3 = 6$, and so forth.) In this scheme, a walk that uses three edges, with weight $1/3! = 1/6$, is 20 times more valuable than a walk along five edges, with weight $1/5! = 1/120$. We can now measure the **connectivity** between v_i and v_j as

$$\text{Con}[i,\ j] = \left(\mathcal{I} + A + \frac{A^2}{2!} + \frac{A^3}{3!} + \frac{A^4}{4!} + \cdots\right)_{i,\ j} . \tag{3.35}$$

Admittedly, it is not clear that the sum of infinitely many matrices implied by the right-hand side of (3.35) makes sense. A full explanation would require an advanced course in calculus. In practice, we will always stop adding after some finite number of terms depending on the total number of edges in the graph or some maximum number of decimal places we want in our numerical calculations.

Remark The identity matrix $\mathcal{I}$ is included in the formula (3.35). That is okay because, when $i \neq j$, the $[i,\ j]$-entry of $\mathcal{I}$ is just 0, so we are not introducing any new connections between different vertices. In the case where $i = j$, then we are saying that there is exactly one (fictitious) walk of length 0 edges connecting v_i to itself. The other terms in the sum for $\text{Con}[i,\ i]$ correspond to loops in the graph that start and end at v_i. Adding $+1$ to this for each vertex will not affect the relative connectedness of the vertices.

The mathematical appeal of the sum in formula (3.35) comes from its relationship to the topic of *power series* from calculus. The power series for the exponential function is given by

$$\exp(x) = e^x = 1 + x + \frac{x^2}{2!} + \frac{x^3}{3!} + \frac{x^4}{4!} + \cdots .$$

When we cut this series off after a certain point, we get a polynomial that *approximates* the exponential function. For instance, we can estimate the cube root of e, or $e^{1/3}$, by evaluating the polynomial $p(x) = 1 + x + x^2/2 + x^3/6$ at $x = 1/3$. This gives us

$$\exp(1/3) \approx 1 + (1/3) + (1/3)^2/2 + (1/3)^3/6 = 1.\overline{395061728} .$$

If we include the terms $(1/3)^4/4! + (1/3)^5/5! + (1/3)^6/6!$, the estimate improves to $e^{1/3} \approx 1.395612330$. As a rule, we get better approximations when we include more terms from the power series. The sum in formula (3.35) is the same as the power series for $\exp(x)$, but with the adjacency matrix A in place of x. We write this new matrix as $\exp(A)$. This allows us to rewrite the formula for connectivity between two vertices in a graph as

$$\text{Con}[i,\ j] = (\exp(A))_{i,\ j} . \tag{3.36}$$

In practice, we would make a decision about the maximum number of edges to be considered for walks between two vertices. If the small-world theory of six degrees of separation is correct, using walks along five edges or fewer is likely enough to distinguish different pairs of vertices. So, we can approximate the connectivity as

$$\mathrm{Con}[i,\ j] \approx \left(\mathcal{I} + A + \frac{A^2}{2!} + \frac{A^3}{3!} + \frac{A^4}{4!} + \frac{A^5}{5!}\right)_{i,\ j}. \tag{3.37}$$

For the case where $i = j$, formulas (3.35) and (3.37) offer a measurement of the connectivity of a vertex with itself. Each included walk is a round trip or loop that uses v_i as the hub. If there are lots of such loops of various lengths, then v_i would seem to be a strong hub. Thus, we will interpret the connectivity of a vertex with itself as a **hub score** for the vertex v_i:

$$\mathrm{Hub}(i) = \mathrm{Con}[i,\ i] = (\exp(A))_{i,\ i}\,. \tag{3.38}$$

One way to measure how well connected a graph is overall is to combine the hub scores of all the different vertices. This total will be higher when there are more hubs with higher hub scores, which would allow for easier and more efficient communication within the graph as a whole. For instance, a system of airports with more hubs connecting to and from other hubs will need fewer intermediate stops. The combined hub score will be lower when there are some relatively weak links, making it harder to communicate between different parts of the graph. This combination of hub scores is called the **Estrada index** of the graph.

Definition 3.22 The **Estrada index** of a graph $\mathcal{G}$ is defined as

$$\mathrm{EI}(\mathcal{G}) := \sum_{i=1}^{N} \mathrm{Hub}(i) = \sum_{i=1}^{N} (\exp(A))_{i,\ i}\,. \tag{3.39}$$

Example 3.23 Look again at the graph with 5 vertices described in Example 3.6. In this example, we get

$$I + A + \frac{A^2}{2!} + \frac{A^3}{3!} + \frac{A^4}{4!} + \frac{A^5}{5!} \approx \begin{bmatrix} 3.45 & 2.5\overline{66} & 1.88\bar{3} & 0.68\bar{3} & 2.5\overline{66} \\ & 2.725 & 0.9\overline{33} & 0.25 & 2.358\bar{3} \\ & & 2.2\overline{66} & 1.38\bar{3} & 0.9\overline{33} \\ & & & 1.58\bar{3} & 0.25 \\ & & & & 2.725 \end{bmatrix}.$$

(The matrix is symmetric, so we only need to record the entries on and above the diagonal.) Using formula (3.37) to estimate the connectivity between vertices, we see that vertices v_1, v_2, and v_5 are well connected to each other. Vertex v_1 is the best hub, with vertices v_2 and v_5 also having good hub scores. Vertex v_4 has the weakest connections within this graph.

In general, we can define the **matrix exponential** $\exp(A)$ for any square matrix. This matrix is quite difficult to compute, however, especially when A is large. For that reason, $\exp(A)$ is mainly of theoretical value.

3.10 Exercises

1. Flags with three vertical or horizontal color bands.
   (a) (Flag of France) Compute the **outer product**
   $$\begin{bmatrix} 1 \\ 1 \\ 1 \end{bmatrix} \begin{bmatrix} \text{blue white red} \end{bmatrix}.$$
   (b) (Flag of Sierra Leone) Compute the **outer product**
   $$\begin{bmatrix} \text{green} \\ \text{white} \\ \text{blue} \end{bmatrix} \begin{bmatrix} 1 & 1 & 1 \end{bmatrix}.$$
   (c) Describe the flags of the Netherlands and Côte D'Ivoire using outer products.
2. Let $\mathbf{u} = \begin{bmatrix} 1 \\ 0 \\ 1 \end{bmatrix}$ and $\mathbf{v} = \begin{bmatrix} 0 \\ 1 \\ 0 \end{bmatrix}$.
   (a) Compute the outer products $\mathbf{u}\mathbf{u}^T$ and $\mathbf{v}\mathbf{v}^T$.
   (b) Show that the sum $\mathbf{u}\mathbf{u}^T + \mathbf{v}\mathbf{v}^T$ produces a 3×3 checkerboard matrix of 1s and 0s.
   (c) Try to form a 4×4 checkerboard pattern of 1s and 0s, starting with a 1 in the upper left corner, as a sum of two outer products.
   (d) Try to form any $M \times M$ checkerboard of 1s and 0s, starting with a 1 in the upper left corner, as a sum of two outer products.
3. Let $\mathbf{u} = \begin{bmatrix} 1 \\ 0 \\ 1 \end{bmatrix}$ and $\mathbf{v} = \begin{bmatrix} 0 \\ 1 \\ 0 \end{bmatrix}$.
   (a) Compute the outer products $\mathbf{u}\mathbf{v}^T$ and $\mathbf{v}\mathbf{u}^T$.
   (b) Show that the sum $\mathbf{u}\mathbf{v}^T + \mathbf{v}\mathbf{u}^T$ produces a 3×3 checkerboard matrix of 0s and 1s that starts with a 0 in the upper left corner. (This reverses the pattern from the previous exercise.)
   (c) Express any $M \times M$ checkerboard of 0s and 1s, starting with a 0 in the top left corner, as a sum of two outer products.
4. Let $\mathbf{u} = \begin{bmatrix} 1 \\ 2 \\ 1 \end{bmatrix}$ and $\mathbf{v} = \begin{bmatrix} 1 \\ -1 \\ 1 \end{bmatrix}$.

(a) Compute the outer products $\mathbf{uu}^T$ and $\mathbf{vv}^T$.
(b) Show that the matrix $(1/3) \cdot \left(\mathbf{uu}^T - \mathbf{vv}^T\right)$ produces a 3×3 "plus sign" matrix with 1s in the middle row and column and 0s in the corners.

5. Refer to the term–document matrix L shown in Table 1.5.

(a) Use the matrix product $L^T L$ to calculate the inner product of the document vectors of each pair of documents.
(b) Compute the matrix L_c, obtained from L by converting each column to a unit vector in the same direction.
(c) Use the matrix product $L_c^T L_c$ to calculate the cosine of the angle between the document vectors of each pair of documents.
(d) Use the matrix/vector product $L_c^T \mathbf{q}/||\mathbf{q}||$ to calculate the cosines of the angles between the document vectors and the given query vector $\mathbf{q}$.

(i) $\mathbf{q} = \begin{bmatrix} 0 & 1 & 1 & 0 & 0 \end{bmatrix}^T$ = geometry + probability.
(ii) $\mathbf{q} = \begin{bmatrix} 0 & 1 & 0 & 0 & 1 \end{bmatrix}^T$ = geometry + symmetry.
(iii) $\mathbf{q} = \begin{bmatrix} 1 & 0 & 0 & 1 & 0 \end{bmatrix}^T$ = dimension + statistics.

6. Refer to the term–document matrix L shown in Table 1.5.

(a) Use the matrix product LL^T to calculate the inner product of the term vectors of each pair of terms in the dictionary.
(b) Compute the matrix L_r, obtained from L by converting each row to a unit vector in the same direction.
(c) Use the matrix product $L_r L_r^T$ to calculate the cosine of the angle between the term vectors of each pair of terms.
(d) Identify those pairs of terms for which the cosine computed in part (c) is

(i) Greater than 0.5
(ii) Greater than 0.7
(iii) Greater than 0.8

7. Let R be the matrix of movie ratings shown in Table 1.4.

(a) Use the matrix product $R^T R$ to calculate the inner product of the rating vectors of each pair of viewers.
(b) Compute the matrix R_c, obtained from R by converting each column to a unit vector in the same direction.
(c) Use the matrix product $R_c^T R_c$ to calculate the cosine of the angle between the rating vectors of each pair of viewers.
(d) Using the answers to parts (a) and (c), discuss which pairs of viewers have tastes that are most compatible and which have tastes that are least compatible.

8. Let R be the matrix of movie ratings shown in Table 1.4.

(a) Use the matrix product RR^T to calculate the inner product of the rating vectors of each different pair of movies. What does each of these inner products represent?
(b) Compute the matrix R_r, obtained from R by converting each row to a unit vector in the same direction.
(c) Use the matrix product $R_r R_r^T$ to calculate the cosine of the angle between the rating vectors of each different pair of movies.
(d) Using the answers to parts (a) and (c), discuss which pairs of movies tend to be rated the same by more viewers and which pairs of movies tend to be rated differently by most viewers.

9. In this exercise, we explore several scenarios involving a metropolitan area consisting of a city and its suburbs. The assumptions are as follows:

- Each year, the given proportion p of city dwellers relocate to the suburbs.
- Each year, the given proportion q of the suburbanites move into the city.
- The metropolitan area has a permanent total population of 150 thousand.
- In year 0, the initial population distribution is $\mathbf{p}_0 = \begin{bmatrix} 2/3 \\ 1/3 \end{bmatrix}$. (That is, $2/3$ of the population, or 100 thousand people, live in the city and $1/3$, or 50 thousand people, live in the suburbs in year 0.)

For each scenario, perform the following steps.

(i) Write down the transition matrix T for the two-state Markov process. Remember that each column of T must sum to 1.
(ii) Compute the population vectors $\mathbf{p}_1 = T\mathbf{p}_0$, $\mathbf{p}_2 = T\mathbf{p}_1$, and $\mathbf{p}_3 = T\mathbf{p}_2$ for each of the next three years.
(iii) Use a computer to find the population vector $\mathbf{p}_k = T^k\mathbf{p}_0$ for $k = 20$, $k = 25$, and $k = 30$. Make a conjecture about the stationary population distribution $\mathbf{p}_*$.
(iv) Compute the vector $\mathbf{p}_\sharp = \left(\frac{1}{p+q}\right) \cdot \begin{bmatrix} q \\ p \end{bmatrix}$ and compare this to your conjecture about $\mathbf{p}_*$. (The numbers p and q are the transition proportions defined in each scenario.)

Here are the transition proportions for the different scenarios:

(a) $p = 0.15$; $q = 0.10$.
(b) $p = 0.10$; $q = 0.10$.
(c) $p = 0.20$; $q = 0.10$.
(d) $p = 0.30$; $q = 0.10$.

10. Consider a two-state Markov process to model the weather. State A represents a *sunny* day, while state B represents a *rainy* day. (In this simplistic model, these are the only types of weather.) A transition matrix $\begin{bmatrix} 1-p & q \\ p & 1-q \end{bmatrix}$ means that,

with probability p, a sunny day is followed by a rainy day and, with probability q, a rainy day is followed by a sunny day.

For each transition matrix T given below, compute the following:

(i) Assume day 0 is sunny and set $\mathbf{x}_0 = \begin{bmatrix} 1 \\ 0 \end{bmatrix}$. Compute the vectors $\mathbf{x}_1 = T\mathbf{x}_0$, $\mathbf{x}_2 = T\mathbf{x}_1$, and $\mathbf{x}_3 = T\mathbf{x}_2$ to determine the probabilities of sun and rain on each of the next three days.

(ii) Use a computer to find the vectors $\mathbf{x}_k = T^k\mathbf{x}_0$ for $k = 20$, $k = 25$, and $k = 30$.

(iii) Make a conjecture about the stationary distribution $\mathbf{x}_*$. This gives the long-term percentages of sunny days and of rainy days.

(iv) Compute the vector $\mathbf{x}_\sharp = \left(\frac{1}{p+q}\right) \cdot \begin{bmatrix} q \\ p \end{bmatrix}$ and compare this to your conjecture about $\mathbf{x}_*$. (The numbers p and q are the transition probabilities defined in each example.)

$$\text{(a)} \begin{bmatrix} 0.8 & 0.5 \\ 0.2 & 0.5 \end{bmatrix}; \text{ (b)} \begin{bmatrix} 0.8 & 0.6 \\ 0.2 & 0.4 \end{bmatrix}; \text{ (c)} \begin{bmatrix} 0.8 & 0.7 \\ 0.2 & 0.3 \end{bmatrix}.$$

11. Consider a "flower" graph consisting of three triangular petals joined at a common central vertex, labeled v_7 here. The adjacency matrix is

$$A = \begin{bmatrix} 0&1&0&0&0&0&1 \\ 1&0&0&0&0&0&1 \\ 0&0&0&1&0&0&1 \\ 0&0&1&0&0&0&1 \\ 0&0&0&0&0&1&1 \\ 0&0&0&0&1&0&1 \\ 1&1&1&1&1&1&0 \end{bmatrix}.$$

(a) Sketch the graph.
(b) Determine the degrees of all seven vertices.
(c) Compute the total number of edges in the graph.
(d) Write down the matrix A^2. Then count the following:

- How many paths along two edges are there from v_1 to v_4?
- How many paths along two edges are there from the central vertex v_7 to any other vertex?

(e) Use Formula (3.12) to determine the number of 4-cycles in this graph.
(f) Write down the matrix A^3. Then use Formula (3.11) to determine the number of triangles in the graph. (Hint: You know the answer already!)

12. Consider a "barbell" graph consisting of two triangles (the "bells") joined by a line segment (the "bar"). The adjacency matrix is

$$A = \begin{bmatrix} 0 & 1 & 1 & 0 & 0 & 0 \\ 1 & 0 & 1 & 0 & 0 & 0 \\ 1 & 1 & 0 & 1 & 0 & 0 \\ 0 & 0 & 1 & 0 & 1 & 1 \\ 0 & 0 & 0 & 1 & 0 & 1 \\ 0 & 0 & 0 & 1 & 1 & 0 \end{bmatrix}.$$

(a) Sketch the graph.
(b) Determine the degrees of all six vertices.
(c) Compute the total number of edges in the graph.
(d) Write down the matrix A^2. Then count the following:

- How many paths along two edges are there from v_1 to v_2?
- How many paths along two edges are there from v_1 to v_4?

(e) Use formula (3.12) to determine the number of 4-cycles in this graph.
(f) Write down the matrix A^3. Then use formula (3.11) to determine the number of triangles in the graph. (Hint: You know the answer already!)

13. Consider a graph consisting of a regular pentagon where each vertex is also connected to a sixth vertex in the middle. This looks like a wheel with five sides and a central hub with spokes. The adjacency matrix is

$$A = \begin{bmatrix} 0 & 1 & 0 & 0 & 1 & 1 \\ 1 & 0 & 1 & 0 & 0 & 1 \\ 0 & 1 & 0 & 1 & 0 & 1 \\ 0 & 0 & 1 & 0 & 1 & 1 \\ 1 & 0 & 0 & 1 & 0 & 1 \\ 1 & 1 & 1 & 1 & 1 & 0 \end{bmatrix}.$$

(a) Sketch the graph.
(b) Determine the degrees of all six vertices.
(c) Compute the total number of edges in the graph.
(d) Write down the matrix A^2. Then count the following:

- How many paths along two edges are there from v_1 to v_3?
- How many paths along two edges are there from the hub v_6 to any other vertex?

(e) Use formula (3.12) to determine the number of 4-cycles in this graph.

(f) Write down the matrix A^3. Then use formula (3.11) to determine the number of triangles in the graph.

14. Consider a graph consisting of a square with one diagonal, connecting vertices v_2 and v_4, say. The corresponding adjacency matrix is

$$A = \begin{bmatrix} 0 & 1 & 0 & 1 \\ 1 & 0 & 1 & 1 \\ 0 & 1 & 0 & 1 \\ 1 & 1 & 1 & 0 \end{bmatrix}.$$

(a) Sketch the graph.
(b) Determine the degrees of all four vertices.
(c) Verify that total number of edges in the graph agrees with the value $(1/2) *$ (sum of all entries of A).
(d) Write down the matrix A^2. Then count the following:

- How many paths along two edges are there from v_1 to v_4?
- How many paths along two edges are there from v_2 to v_4?

(e) Use formula (3.12) to determine the number of 4-cycles in the graph.
(f) Write down the matrix A^3. Then count the following:

- How many triangles in the graph include vertex v_3?
- How many triangles in the graph include v_4?

(g) Use formula (3.18) to calculate *hub scores* for vertices v_1 and v_4.
(h) Use formula (3.37) to approximate the connectivity scores for the pairs of vertices v_1 and v_2 and v_1 and v_3; that is, estimate Con[1, 2] and Con[1, 3].
(i) Use formula (3.38) and the degree 5 approximation to $\exp(A)$ to approximate the hub scores Hub(1) and Hub(4). Compare these to the values computed in part (g) of this exercise.

15. The table shows the population of the United States, in millions, every ten years from 1950 to 2020, according to the official US Census.

Year (since 1950):	0	10	20	30
US pop. (millions):	150.7	179.3	203.2	226.5
Year (since 1950):	40	50	60	70
US pop. (millions):	248.7	281.4	308.7	331.4

(a) Create a scatter plot showing the year (0 through 70) on the horizontal axis and the US population on the vertical axis. Does it seem possible that there is a linear relationship between the two quantities?

(b) Create the 8×2 matrix A that has all 1s in the first column and the year data in the second column.

(c) Referring to the **normal equation** (3.22), compute the matrix $A^T A$ and the vector $A^T \mathbf{y}$ where $\mathbf{y}$ is the vector of US population data.

(d) Solve the normal equation, as in (3.23), to find the regression parameters

$$\begin{bmatrix} \alpha \\ \beta \end{bmatrix} = \left(A^T A\right)^{-1} A^T \mathbf{y}.$$

(Use a computer!)

(e) What does the intercept value α mean in this context? What does the slope parameter β mean in this context?

(f) Create a plot showing the scatter plot from part (a) along with the regression line $y = \alpha + \beta x$ in the same picture. Does the line look like a good fit for the data?

(g) Compute the value R^2, as in (3.25). Remember that the closer this is to 1, the better the line explains the data.

16. In Chapter 1, Exercise 7, we computed the correlation coefficient for team home runs (HR) and team wins for the 2019 season for six teams in the American League of Major League Baseball: New York Yankees (NYY), Baltimore Orioles (Balt), Cleveland Indians (Cleve), Chicago White Sox (CWS), Texas Rangers (TX), and Los Angeles Angels (LAA). The data are shown in the table.

Team	NYY	Balt	Cleve	CWS	TX	LAA
HR	306	213	223	182	223	220
Wins	103	54	93	72	78	72

(a) Create a scatter plot showing team HR on the horizontal axis and team wins on the vertical axis. Does it seem possible that there is a linear relationship between the two quantities?

(b) Create the 6×2 matrix A that has all 1s in the first column and the team HR data in the second column.

(c) Referring to the **normal equation** (3.22), compute the matrix $A^T A$ and the vector $A^T \mathbf{y}$ where $\mathbf{y}$ is the vector of team wins data.

(d) Solve the normal equation, as in (3.23), to find the regression parameters

$$\begin{bmatrix} \alpha \\ \beta \end{bmatrix} = \left(A^T A\right)^{-1} A^T \mathbf{y}.$$

(Use a computer!)

(e) What does the intercept value α mean in this context? What does the slope parameter β mean in this context?

(f) Create a plot showing the scatter plot from part (a) along with the regression line $y = \alpha + \beta x$ in the same picture. Does the line look like a good fit for the data?

(g) Compute the value R^2, as in (3.25). Remember that the closer this is to 1, the better the line explains the data.

17. In Chapter 1, Exercise 8, we computed the correlation coefficient for team earned run average (ERA) and team wins for the 2019 season for six teams in the National League of Major League Baseball: Washington Nationals (Wash), Philadelphia Phillies (Phil), Milwaukee Brewers (Mil), Pittsburgh Pirates (Pitt), Los Angeles Dodgers (LAD), and San Francisco Giants (SFG). The data are shown in the table.

Team	Wash	Phil	Mil	Pitt	LAD	SFG
ERA	4.27	4.53	4.40	5.18	3.37	4.38
Wins	93	81	89	69	106	77

(a) Create a scatter plot showing team ERA on the horizontal axis and team wins on the vertical axis. Does it seem possible that there is a linear relationship between the two quantities?

(b) Create the 6×2 matrix A that has all 1s in the first column and the team ERA data in the second column.

(c) Referring to the **normal equation** (3.22), compute the matrix $A^T A$ and the vector $A^T \mathbf{y}$ where $\mathbf{y}$ is the vector of team wins data.

(d) Solve the normal equation, as in (3.23), to find the regression parameters

$$\begin{bmatrix} \alpha \\ \beta \end{bmatrix} = \left(A^T A\right)^{-1} A^T \mathbf{y}.$$

(Use a computer!)

(e) What does the intercept value α mean in this context? What does the slope parameter β mean in this context?

(f) Create a plot showing the scatter plot from part (a) along with the regression line $y = \alpha + \beta x$ in the same picture. Does the line look like a good fit for the data?

(g) Compute the value R^2, as in (3.25). Remember that the closer this is to 1, the better the line explains the data.

18. *(Proof Problem)* For the optimal solution $\begin{bmatrix} \alpha \\ \beta \end{bmatrix}$ to the normal equation shown in equation (3.29), **show that** α satisfies

$$\alpha = \bar{y} - \beta \bar{x},$$

where $\bar{y} = \sum y_i/N$ and $\bar{x} = \sum x_i/N$ are the average values of the sample observations of Y and X, respectively.

19. For each vector $\mathbf{a}$ given below, compute the following:

    (i) $P_{\mathbf{a}}$, the matrix for projection along the line generated by $\mathbf{a}$
    (ii) $R_{\mathbf{a}}$, the matrix for reflection across the line generated by $\mathbf{a}$
    (iii) $P_{\mathbf{a}}\begin{bmatrix} 3 \\ 4 \end{bmatrix}$ and $R_{\mathbf{a}}\begin{bmatrix} 3 \\ 4 \end{bmatrix}$, the projection and reflection, respectively, of the vector $\begin{bmatrix} 3 \\ 4 \end{bmatrix}$
    (iv) $P_{\mathbf{a}}\begin{bmatrix} x \\ y \end{bmatrix}$ and $R_{\mathbf{a}}\begin{bmatrix} x \\ y \end{bmatrix}$, the projection and reflection, respectively, of the generic vector $\begin{bmatrix} x \\ y \end{bmatrix}$

    (a) $\mathbf{a} = \begin{bmatrix} \sqrt{3} \\ 1 \end{bmatrix}$
    (b) $\mathbf{a} = \begin{bmatrix} \sqrt{3} \\ -1 \end{bmatrix}$
    (c) $\mathbf{a} = \begin{bmatrix} -1 \\ \sqrt{3} \end{bmatrix}$
    (d) $\mathbf{a} = \begin{bmatrix} -1 \\ -\sqrt{3} \end{bmatrix}$

20. Let $\mathbf{a} = \begin{bmatrix} 1 \\ 1 \end{bmatrix}$ and $\mathbf{b} = \begin{bmatrix} 1 \\ -1 \end{bmatrix}$.

    (a) Write down the matrices for the projections $P_{\mathbf{a}}$ and $P_{\mathbf{b}}$.
    (b) Compute the matrix products $P_{\mathbf{b}}P_{\mathbf{a}}$, $P_{\mathbf{a}}P_{\mathbf{b}}$, and $P_{\mathbf{a}}^2 = P_{\mathbf{a}}P_{\mathbf{a}}$.

21. Let $\mathbf{a} = \begin{bmatrix} 1 \\ s \end{bmatrix}$, for some real number s.

    (a) Write down the matrices for the projection $P_{\mathbf{a}}$ and the reflection $R_{\mathbf{a}}$.
    (b) *(Proof Problem)* Show that, for $\mathbf{a} = \begin{bmatrix} 1 \\ s \end{bmatrix}$ and $\mathbf{b} = \begin{bmatrix} 1 \\ t \end{bmatrix}$, with $s \neq t$, then $R_{\mathbf{a}} \neq R_{\mathbf{b}}$. (This confirms the claim made in Example 3.20 that the identity matrix has infinitely many square roots.)

22. For each given multistep transformation of $\mathbb{R}^2$, express the matrix for the transformation as a product of basic transformations listed in Section 3.8. Sketch the effect of each given transformation on the unit square. (The unit square has corners at (0, 0), (1, 0), (1, 1), and (0, 1).)

    (a) Stretch by a factor of 3 in the x-direction and shrink by a factor of 1/2 in the y-direction.

(b) Rotate counterclockwise by $\pi/4$ (or 45°); then reflect across the line $y = x$; then rotate clockwise by $\pi/4$ (or 45°). Which of the basic transformations do we get?
(c) Shear one unit to the right in the x-direction; then rotate counterclockwise by $\pi/2$ (or 90°); then reflect across the line $y = x$.
(d) Shear one unit up in the y-direction; then reflect across the y-axis; then reflect across the x-axis.

23. *(Proof Problem)* Suppose $A = \left[\mathbf{a}_1 \middle| \mathbf{a}_2 \middle| \mathbf{a}_3\right]$ is any 3×3 matrix whose columns are mutually orthogonal unit vectors in $\mathbb{R}^3$. Show that $A^T A = \mathcal{I}_3$.
24. *(Proof Problem)* Prove the following statement: If the $N \times N$ matrix A is idempotent, then so is the matrix $\mathcal{I}_N - A$.
25. *(Proof Problem)* Let T be the transition matrix of a Markov process. Show that, for any vector $\mathbf{x}$, the coordinates of the vector $T\mathbf{x}$ have the same sum as those of $\mathbf{x}$. (*Hint:* The entries in each column of T sum to 1.)

3.11 Projects

Project 3.1 (More Images of Flags) Suppose we wish to produce a flag that has a cross of one color against a background of a different color. For example, the flag of Denmark has a white cross against a red background. The flags of England and Georgia show a red cross against a white background, called St. George's cross. In this project, we imagine such a flag as an $M \times N$ matrix, with $M \leq N$. The background and the cross will have different fixed color values. Using outer products, we can compute all $M \cdot N$ entries in the flag matrix from the coordinates of a relatively small number of vectors from $\mathbb{R}^M$ and $\mathbb{R}^N$. The total number of coordinates of these vectors has the form $c \cdot (M + N)$, depending on the width of the cross. Usually, we have $c \cdot (M + N) < M \cdot N$, resulting in a savings in data storage.

1. For any positive integers k and m, with $k \leq m$, let $\mathbf{e}_{k,\,m}$ denote the kth column vector of the $m \times m$ identity matrix, $\mathcal{I}_m$. Now, for positive integers $i \leq M$ and $j \leq N$, describe the entries of the following $M \times N$ outer product matrices:

   (a) $\mathbf{e}_{i,\,M}\mathbf{1}_N^T$
   (b) $\mathbf{1}_M\mathbf{e}_{j,\,N}^T$
   (c) $\mathbf{e}_{i,\,M}\mathbf{e}_{j,\,N}^T$

2. Thus, describe the entries of the $M \times N$ matrix

$$\mathbf{e}_{i,\,M}\mathbf{1}_N^T + \mathbf{1}_M\mathbf{e}_{j,\,N}^T - \mathbf{e}_{i,\,M}\mathbf{e}_{j,\,N}^T .$$

3. Describe the entries of the $M \times N$ matrix $\mathbf{1}_M \, \mathbf{1}_N^T$.
4. Describe how to produce a flag where the background color value is 1 and the color value of the cross is 0.
5. Write down a combination of outer products to represent the flag of Denmark as a 5×9 matrix with the cross occupying the 3rd row and 3rd column.
6. The official proportions of the Danish flag are 56:107. Find a combination of outer products to design a 56×107 matrix with a cross of 1s in the eight rows and eight columns numbered 25 through 32 and all other entries equal to 0.
7. *(Challenge)* Use outer products to model the flag of Norway in which the blue cross has a white border that separates it from the red background.

Project 3.2 (Information Retrieval)

Part I. Use the library of documents and dictionary of terms shown in Table 1.3, in Section 1.6 above.

1. Create a query vector using terms 3, 8, 12, and 18. That is,

$$\mathbf{q} = \begin{bmatrix} 0\,0\,1\,0\,0\,0\,0\,1\,0\,0\,0\,1\,0\,0\,0\,0\,0\,1 \end{bmatrix}^T .$$

2. Compute the cosine of the angle between this query and the vector for each of the eight documents in the table.
3. Using a threshold of $cosine > 0.5$, determine which documents are relevant to the query.
4. Using a threshold of $cosine > 0.7$, determine which documents are relevant to the query.
5. Create two additional queries of your own and repeat the analysis for each one.

Part II. Create a new library of documents and dictionary of terms. Use at least 8 documents and 15 terms. Create two separate queries and analyze each one as above using two different thresholds for the cosine value.

Project 3.3 (Make Your Own Movie Rating Service) Revisit Project 1.3. Generate a list of at least ten films and have at least ten friends rate them using the system $+1 =$ like, $-1 =$ do not like, and $0 =$ did not see, as in Table 1.4, in Section 1.6. This time, form a matrix R whose columns are the normalized rating vectors of the different viewers. Then use $R^T R$ to compute the cosines of the angles between everyone's rating vectors. Find out which friends have similar tastes and which friends can help each other to expand their horizons.

Project 3.4 (Connections in a Graph)

Part I. A particular graph with nine vertices has the following *adjacency matrix*:

$$A = \begin{bmatrix} 0&0&0&1&0&0&1&0&0 \\ 0&0&1&0&0&1&0&0&0 \\ 0&1&0&1&1&1&0&1&0 \\ 1&0&1&0&0&0&1&0&0 \\ 0&0&1&0&0&0&0&0&1 \\ 0&1&1&0&0&0&0&0&0 \\ 1&0&0&1&0&0&0&1&0 \\ 0&0&1&0&0&0&1&0&0 \\ 0&0&0&0&1&0&0&0&0 \end{bmatrix}.$$

1. Sketch a diagram of this graph. For example, place nine dots around a ring to represent the vertices. Then join them according to the entries of the adjacency matrix.
2. Use the adjacency matrix A to answer the following:
   (a) Compute the degrees of all nine vertices.
   (b) Find the total number of edges in this graph.
3. Compute the matrix A^2; then answer the following:
   (a) Explain why each diagonal entry of A^2 represents the degree of the corresponding vertex.
   (b) How many common neighbors do vertices v_1 and v_8 have? Explain your reasoning.
   (c) How many common neighbors do vertices v_3 and v_8 have? Explain your reasoning.
   (d) Determine the number of 4-cycles in this graph. Which vertices are involved? Explain your reasoning. (formula (3.12) may be useful.)
4. Compute the matrix A^3; then answer the following:
   (a) Write down the number of different ways to "walk" from vertex v_1 to vertex v_3 along 3 edges. Explain your reasoning.
   (b) Write down the number of different ways to "walk" from vertex v_1 to vertex v_7 along 3 edges. Explain your reasoning.
   (c) Find an example of a walk along 3 edges between two different vertices in the graph that is *not the shortest path* between these vertices.
   (d) Compute the total number of triangles in this graph. Which vertices are involved? Explain your reasoning. (formula (3.11) may be useful.)
5. (Optional) Compute the approximate matrix exponential of A using powers up to A^5:

$$\exp(A) \approx \mathcal{I} + A + A^2/2 + A^3/3! + A^4/4! + A^5/5!\,.$$

Use this approximation along with formulas (3.36) and (3.38) to answer the following:

(a) Find the pair of vertices whose connection score Con$[i, j]$ is the largest.
(b) Find the vertex that has the highest hub score Hub(i). (This might be considered to be the best connected hub in this particular graph.)
(c) List all the vertices in decreasing order of Hub(i).

Part II.

- Create a graph of your own with at least ten vertices, at least two triangles, and at least one 4-cycle.
- Form its adjacency matrix.
- Answer all of the questions from Part I for this graph.

Project 3.5 (Simple Linear Regression: Le Tour de France) Apply the ideas of Section 3.5 to explore the relationship between the distance (in kilometers) and the average percent gradient for all climbs within a given category from the 2018 Tour de France. The following table shows the relevant data for Category 3 and Category 2 climbs from the 2018 Tour de France. Note that Category 2 climbs are considered to be more difficult than Category 3 climbs.

2018 Tour de France							
Category 3 climbs							
Distance (km)	1.9	3	2.2	1.5	1.9	2.4	3.1
Average % gradient	6.6	6.3	5.9	7	6.9	6.9	5.9
Category 2 climbs							
Distance (km)	8.6	5.4	3.0	9.1	3.4	5.7	
Average % gradient	5.8	7.1	10.2	5.3	8.2	6.5	

For each of these two categories of climbs, do the following:

(a) Create the matrix A that has all 1s in the first column and the distance data for the category in the second column.
(b) Form the vector $\mathbf{y}$ that contains the category's gradient data.
(c) Compute the solution to the normal equation, as in (3.23) and (3.29).
(d) Write down the equation of the regression line that best describes the average percent gradient as a function of the distance for the given category.
(e) Briefly explain the meaning of the slope coefficient.
(f) Create a plot showing the data points and the regression line in the same plot window. (Show *distance* on the horizontal axis and *average gradient* on the vertical axis.)
(g) Compute the value of R-squared for the regression line computed above. Use formula (3.25). Based on the value of R-squared, discuss whether the distance of a climb does a good job of explaining the average gradient of the climb for the given category.

Project 3.6 (Simple Linear Regression: Child Mortality Rates) Refer to Examples 3.15 and 1.16 for context and to Table 1.2 for the data.

For each comparison described below, do the following:

(a) Compute the least squares regression line.
(b) Briefly explain the meaning of the slope coefficient.
(c) Compute the corresponding value of R^2.
(d) Create a plot showing the scatter plot of the data along with the regression line:

1. Child mortality rate (CM) as a function of the female literacy rate (FLR)
2. Child mortality rate (CM) as a function of per capita gross national product (pcGNP)
3. Female literacy rate (FLR) as a function of per capita gross national product (pcGNP)

Project 3.7 (Simple Linear Regression: Disposable Personal Income and Savings) In Project 1.1, we looked at the correlation between aggregate savings and disposable personal income (DPI) in the United States for the years 1970–1993. The relevant data are shown in Table 1.6. As discussed, there was a significant economic recession in 1982.

1. For each of the time periods 1970–1993, 1970–1981, and 1982–1993, do the following:

   (a) Compute the least squares regression line for savings as a function of DPI.
   (b) Briefly explain the meaning of the slope coefficient.
   (c) Compute the corresponding value of R^2.
   (d) Create a scatter plot of the data for the time interval. Include the regression line in the same plot window.

2. Discuss the results of the regression analysis in #1. In particular, discuss whether or not the results support the idea that the relationship between savings and DPI changed after the recession.

Project 3.8 (Information Retrieval Using k-Means) Referring to the term–document array shown in Table 1.3, in Section 1.6, apply the method of k-means to form clusters of documents that tend to have similar term contents. Do this for both $k = 2$ and $k = 3$ and compare the results. See Example 3.17 for guidance.

Chapter 4
Linear Systems

4.1 Linear Equations

The projection of any vector $\mathbf{x}$ along a fixed nonzero vector $\mathbf{a}$ is defined by $\text{proj}_{\mathbf{a}}\mathbf{x} = \left(\mathbf{a}^T\mathbf{x}/||\mathbf{a}||^2\right)\mathbf{a}$, as in (1.37). So, two vectors $\mathbf{x}$ and $\mathbf{y}$ have *the same projection* along $\mathbf{a}$ when $\mathbf{a}^T\mathbf{x} = \mathbf{a}^T\mathbf{y}$. Geometrically, $\mathbf{x}$ and $\mathbf{y}$ "cast the same shadow" along the line generated by $\mathbf{a}$. The difference vector $\mathbf{x} - \mathbf{y}$ satisfies

$$\mathbf{a}^T(\mathbf{x} - \mathbf{y}) = \mathbf{a}^T\mathbf{x} - \mathbf{a}^T\mathbf{y} = 0\,.$$

That is, $\mathbf{x} - \mathbf{y}$ is *orthogonal* to $\mathbf{a}$.

Now fix a number b and consider the set of all vectors $\mathbf{x}$ for which $\mathbf{a}^T\mathbf{x} = b$. One such vector is $\mathbf{x} = (b/||\mathbf{a}||^2)\cdot\mathbf{a}$. All solutions have the same projection along $\mathbf{a}$, and the difference between any two solutions is orthogonal to $\mathbf{a}$. If $\mathbf{x}$ and $\mathbf{y}$ are any two solutions, then the line segment connecting their terminal points is parallel to $\mathbf{x} - \mathbf{y}$ and, therefore, is orthogonal to $\mathbf{a}$. Thus, the terminal points of all solutions to $\mathbf{a}^T\mathbf{x} = b$ collectively form a set that is orthogonal to $\mathbf{a}$ in the sense that $\mathbf{a}$ is orthogonal to every line in this set. We say that $\mathbf{a}$ is a **normal vector** to this set of terminal points. When $b = 0$, the equation $\mathbf{a}^T\mathbf{x} = 0$ is called **homogeneous**. The set of solutions to the homogeneous equation consists of all vectors $\mathbf{x}$ that are orthogonal to $\mathbf{a}$, including the zero vector $\mathbf{x} = \mathbf{0}$.

Geometrically, when $N = 2$ and $\mathbf{a}$ is a nonzero vector in $\mathbb{R}^2$, the solutions to $\mathbf{a}^T\mathbf{x} = b$ determine a line in $\mathbb{R}^2$ that is perpendicular to $\mathbf{a}$. When $N = 3$, the terminal points of the solutions form a 2-dimensional plane in $\mathbb{R}^3$ orthogonal to the nonzero vector $\mathbf{a}$. More generally, for $N > 3$, the set of terminal points of all vectors $\mathbf{x}$ in $\mathbb{R}^N$

Supplementary Information The online version contains supplementary material available at https://doi.org/10.1007/978-3-031-39562-8_4.

T. G. Feeman, *Applied Linear Algebra and Matrix Methods*,
Springer Undergraduate Texts in Mathematics and Technology,
https://doi.org/10.1007/978-3-031-39562-8_4

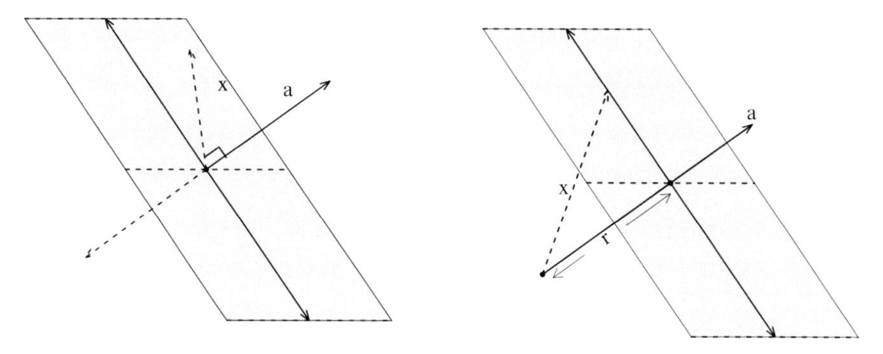

Fig. 4.1 For a given nonzero vector **a** and a number b, the set of terminal points of all vectors **x** for which $\mathbf{a}^T\mathbf{x} = b$ forms a *hyperplane* orthogonal to **a**. The figure on the left shows the homogeneous case, where $b = 0$. The figure on the right shows the case where $b \neq 0$, with $r = |b|/||\mathbf{a}||$

for which $\mathbf{a}^T\mathbf{x} = b$ is called a **hyperplane** in $\mathbb{R}^N$, a term meant to evoke a higher-dimensional plane. The hyperplane contains the terminal point of $(b/||\mathbf{a}||^2) \cdot \mathbf{a}$ and has **a** as a normal vector. Figure 4.1 shows the $\mathbb{R}^3$ case for both $b = 0$ and $b \neq 0$.

Definition 4.1 A **linear equation** with N unknowns, labeled x_1 through x_N, is any equation of the form

$$a_1x_1 + a_2x_2 + \cdots a_Nx_N = b\,, \tag{4.1}$$

where the coefficients a_1 through a_N and the value b on the right-hand side are real numbers.

- The linear equation (4.1) is the same as

$$\mathbf{a}^T\mathbf{x} = \begin{bmatrix} a_1 \; a_2 \cdots a_N \end{bmatrix} \begin{bmatrix} x_1 \\ x_2 \\ \vdots \\ x_N \end{bmatrix} = b\,. \tag{4.2}$$

  The vector $\mathbf{a} = [a_1\; a_2\; \ldots\; a_N]^T$ is called the **coefficient vector** and $\mathbf{x} = [x_1\; x_2\; \ldots\; x_N]^T$ is the vector of unknowns.
- A **solution** to the linear equation $a_1x_1 + a_2x_2 + \cdots a_Nx_N = b$ consists of a vector of real number values for the unknowns $x_1, \ldots, x_N$ that yields a true equation when plugged in.
- For $\mathbf{a} \neq \mathbf{0}$, the set of terminal points of all solutions to (4.1) forms a *hyperplane* in $\mathbb{R}^N$. The coefficient vector **a** is orthogonal to every line in the hyperplane and is called a **normal vector** to the hyperplane.
- For $\mathbf{a} \neq \mathbf{0}$, the terminal point of the vector $(b/||\mathbf{a}||^2) \cdot \mathbf{a}$ is the point closest to the origin in the hyperplane determined by (4.1). The length of this vector is

$r = |b|/||\mathbf{a}||$, which is, therefore, the orthogonal distance to the origin from the hyperplane.

Example 4.2 The equation $3x_1 + 4x_2 - 5x_3 = 12$ is a linear equation in three unknowns. It has the form $\mathbf{a}^T\mathbf{x} = b$, where $\mathbf{a} = \begin{bmatrix} 3 & 4 & -5 \end{bmatrix}^T$ and $b = 12$. Thus, one solution is

$$\mathbf{x} = (b/||\mathbf{a}||^2) \cdot \mathbf{a} = \begin{bmatrix} 0.72 \\ 0.96 \\ -1.20 \end{bmatrix}.$$

The values $x_1 = 3$, $x_2 = 2$, and $x_3 = 1$ provide another solution, since $3 \cdot 3 + 4 \cdot 2 - 5 \cdot 1 = 12$. Taking $x_1 = 2$, $x_2 = -1$, and $x_3 = -2$ yields a third solution, since $3 \cdot 2 + 4 \cdot (-1) - 5 \cdot (-2) = 12$. Collectively, the set of all solutions generates a plane in $\mathbb{R}^3$ that is orthogonal to the vector $\begin{bmatrix} 3 & 4 & -5 \end{bmatrix}^T$ and contains the terminal point of each solution vector. For example, this plane passes through the points $(0.72,\ 0.96,\ -1.20)$, $(3,\ 2,\ 1)$, and $(2,\ -1,\ -2)$.

Example 4.3 Consider the linear equation $\mathbf{a}^T\mathbf{x} = b$ in the case where the coefficient vector is $\mathbf{a} = \mathbf{0}$. Then, for every vector $\mathbf{x}$, we have $\mathbf{a}^T\mathbf{x} = \mathbf{0}^T\mathbf{x} = 0$. If the right-hand side is $b = 0$, then *every vector* $\mathbf{x}$ in $\mathbb{R}^N$ will satisfy $\mathbf{a}^T\mathbf{x} = b$. On the other hand, if $b \neq 0$, then the equation has *no solutions*.

Example 4.4 In $\mathbb{R}^2$, the vector $\mathbf{a} = \begin{bmatrix} a_1 \\ a_2 \end{bmatrix}$ lies on a line through the origin with slope a_2/a_1, or on the vertical coordinate axis if $a_1 = 0$. Every vector orthogonal to $\mathbf{a}$ has the form $\mathbf{x} = t \cdot \begin{bmatrix} -a_2 \\ a_1 \end{bmatrix}$, where t can be any real number. These vectors generate the line of slope $-a_1/a_2$ through the origin, or the vertical coordinate axis if $a_2 = 0$. One particular solution to $\mathbf{a}^T\mathbf{x} = b$ is $\mathbf{x}_* = \begin{bmatrix} a_1 \cdot b/(a_1^2 + a_2^2) \\ a_2 \cdot b/(a_1^2 + a_2^2) \end{bmatrix}$. If $\mathbf{x}$ is any other solution, then $\mathbf{x} - \mathbf{x}_*$ is orthogonal to $\mathbf{a}$. Thus, every solution to $\mathbf{a}^T\mathbf{x} = b$ looks like

$$\mathbf{x} = \mathbf{x}_* + t \cdot \begin{bmatrix} -a_2 \\ a_1 \end{bmatrix}.$$

The terminal points of these solutions form a line of slope $-a_1/a_2$ at a distance $|b|/\sqrt{a_1^2 + a_2^2}$ from the origin.

Example 4.5 In medical imaging, a CAT scan is created by analyzing information from a large family of X-rays traveling through the patient. The path of each X-ray is a line in a plane, which we catalog, like so. Choose an angle θ and a real number b. Set $\mathbf{a} = [\cos(\theta)\ \ \sin(\theta)]^T$ and define the line $\ell(b,\theta) = \left\{\mathbf{x} \in \mathbb{R}^2 \,|\, \mathbf{a}^T\mathbf{x} = b\right\}$.

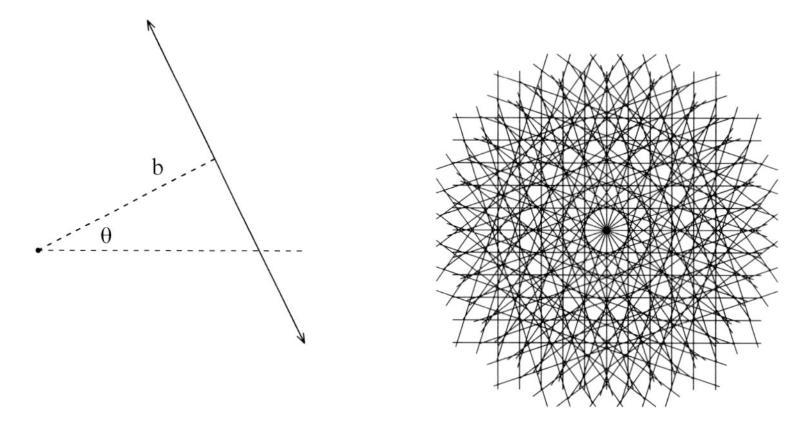

Fig. 4.2 *Left,* a single line $\ell_{b,\theta}$; *right,* a configuration of lines $\ell_{b,\theta}$ for 11 values of b with θ in increments of $\pi/10$

Thus, $\ell(b, \theta)$ passes through the point with coordinates $(b\cos(\theta), b\sin(\theta))$ and is orthogonal to **a**. Every line in the plane is uniquely represented as $\ell(b, \theta)$ for some real number b and some θ with $0 \le \theta < \pi$. The points on $\ell(b, \theta)$ are the solutions to the linear equation $\cos(\theta) \cdot x_1 + \sin(\theta) \cdot x_2 = b$. Figure 4.2 depicts a configuration of X-ray lines.

4.2 Systems of Linear Equations

A **system of linear equations** is any collection of two or more linear equations with the same number of unknowns. A system of M equations with N unknowns, labeled $x_1, x_2, \ldots, x_N$, has the form

$$\begin{cases} a_{1,1}x_1 + a_{1,2}x_2 + \cdots + a_{1,N}x_N = b_1 \\ a_{2,1}x_1 + a_{2,2}x_2 + \cdots + a_{2,N}x_N = b_2 \\ \quad\vdots \\ a_{M,1}x_1 + a_{M,2}x_2 + \cdots + a_{M,N}x_N = b_M \,. \end{cases} \tag{4.3}$$

Each coefficient $a_{i,j}$ has two subscripts, the first of which tells us which equation we are looking at and the second of which matches the unknown.

A **solution** to a system of linear equations consists of specific numerical values for the unknowns that simultaneously satisfy *all equations in the system* when these values are plugged in. Usually, our goal will be to find all solutions to a given system of linear equations. When each equation in the system represents a hyperplane in $\mathbb{R}^N$, our goal is to locate and describe the common intersection of these hyperplanes. For example, the set of solutions to a system of three linear equations in three unknowns typically represents the intersection of three planes in $\mathbb{R}^3$. Usually this is a single point, but it could be empty, or form a line or a plane.

A system of linear equations that has no solutions is said to be **inconsistent**.

Every system of linear equations can be formulated using matrices and vectors. For the generic system shown in (4.3), we form the $M \times N$ matrix $A = \left[a_{i,j} \right]$, called the **coefficient matrix**, or **matrix of coefficients**, for the system. We also form the vector of unknowns $\mathbf{x} = [x_1\ x_2\ \ldots\ x_N]^T$ and the right-hand side vector $\mathbf{b} = [b_1\ b_2\ \ldots\ b_M]^T$. The equation $A\mathbf{x} = \mathbf{b}$ is a concise expression of our system. That is,

$$A\mathbf{x} = \begin{bmatrix} a_{1,1} & a_{1,2} & \cdots & a_{1,N} \\ a_{2,1} & a_{2,2} & \cdots & a_{2,N} \\ \vdots & \vdots & \ddots & \vdots \\ a_{M,1} & a_{M,2} & \cdots & a_{M,N} \end{bmatrix} \begin{bmatrix} x_1 \\ x_2 \\ \vdots \\ x_N \end{bmatrix} = \begin{bmatrix} b_1 \\ b_2 \\ \vdots \\ b_M \end{bmatrix} = \mathbf{b}\,. \tag{4.4}$$

In this framework, the solution set for the system comprises all vectors $\mathbf{x}$ for which $A\mathbf{x} = \mathbf{b}$.

Example 4.6

(i) The values $x = 4$, $y = -1$, and $z = 3$ satisfy the system

$$\begin{cases} x + 5y + z = 2 \\ 2x + y - 2z = 1 \\ x + 7y + 2z = 3\,, \end{cases}$$

since the statements $4 - 5 + 3 = 2$, $8 - 1 - 6 = 1$, and $4 - 7 + 6 = 3$ are all true. In matrix form, the system is given by

$$\begin{bmatrix} 1 & 5 & 1 \\ 2 & 1 & -2 \\ 1 & 7 & 2 \end{bmatrix} \begin{bmatrix} x \\ y \\ z \end{bmatrix} = \begin{bmatrix} 2 \\ 1 \\ 3 \end{bmatrix}.$$

The vector $\mathbf{x} = \begin{bmatrix} 4 \\ -1 \\ 3 \end{bmatrix}$ is a solution to the system because

$$\begin{bmatrix} 1 & 5 & 1 \\ 2 & 1 & -2 \\ 1 & 7 & 2 \end{bmatrix} \begin{bmatrix} 4 \\ -1 \\ 3 \end{bmatrix} = \begin{bmatrix} 2 \\ 1 \\ 3 \end{bmatrix}.$$

In fact, this is the unique solution to this system, as we verify later.

(ii) The system

$$\begin{cases} x_1 + 2x_2 + 3x_3 = 1 \\ x_1 + 2x_2 + 3x_3 = 3 \end{cases}$$

has no solutions. The two equations have the same left-hand side, but the right-hand sides are different. These equations describe two parallel planes in $\mathbb{R}^3$. Geometrically, a vector $\mathbf{x}$ cannot have different projections along the same vector $\begin{bmatrix} 1 & 2 & 3 \end{bmatrix}^T$.

(iii) For every real number t, the values $x_1 = 4 + 3t$, $x_2 = 3 - 2t$, and $x_3 = t$ give a solution to the system of equations

$$\begin{cases} x_1 + 3x_2 + 3x_3 = 13 \\ 2x_1 + 5x_2 + 4x_3 = 23 \\ 2x_1 + 7x_2 + 8x_3 = 29 . \end{cases}$$

Equivalently, the vector $\mathbf{x} = \begin{bmatrix} 4+3t & 3-2t & t \end{bmatrix}^T$ satisfies

$$\begin{bmatrix} 1 & 3 & 3 \\ 2 & 5 & 4 \\ 2 & 7 & 8 \end{bmatrix} \mathbf{x} = \begin{bmatrix} 1 & 3 & 3 \\ 2 & 5 & 4 \\ 2 & 7 & 8 \end{bmatrix} \begin{bmatrix} 4+3t \\ 3-2t \\ t \end{bmatrix} = \begin{bmatrix} 13 \\ 23 \\ 29 \end{bmatrix} .$$

We write this general solution vector as

$$\mathbf{x} = \begin{bmatrix} 4 \\ 3 \\ 0 \end{bmatrix} + t \cdot \begin{bmatrix} 3 \\ -2 \\ 1 \end{bmatrix} .$$

Geometrically, the equations in the system describe three planes in $\mathbb{R}^3$ whose intersection is the line passing through the point (4, 3, 0) and parallel to the vector $\begin{bmatrix} 3 & -2 & 1 \end{bmatrix}^T$.

4.3 Row Reduction

To help us find a method for solving linear systems, think about where we would like to end up. If the unknowns are x_1 through x_N, then a solution should look like

$$x_1 = \widehat{b}_1 \, , \; x_2 = \widehat{b}_2 \, , \; \ldots \, , \; x_N = \widehat{b}_N \, .$$

This corresponds to the system

$$\begin{bmatrix} 1 & 0 & \cdots & 0 \\ 0 & 1 & \cdots & 0 \\ \vdots & \vdots & \ddots & \vdots \\ 0 & 0 & \cdots & 1 \end{bmatrix} \begin{bmatrix} x_1 \\ x_2 \\ \vdots \\ x_N \end{bmatrix} = \begin{bmatrix} \widehat{b}_1 \\ \widehat{b}_2 \\ \vdots \\ \widehat{b}_N \end{bmatrix} .$$

In this ideal system, the coefficient matrix is the identity matrix. Thus, our goal is to modify the original coefficient matrix in such a way that it comes to resemble the identity matrix. This is equivalent to modifying the original equations in order to eliminate as many unknowns from as many equations as possible. In the nineteenth century, a methodical procedure for achieving these goals was prominently used by Carl Friedrich Gauss and is now known as **Gaussian elimination**. This procedure relies on two basic algebraic facts:

(i) An equation $P = Q$ is true if, and only if, the equation $c \cdot P = c \cdot Q$ is true for every nonzero real number c. This means that **we can multiply both sides of any equation in our system by the same nonzero constant** without altering the set of solutions.

(ii) Two equations $P = Q$ and $R = S$ are both true if, and only if, the equations $P = Q$ and $R + c \cdot P = S + c \cdot Q$ are both true for every constant c. In other words, **we can modify any equation in our system by adding a constant multiple of another equation to it** without changing the solution set for the system.

To these two modifications, we add a third.

(iii) **We may change the order in which we write the equations.** This is cosmetic and does not alter the solution set for the system.

Example 4.7 To illustrate, we use Gaussian elimination to solve the system

$$\begin{aligned} 5x + y &= 2 \\ 7x + 2y &= 3\,. \end{aligned}$$

First, make the leading coefficient in the first equation equal to 1 by multiplying the equation by $1/5$.

$$\begin{aligned} x + (1/5)y &= 2/5 \text{ (New eqn \#1 } = (1/5) \cdot \text{ (old eqn \#1).)} \\ 7x + 2y &= 3\,. \end{aligned}$$

Next, eliminate the x term from the second equation by adding (-7) times the new first equation to it.

$$\begin{aligned} x + (1/5)y &= 2/5 \\ (3/5)y &= 1/5 \text{ (New eqn \#2 } = \text{ (old eqn \#2) } - 7 \cdot \text{ (eqn \#1).)}\,. \end{aligned}$$

Next, multiply the new second equation by $(5/3)$ to get a coefficient of 1 on the y term, which is now the leading term of that equation.

$$\begin{aligned} x + (1/5)y &= 2/5 \\ y &= 1/3 \text{ (New eqn \#2 } = (5/3) \cdot \text{ (old eqn \#2).)}\,. \end{aligned}$$

Finally, eliminate the y term from the first equation by adding $-(1/5)$ times the new second equation to it.

$$\begin{aligned} x \quad &= 1/3 \text{ (New eqn \#1} = \text{(old eqn \#1)} - (1/5)\cdot \text{(eqn \#2).)} \\ y &= 1/3\,. \end{aligned}$$

Our new, modified system of equations explicitly displays the solution $x = 1/3$ and $y = 1/3$. None of the algebra we did changes the solution set in any way, so this is the unique solution to the original system of equations. Geometrically, at each step along the way, we have the equations of two lines in $\mathbb{R}^2$. Each pair of lines has the same intersection as the previous pair. In some steps, we rewrite the equation of one of the lines by multiplying both sides by a constant. In the other steps, we replace one of the lines with a different line that is parallel to a coordinate axis by rotating or **pivoting** the existing line about the intersection point we want to locate. In the end, we have one vertical line and one horizontal line whose intersection point is easy to identify.

The modifications used in Gaussian elimination affect only the coefficients and right-hand sides of the various equations. We can keep track of our changes by forming the **augmented matrix** for the system. For the generic system shown in (4.4), the augmented matrix is

$$\left[\,A\middle|\mathbf{b}\,\right] = \left[\begin{array}{cccc|c} a_{11} & a_{12} & \cdots & a_{1N} & b_1 \\ a_{21} & a_{22} & \cdots & a_{2N} & b_2 \\ \vdots & \vdots & \ddots & \vdots & \vdots \\ a_{M1} & a_{M2} & \cdots & a_{MN} & b_M \end{array}\right]. \tag{4.5}$$

This is a **partitioned matrix**, where the coefficient matrix A fills up the first N columns and the right-hand side vector occupies the remaining column. The vertical partition line does not have to be there, but its presence can help to remind us that we are looking at an augmented matrix. Also, when we translate the augmented matrix back into a system of equations, the partition indicates where the "=" sign goes.

Each of the modifications allowed by Gaussian elimination corresponds to an adjustment of the augmented matrix for the system, as follows.

(i) When we multiply both sides of an equation by the same nonzero constant $c \neq 0$, we multiply every coefficient and also the right-hand side by that constant. In other words, we **multiply all entries in one row of the augmented matrix by a nonzero constant** $c \neq 0$**.**

$$\left[\begin{array}{cccc|c} \vdots & \vdots & \ddots & \vdots & \vdots \\ a_{i,1} & a_{i,2} & \cdots & a_{i,N} & b_i \\ \vdots & \vdots & \ddots & \vdots & \vdots \end{array}\right] \mapsto \left[\begin{array}{cccc|c} \vdots & \vdots & \ddots & \vdots & \vdots \\ ca_{i,1} & ca_{i,2} & \cdots & ca_{i,N} & cb_i \\ \vdots & \vdots & \ddots & \vdots & \vdots \end{array}\right].$$

(ii) When we modify equation k by adding a constant multiple of equation i to it, the effect is to replace each coefficient $a_{k,j}$ with $(a_{k,j} + c \cdot a_{i,j})$ and the right-hand side b_k with $(b_k + c \cdot b_i)$. Thus, we **add c times the entries in row i of the augmented matrix to the corresponding entries in row k.** Row k of the new augmented matrix is

$$\left[\begin{array}{cccc|c} \vdots & \vdots & \ddots & \vdots & \vdots \\ (a_{k,1} + ca_{i,1}) & (a_{k,2} + ca_{i,2}) & \cdots & (a_{k,N} + ca_{i,N}) & (b_k + cb_i) \\ \vdots & \vdots & \ddots & \vdots & \vdots \end{array}\right].$$

As we go, we will choose the constant c in order to eliminate one or more unknowns from equation k, thus making that equation simpler.

(iii) The cosmetic step of changing the order of the equations allows us now to **switch two rows of the augmented matrix.**

$$\left[\begin{array}{cccc|c} \vdots & \vdots & \ddots & \vdots & \vdots \\ a_{i,1} & a_{i,2} & \cdots & a_{i,N} & b_i \\ \vdots & \vdots & \ddots & \vdots & \vdots \\ a_{k,1} & a_{k,2} & \cdots & a_{k,N} & b_k \\ \vdots & \vdots & \ddots & \vdots & \vdots \end{array}\right] \mapsto \left[\begin{array}{cccc|c} \vdots & \vdots & \ddots & \vdots & \vdots \\ a_{k,1} & a_{k,2} & \cdots & a_{k,N} & b_k \\ \vdots & \vdots & \ddots & \vdots & \vdots \\ a_{i,1} & a_{i,2} & \cdots & a_{i,N} & b_i \\ \vdots & \vdots & \ddots & \vdots & \vdots \end{array}\right].$$

Definition 4.8 We have identified three ways of modifying the rows of an augmented matrix such that the set of solutions to the corresponding system of linear equations is preserved. These modifications are called **elementary row operations**. The process of carrying out Gaussian elimination using an augmented matrix and elementary row operations is called **row reduction**. The elementary row operations are as follows:

(i) **Multiply the entries of a row by a nonzero constant.**
(ii) **Modify a row by adding a constant multiple of another row to it.**
(iii) **Switch the positions of two rows.**

Example 4.9 The system $\begin{cases} 5x + y = 2 \\ 7x + 2y = 3 \end{cases}$, from Example 4.7, has the augmented matrix $\left[\begin{array}{cc|c} 5 & 1 & 2 \\ 7 & 2 & 3 \end{array}\right]$. We now apply the elementary row operations that correspond to the same steps we used above.

$$\left[\begin{array}{cc|c} 5 & 1 & 2 \\ 7 & 2 & 3 \end{array}\right] \quad \overset{(1/5)\cdot(\text{Row \#1})}{\longmapsto} \quad \left[\begin{array}{cc|c} 1 & 1/5 & 2/5 \\ 7 & 2 & 3 \end{array}\right]$$

$$\overset{(\text{Row }\#2)-7\cdot(\text{Row }\#1)}{\longmapsto} \left[\begin{array}{cc|c} 1 & 1/5 & 2/5 \\ 0 & 3/5 & 1/5 \end{array}\right]$$

$$\overset{(5/3)\cdot(\text{Row }\#2)}{\longmapsto} \left[\begin{array}{cc|c} 1 & 1/5 & 2/5 \\ 0 & 1 & 1/3 \end{array}\right]$$

$$\overset{(\text{Row }\#1)-(1/5)\cdot(\text{Row }\#1)}{\longmapsto} \left[\begin{array}{cc|c} 1 & 0 & 1/3 \\ 0 & 1 & 1/3 \end{array}\right].$$

Translating the last version of the augmented matrix back into equations yields the unique solution $\begin{cases} x = 1/3 \\ y = 1/3 \end{cases}$. We can, and should, check our answer by plugging back into the original system. Thus,

$$5(1/3) + (1/3) = 6/3 = 2 \text{ and } 7(1/3) + 2(1/3) = 9/3 = 3\,, \text{ as desired.}$$

In matrix form,

$$\begin{bmatrix} 5 & 1 \\ 7 & 2 \end{bmatrix} \begin{bmatrix} 1/3 \\ 1/3 \end{bmatrix} = \begin{bmatrix} 5(1/3) + (1/3) \\ 7(1/3) + 2(1/3) \end{bmatrix} = \begin{bmatrix} 2 \\ 3 \end{bmatrix}.$$

Example 4.10 As promised in Example 4.6 (i), we show that the values $x = 4$, $y = -1$, and $z = 3$ are the only solution to the system

$$\begin{cases} x + 5y + z = 2 \\ 2x + y - 2z = 1 \\ x + 7y + 2z = 3 \end{cases}.$$

The augmented matrix is $\left[\begin{array}{ccc|c} 1 & 5 & 1 & 2 \\ 2 & 1 & -2 & 1 \\ 1 & 7 & 2 & 3 \end{array}\right]$. Row reduction proceeds as follows.

$$\left[\begin{array}{ccc|c} 1 & 5 & 1 & 2 \\ 2 & 1 & -2 & 1 \\ 1 & 7 & 2 & 3 \end{array}\right] \overset{\substack{(\text{Row }\#2)-2\cdot(\text{Row }\#1) \\ (\text{Row }\#3)-(\text{Row }\#1)}}{\longmapsto} \left[\begin{array}{ccc|c} 1 & 5 & 1 & 2 \\ 0 & -9 & -4 & -3 \\ 0 & 2 & 1 & 1 \end{array}\right]$$

$$\overset{-(1/9)\cdot(\text{Row }\#2)}{\longmapsto} \left[\begin{array}{ccc|c} 1 & 5 & 1 & 2 \\ 0 & 1 & 4/9 & 1/3 \\ 0 & 2 & 1 & 1 \end{array}\right]$$

$$\overset{\substack{(\text{Row }\#1)-5\cdot(\text{Row }\#2) \\ (\text{Row }\#3)-2\cdot(\text{Row }\#2)}}{\longmapsto} \left[\begin{array}{ccc|c} 1 & 0 & -11/9 & 1/3 \\ 0 & 1 & 4/9 & 1/3 \\ 0 & 0 & 1/9 & 1/3 \end{array}\right]$$

$$\overset{9\cdot(\text{Row \#3})}{\longmapsto} \quad \left[\begin{array}{ccc|c} 1 & 0 & -11/9 & 1/3 \\ 0 & 1 & 4/9 & 1/3 \\ 0 & 0 & 1 & 3 \end{array}\right]$$

$$\overset{\substack{(\text{Row \#1})+(11/9)\cdot(\text{Row \#3}) \\ (\text{Row \#2})-(4/9)\cdot(\text{Row \#3})}}{\longmapsto} \quad \left[\begin{array}{ccc|c} 1 & 0 & 0 & 4 \\ 0 & 1 & 0 & -1 \\ 0 & 0 & 1 & 3 \end{array}\right]$$

From this version of the augmented matrix, we see that the unique solution is $x = 4$, $y = -1$, and $z = 3$, as claimed.

There was a stage along the way when we could have stopped with the upper triangular augmented matrix $\left[\begin{array}{ccc|c} 1 & 5 & 1 & 2 \\ 0 & 1 & 4/9 & 1/3 \\ 0 & 0 & 1 & 3 \end{array}\right]$. The bottom row here implies that $z = 3$. Then, **backward substitution** in the second row yields $y = 1/3 - (4/9)z = -1$. Further substitution in the first row yields $x = 2 - 5y - z = 4$. The arithmetic required for the backward substitution is the same as that used in the additional elementary row operations.

Example 4.11 Three planes can meet in a common line, as with pages of a book where they meet along the binding. For example, revisit the system

$$\begin{cases} x_1 + 3x_2 + 3x_3 = 13 \\ 2x_1 + 5x_2 + 4x_3 = 23 \\ 2x_1 + 7x_2 + 8x_3 = 29 \end{cases},$$

from Example 4.6(iii). Elementary row operations applied to the augmented matrix produce the following result.

$$\left[\begin{array}{ccc|c} 1 & 3 & 3 & 13 \\ 2 & 5 & 4 & 23 \\ 2 & 7 & 8 & 29 \end{array}\right] \overset{\substack{(\text{Row \#2})-2\cdot(\text{Row \#1}) \\ (\text{Row \#3})-2(\text{Row \#1})}}{\longmapsto} \left[\begin{array}{ccc|c} 1 & 3 & 3 & 13 \\ 0 & -1 & -2 & -3 \\ 0 & 1 & 2 & 3 \end{array}\right]$$

$$\overset{\text{switch Rows \#2 and \#3}}{\longmapsto} \left[\begin{array}{ccc|c} 1 & 3 & 3 & 13 \\ 0 & 1 & 2 & 3 \\ 0 & -1 & -2 & -3 \end{array}\right]$$

$$\overset{\substack{(\text{Row \#1})-3\cdot(\text{Row \#2}) \\ (\text{Row \#3})+(\text{Row \#2})}}{\longmapsto} \left[\begin{array}{ccc|c} 1 & 0 & -3 & 4 \\ 0 & 1 & 2 & 3 \\ 0 & 0 & 0 & 0 \end{array}\right].$$

Look what happened to the bottom row! In the next to last step, the third row of the augmented matrix was the negative of the second row. When we added the second row to it, the third row became a row of all 0s. It may be tempting to think that this row no longer matters, but it is *crucial* that the corresponding equation $0 = 0$ is *true*. Therefore, any values for the unknowns that satisfy the first two equations will automatically satisfy the third. The other consequence of this row of all 0s is that there is no equation that solves for the unknown x_3. Accordingly, x_3 is a **free variable**—x_3 can have potentially any value. It is customary to rename a free variable as a parameter, say $x_3 = t$. Then the first row of the final augmented matrix yields $x_1 - 3t = 4$, or, equivalently, $x_1 = 4 + 3t$. The second row of the final augmented matrix corresponds to the equation $x_2 + 2t = 3$, so that $x_2 = 3 - 2t$. In vector form, the general solution to the system is given by

$$\mathbf{x} = \begin{bmatrix} x_1 \\ x_2 \\ x_3 \end{bmatrix} = \begin{bmatrix} 4 + 3t \\ 3 - 2t \\ t \end{bmatrix} = \begin{bmatrix} 4 \\ 3 \\ 0 \end{bmatrix} + t \cdot \begin{bmatrix} 3 \\ -2 \\ 1 \end{bmatrix}$$

where t can have any real number value. From this description, we see that the solution set forms a line in $\mathbb{R}^3$ through the point with coordinates (4, 3, 0) and parallel to the vector $\begin{bmatrix} 3 & -2 & 1 \end{bmatrix}^T$.

4.4 Row Echelon Forms

The examples above raise the question, *"When we perform elementary row operations on an augmented matrix, how do we know when to stop?"* There are two standard answers to this question. Creating a modified augmented matrix that is in **row echelon form** allows us to produce the solution via backward substitution. Putting the augmented matrix into **reduced row echelon form** allows us more or less to read off the solution without additional work. In fact, these two approaches require the same amount of calculation overall. Depending on the context, one or the other approach may seem more convenient.

Consider reduced row echelon form first. At the end of our work in Example 4.9 above, each row of the modified augmented matrix has a 1 as the leftmost nonzero entry. These are called **leading 1s**. Moreover, each column contains only one of these leading 1s and the remaining entries in the column are all 0s. Thus, the modified augmented matrix has one row for each unknown giving the value of that unknown and not involving any other unknowns. Conveniently, as we move down through the rows, the columns containing successive leading 1s move to the right, as if descending a staircase. This is the reason for using the word *echelon*. Thanks to the echelon formation of the leading 1s, we solve for the various unknowns in their original order. The last modified augmented matrix in Example 4.11 has some of these features, but the third column looks different, reflecting the fact that x_3 is a free variable. Also, this matrix has a row of all 0s located at the bottom. Both of these augmented matrices are in reduced row echelon form, which we now define.

Definition 4.12 A matrix is said to be in **reduced row echelon form**, or **RREF**, if the following conditions all apply:

(i) The leftmost nonzero entry in every row is a 1. These entries are called **leading 1s**.
(ii) In each column that contains the leading 1 of some row, all other entries are 0s.
(iii) The leading 1s are arranged in a "down and to the right" echelon in which the leading 1 in a lower row is in a column farther to the right.
(iv) Any rows of all 0s are located at the bottom.

Features (i) and (ii) are essential and come from eliminating as many unknowns from as many equations as possible. Features (iii) and (iv) both make the array easier to read. They also have important implications for computing matrix inverses, in Section 4.5.

When a matrix A has been transformed, using elementary row operations, into a matrix that is in reduced row echelon form, we will denote the result by RREF(A). If column j of RREF(A) contains the leading 1 of some row and is otherwise all 0, then the corresponding unknown (e.g., x_j) is called a **leading variable**. Unknowns that are not leading variables are called **free variables**. Notice that a system with fewer equations than unknowns *must have free variables*.

Remark It can happen that a row of RREF(A) will have all 0s in the columns corresponding to the unknowns but a 1 in the right-hand side column. That row is then equivalent to the false statement "$0 = 1$". There is no possible choice of values for the unknowns that can make that statement true, so the corresponding system of equations **has no solutions!** Such a system is said to be **inconsistent**. A system that has more equations than unknowns is either inconsistent or contains redundant equations, resulting in one or more rows of all 0s in the reduced row echelon form of the augmented matrix.

Example 4.13 The system

$$\begin{aligned} 2x + \ y + \ z + 5w &= 3 \\ -x + 9y + 3z \qquad &= 1 \\ 3y + \ z + \ w &= 5 \\ 2x - 7y \qquad - w &= 7. \end{aligned}$$

is *inconsistent* because

$$\text{RREF}\left(\left[\begin{array}{cccc|c} 2 & 1 & 1 & 5 & 3 \\ -1 & 9 & 3 & 0 & 1 \\ 0 & 3 & 1 & 1 & 5 \\ 2 & -7 & 0 & -1 & 7 \end{array}\right]\right) = \left[\begin{array}{cccc|c} 1 & 0 & 0 & 3 & 0 \\ 0 & 1 & 0 & 1 & 0 \\ 0 & 0 & 1 & -2 & 0 \\ 0 & 0 & 0 & 0 & 1 \end{array}\right].$$

The bottom row implies the impossible requirement that "$0 = 1$".

Algorithm 4.14 Reduced row echelon form via row reduction Here is a systematic approach for transforming a matrix $A = [a_{i,j}]$ into reduced row echelon form using elementary row operations.

Step 1. (a) Multiply the first row of A by $1/a_{1,1}$. This creates a leading 1 in the top left corner. (b) For each $i > 1$, add $-a_{i,1}/a_{1,1}$ times row 1 of A to row i. This eliminates the unknown x_1 from all but the first row.

Fine print: If $a_{1,1} = 0$, then switch the first row of A with a different row that starts with a nonzero entry. If all coefficients in the first column are 0s, then x_1 is a free variable and will not effect the values of any of the other unknowns.

Step k for $k \geq 2$. Assume that each of the first $k - 1$ rows has a leading 1 as the leftmost nonzero entry and that the remaining entries in the corresponding columns are all 0s. From the position of the leading 1 created in the previous step, move down by one row and to the right by one column. Call the entry located there $a_{k,\ell}$.

(a) Multiply row k by $1/a_{k,\ell}$ to create a leading 1 in row k.
(b) For each $i \neq k$, add $-a_{i,\ell}/a_{k,\ell}$ times row k to row i. This eliminates the unknown x_ℓ from all rows except row k.

Fine print: If $a_{k,\ell} = 0$, then switch row k with a *lower* row that starts with a nonzero entry. If there are no such lower rows, then x_ℓ is a *free variable*. Move one column to the right along row k to the entry $a_{k,\ell+1}$ and try again with Step k.

Final step. Repeat Step k until every row either has a leading 1 as its leftmost nonzero entry or is a row of all 0s. The procedure described in Step k ensures that the leading 1s form a "down and to the right" echelon, while the *fine print* guarantees that any rows of all 0s are located at the bottom.

Remark Algorithm 4.14 shows how to transform any matrix into a matrix that is in reduced row echelon. In fact, the end result is the *only* such matrix. That is, RREF(A) is unique. We do not attempt to prove this subtlety here.

With fewer elementary row operations than are required by RREF, we can produce a matrix that has just enough of the essential features of RREF to allow us to identify the solutions to our system of equations. This version is called **row echelon form**.

Definition 4.15 A matrix is said to be in **row echelon form**, abbreviated as **REF**, if the following conditions all apply:

(i) In each column that contains the leftmost nonzero entry of some row, all entries below that are 0s.
(ii) The leftmost nonzero entry of a lower row is in a column farther to the right of the leftmost nonzero entry of a higher row. (This is the same "down and to the right" echelon as RREF.)
(iii) Any rows of all 0s are located at the bottom.

With row echelon form, the leading nonzero entry of each row is not required to be a 1 and only the entries below such a leading nonzero entry are required to be 0s. As with RREF, if column j of an augmented matrix in row echelon form contains the leftmost nonzero entry of some row, then the corresponding unknown is a **leading variable**. All other unknowns are **free variables**. When the augmented matrix of a system of linear equations has been transformed into row echelon form, the solution to the system is found by using **backward substitution**, like so.

Example 4.16 The augmented matrix for the system

$$\begin{cases} x + \ y + \ z + 5w = 3 \\ x + 9y + 3z \qquad = 1 \\ \quad 8y + \ z + \ w = 5 \\ 2x + 2y + 3z - w = 7 \end{cases}$$

transforms into row echelon form as

$$\left[\begin{array}{cccc|c} 1 & 1 & 1 & 5 & 3 \\ 0 & 8 & 2 & -5 & -2 \\ 0 & 0 & -1 & 6 & 7 \\ 0 & 0 & 0 & -5 & 8 \end{array}\right].$$

Starting at the bottom, we see that $-5w = 8$, so $w = -8/5$. Moving up one row, we find the equation $-z+6w = 7$; thus, $z = -(7-6w) = -(7+48/5) = -83/5$. The second row shows that $8y + 2z - 5w = -2$. So $y = (-2 - 2z + 5w)/8 = 29/10$. Finally, the top row gives us $x = 3 - y - z - 5w = 247/10$. The arithmetic needed to solve for each unknown from the later ones is the same as what would have been required to further transform the augmented matrix into reduced row echelon form.

Remark A matrix in row echelon form is **upper triangular** in the sense that the $[i, j]$ entry is 0 whenever $i > j$. If we change the first requirement of row echelon form to read "In each column that contains the leftmost nonzero entry of some row, all entries *above* that are 0s" then such a matrix is **lower triangular**. A system whose augmented matrix is in lower triangular form can be solved via **forward substitution**, where we first solve for x_1 and then solve for each successive unknown in terms of the previous ones.

Algorithm 4.17 Row echelon form via row reduction Any matrix $A = \left[a_{i,j}\right]$ can be transformed into row echelon form as follows:

Step 1. Assuming that $a_{1,1} \neq 0$, then, for each $i > 1$, add $-a_{i,1}/a_{1,1}$ times row 1 of A to row i. This eliminates the unknown x_1 from all but the first row. *Fine print:* If $a_{1,1} = 0$, then switch the first row of A with a different row that starts with a nonzero entry.

Step k for $k \geq 2$. Assume that the entries below the leftmost nonzero entry in each of the first $k - 1$ rows are all 0s. From the position of the leading nonzero

entry in row $k - 1$, move down by one row and to the right by one column. Call the entry located there $a_{k,\ell}$. For each $i > k$, add $-a_{i,\ell}/a_{k,\ell}$ times row k to row i. This eliminates the unknown x_ℓ from all rows below row k. *Fine print:* If $a_{k,\ell} = 0$, then switch row k with a *lower* row that starts with a nonzero entry. If there are no such lower rows, then x_ℓ is a *free variable*. Move one column to the right along row k to the entry $a_{k,\ell+1}$ and try again with Step k.
Final step. Repeat Step k until every row of A has been treated. The procedure described in Step k ensures that the leading nonzero entries of the different rows are arranged in "down and to the right" echelon, while the *fine print* guarantees that any rows of all 0s are located at the bottom.

Remark Unlike with reduced row echelon form, a given matrix can be transformed into a variety of matrices in row echelon form. This is so, in particular, because REF does not require the leading nonzero entries of the rows to be 1s.

4.5 Matrix Inverses (And How to Find Them)

As we discussed in Chapter 2, a square matrix A is said to be invertible if there is a matrix B, of the same size as A, for which $AB = BA = \mathcal{I}$, where $\mathcal{I}$ denotes the identity matrix. If such a matrix B exists, then we call it the **inverse of** A and denote it by A^{-1}.

To determine whether or not a given $N \times N$ matrix is invertible, we think of the potential inverse as an unknown matrix X satisfying $AX = \mathcal{I}_N$. This single matrix equation is a compilation of N systems of linear equations, all having the same coefficient matrix A. For each j between 1 and N, the jth system is given by $A\mathbf{x}_j = \mathbf{e}_j$, where $\mathbf{x}_j$ is the jth column vector of the unknown matrix X and $\mathbf{e}_j$ is the jth column vector of the identity matrix $\mathcal{I}_N$. Each system has an augmented matrix to which we can apply row reduction. The elementary row operations needed will be the same in every case, though, because of the common coefficient matrix. Thus, we can form one combined augmented matrix $[A \mid \mathcal{I}]$ and row reduce the whole thing at once. We stop when we have a matrix in *reduced row echelon form*. There are two possibilities that matter for us now.

(1) Suppose the augmented matrix $[A \mid \mathcal{I}]$ becomes the matrix $[\mathcal{I} \mid B]$ in reduced row echelon form. In other words, suppose the "A side" of the original augmented matrix transforms into the *identity matrix*. Then there are no free variables in the system, and the matrix B on the other side must be the solution we seek. That is,

$$A\,B = \mathcal{I}\,.$$

Conveniently, had we started with the augmented matrix $[B \mid \mathcal{I}]$ and gone backwards, reversing all of our elementary row operations, we would have ended up with $[\mathcal{I} \mid A]$. That means we also get $BA = \mathcal{I}$. In other words, $B = A^{-1}$, the inverse of A.

(2) The other possibility that can occur is that there are free variables in the system $AX = \mathcal{I}$. In that case, when we do row reduction on the augmented matrix $[A \mid \mathcal{I}]$, the "A side" will turn into something that has a row of all 0s at the bottom. But the "$\mathcal{I}$ side" will never have a row of all 0s because every row has a 1 in a column where the other rows have only 0s. So, adding a multiple of one row of $\mathcal{I}$ to another row can never zero out an entire row. This means that our system $AX = \mathcal{I}$ has an impossible "$0 = 1$" contradiction hiding within it. In this case, there is no matrix X that satisfies $AX = \mathcal{I}$, so A *does not have* a matrix inverse.

Algorithm 4.18 Finding A^{-1}

Given a square matrix A, form the augmented matrix $[A \mid \mathcal{I}]$. Transform this into reduced row echelon form. That is, look at $\text{RREF}([A|\mathcal{I}]) = \left[\widehat{A} \mid B\right]$.

(1) If $\widehat{A} = \mathcal{I}$, then $B = A^{-1}$. That is, if the "A side" turns into $\mathcal{I}$, then the "$\mathcal{I}$ side" turns into the matrix inverse A^{-1}, which satisfies $A^{-1}A = AA^{-1} = \mathcal{I}$.

(2) If $\widehat{A} \neq \mathcal{I}$, then A does not have a matrix inverse. That is, if the "A side" turns into anything other than $\mathcal{I}$, then A is not invertible.

Example 4.19 Let $A = \begin{bmatrix} 1 & 4 & 1 \\ 2 & 8 & 3 \\ 2 & 7 & 4 \end{bmatrix}$. Form the augmented matrix $[A \mid \mathcal{I}_3]$ and carry out row reduction:

$$[A \mid \mathcal{I}_3] = \left[\begin{array}{ccc|ccc} 1 & 4 & 1 & 1 & 0 & 0 \\ 2 & 8 & 3 & 0 & 1 & 0 \\ 2 & 7 & 4 & 0 & 0 & 1 \end{array}\right]$$

$$\overset{\text{RREF}}{\Longrightarrow} \left[\begin{array}{ccc|ccc} 1 & 0 & 0 & 11 & -9 & 4 \\ 0 & 1 & 0 & -2 & 2 & -1 \\ 0 & 0 & 1 & -2 & 1 & 0 \end{array}\right] = \left[\mathcal{I}_3 \mid A^{-1}\right].$$

Since the "A side" turned into $\mathcal{I}_3$, the other side must be the inverse A^{-1}. We can check by matrix multiplication that

$$AA^{-1} = \begin{bmatrix} 1 & 4 & 1 \\ 2 & 8 & 3 \\ 2 & 7 & 4 \end{bmatrix} \begin{bmatrix} 11 & -9 & 4 \\ -2 & 2 & -1 \\ -2 & 1 & 0 \end{bmatrix} = \begin{bmatrix} 1 & 0 & 0 \\ 0 & 1 & 0 \\ 0 & 0 & 1 \end{bmatrix} = \mathcal{I}_3,$$

and

$$A^{-1}A = \begin{bmatrix} 11 & -9 & 4 \\ -2 & 2 & -1 \\ -2 & 1 & 0 \end{bmatrix} \begin{bmatrix} 1 & 4 & 1 \\ 2 & 8 & 3 \\ 2 & 7 & 4 \end{bmatrix} = \begin{bmatrix} 1 & 0 & 0 \\ 0 & 1 & 0 \\ 0 & 0 & 1 \end{bmatrix} = \mathcal{I}_3.$$

Example 4.20 Consider $A = \begin{bmatrix} 1 & 5 & -1 \\ 1 & 4 & 1 \\ 2 & 7 & 4 \end{bmatrix}$. Form the augmented matrix $[A \,|\, \mathcal{I}]$ and row reduce. We get

$$\left[\begin{array}{ccc|ccc} 1 & 5 & -1 & 1 & 0 & 0 \\ 1 & 4 & 1 & 0 & 1 & 0 \\ 2 & 7 & 4 & 0 & 0 & 1 \end{array}\right] \overset{\text{RREF}}{\Longrightarrow} \left[\begin{array}{ccc|ccc} 1 & 0 & 9 & -4 & 5 & 0 \\ 0 & 1 & -2 & 1 & -1 & 0 \\ 0 & 0 & 0 & 1 & -3 & 1 \end{array}\right].$$

In this case, the "A side" did not turn into $\mathcal{I}$ and we can see that the bottom row has all 0s on the left half but not all 0s on the right half. This means that the system $AX = \mathcal{I}$ is *inconsistent* and has *no solution*. The matrix A in this example *does not have an inverse*.

Example 4.21 To see one thing we can do with an inverse matrix, look again at $A = \begin{bmatrix} 1 & 4 & 1 \\ 2 & 8 & 3 \\ 2 & 7 & 4 \end{bmatrix}$. Suppose we want to find a matrix X that satisfies $AX = \begin{bmatrix} 1 & 0 & 3 \\ 0 & 2 & 2 \\ -1 & 1 & 0 \end{bmatrix}$. We know that the inverse of A is $A^{-1} = \begin{bmatrix} 11 & -9 & 4 \\ -2 & 2 & -1 \\ -2 & 1 & 0 \end{bmatrix}$. So now we get:

$$AX = \begin{bmatrix} 1 & 0 & 3 \\ 0 & 2 & 2 \\ -1 & 1 & 0 \end{bmatrix} \Leftrightarrow A^{-1}AX = A^{-1}\begin{bmatrix} 1 & 0 & 3 \\ 0 & 2 & 2 \\ -1 & 1 & 0 \end{bmatrix}.$$

Notice that A^{-1} must go to the left on each side! Now, since $A^{-1}A = \mathcal{I}$, it follows that $A^{-1}AX = \mathcal{I}X = X$. Therefore,

$$\begin{aligned} X &= A^{-1}\begin{bmatrix} 1 & 0 & 3 \\ 0 & 2 & 2 \\ -1 & 1 & 0 \end{bmatrix} \\ &= \begin{bmatrix} 11 & -9 & 4 \\ -2 & 2 & -1 \\ -2 & 1 & 0 \end{bmatrix}\begin{bmatrix} 1 & 0 & 3 \\ 0 & 2 & 2 \\ -1 & 1 & 0 \end{bmatrix} = \begin{bmatrix} 7 & -14 & 15 \\ -1 & 3 & -2 \\ -2 & 2 & -4 \end{bmatrix}. \end{aligned}$$

Check our answer with matrix multiplication:

$$AX = \begin{bmatrix} 1 & 4 & 1 \\ 2 & 8 & 3 \\ 2 & 7 & 4 \end{bmatrix}\begin{bmatrix} 7 & -14 & 15 \\ -1 & 3 & -2 \\ -2 & 2 & -4 \end{bmatrix} = \begin{bmatrix} 1 & 0 & 3 \\ 0 & 2 & 2 \\ -1 & 1 & 0 \end{bmatrix},$$

as desired. In general, this example shows that the equation $AX = B$ has the unique solution $X = A^{-1}B$ whenever the inverse A^{-1} is available. This is the same as saying that, when we row reduce the augmented matrix $\left[\, A \middle| B \,\right]$, we get $\left[\, \mathcal{I} \middle| A^{-1}B \,\right]$ in reduced row echelon form.

Example 4.22 Only four elementary row operations are required to find the inverse of a 2×2 matrix, so the end result is relatively easy to remember. For a generic matrix $A = \begin{bmatrix} a & b \\ c & d \end{bmatrix}$, we proceed like so.

$$\begin{aligned} \left[\begin{array}{cc|cc} a & b & 1 & 0 \\ c & d & 0 & 1 \end{array}\right] &\to \left[\begin{array}{cc|cc} 1 & \frac{b}{a} & \frac{1}{a} & 0 \\ c & d & 0 & 1 \end{array}\right] \\ &\to \left[\begin{array}{cc|cc} 1 & \frac{b}{a} & \frac{1}{a} & 0 \\ 0 & \frac{ad-bc}{a} & \frac{-c}{a} & 1 \end{array}\right] \\ &\to \left[\begin{array}{cc|cc} 1 & \frac{b}{a} & \frac{1}{a} & 0 \\ 0 & 1 & \frac{-c}{ad-bc} & \frac{a}{ad-bc} \end{array}\right] \\ &\to \left[\begin{array}{cc|cc} 1 & 0 & \frac{d}{ad-bc} & \frac{-b}{ad-bc} \\ 0 & 1 & \frac{-c}{ad-bc} & \frac{a}{ad-bc} \end{array}\right]. \end{aligned} \tag{4.6}$$

If $ad - bc = 0$, this process breaks down and the matrix A has no inverse. See Example 2.17. If, however, $ad - bc \neq 0$, then the last matrix in (4.6) will be $\left[\, \mathcal{I}_2 \middle| A^{-1} \,\right]$. We sum up our findings in the following formula.

Formula 4.23 (Formula 2.18 Redux) The 2×2 matrix $\begin{bmatrix} a & b \\ c & d \end{bmatrix}$ is invertible if, and only if, $ad - bc \neq 0$. In that case, the inverse is given by

$$\begin{bmatrix} a & b \\ c & d \end{bmatrix}^{-1} = \frac{1}{ad - bc} \cdot \begin{bmatrix} d & -b \\ -c & a \end{bmatrix}. \tag{4.7}$$

So, we switch the two diagonal entries (a and d), change the signs of the other entries (b and c), and divide the whole thing by $ad - bc$.

4.6 Leontief Input–Output Matrices

Wassily Leontief was awarded the Nobel Medal in Economics, in 1973. Among Leontief's many contributions is the concept of the *input–output matrix* for an economy with multiple interdependent sectors. The basic notion is that each productive sector of the economy, in order to function, requires as input some of the output from the other sectors and possibly some of its own output as well. For instance, the agricultural sector requires farm machinery and electrical power, while the electrical power industry requires turbines, solar panels, and also electricity. To quantify the interdependencies, we list the different productive sectors of the economy in some order and label them 1 through N. Then create an $N \times N$ matrix A with one row and one column for each sector of the economy. The matrix entry $A_{[i,\,j]}$, in the ith row and jth column of A, is defined to be the number of dollars' worth of output from sector i needed to produce one dollar of output from sector j. The matrix A is a called an *input–output matrix* for the given economy.

Suppose now that the overall output of sector j is equal to x_j dollars, for each j between 1 and N. Store this information in the output vector $\mathbf{x} = \begin{bmatrix} x_1 \; x_2 \; \cdots \; x_N \end{bmatrix}^T$. With this notation, for each i and j, the product $A_{[i,\,j]}x_j$ equals the number of dollars of output from sector i needed to produce \$1 of output from sector j times the total output x_j from sector j. That is the same as the total amount of output from sector i that goes into producing the total output of sector j. When we fix the value of i and add these products up over all values of j, we get the sum

$$A_{[i,\,1]}x_1 + A_{[i,\,2]}x_2 + \cdots + A_{[i,\,N]}x_N \,.$$

This measures the total amount of output from sector i needed to generate the total output of all sectors combined. Notice that this sum is also the ith coordinate of the vector $A\mathbf{x}$ obtained by multiplying the input–output matrix A by the output vector $\mathbf{x}$.

Example 4.24 Consider an economy with three productive sectors: (1) agriculture, (2) manufacturing, and (3) electrical power. Suppose we have the following input–output matrix.

$A =$

↦	Agriculture	Manufacturing	Power
Agriculture	0.40	0.15	0.20
Manufacturing	0.15	0.30	0.25
Power	0.20	0.25	0.30

.

The entries in the first row indicate that \$0.40 of agricultural output is needed to produce \$1 of agricultural output, \$0.15 of agricultural output is needed to produce \$1 of manufacturing output, and \$0.20 of agricultural output is needed to produce \$1 worth of electrical power. The entries in the third column tell us that, to produce \$1 of output, the electrical power sector requires \$0.20 of agricultural output, \$0.25 of manufacturing output, and \$0.30 of its own output. (Caution: These numbers may not be realistic.)

Suppose now that the output levels of the three sectors are x_1 for agriculture, x_2 for manufacturing, and x_3 for electrical power, where these levels are measured in dollars' worth of output. How much agricultural output is needed to produce this total output? The answer is

$$(0.40)x_1 + (0.15)x_2 + (0.20)x_3 = A_{[1,1]}x_1 + A_{[1,2]}x_2 + A_{[1,3]}x_3 .$$

Each summand $A_{[1,j]}x_j$ equals the number of dollars of agricultural output (sector 1) needed to produce \$1 of output from sector j times the total output x_j from sector j. The sum of the three terms is the matrix product of the first row of the input–output matrix A with the output vector $\mathbf{x}$. Similarly, the matrix product of the second row of A with $\mathbf{x}$ is

$$(0.15)x_1 + (0.30)x_2 + (0.25)x_3 = A_{[2,1]}x_1 + A_{[2,2]}x_2 + A_{[2,3]}x_3 .$$

This measures the total amount of manufacturing output needed to create the combined output of the three sectors. Finally,

$$(0.20)x_1 + (0.25)x_2 + (0.30)x_3 = A_{[3,1]}x_1 + A_{[3,2]}x_2 + A_{[3,3]}x_3$$

indicates the amount of electrical power required by the three sectors put together. To record these computations in one tableau, we write

$$\begin{aligned} A\mathbf{x} &= \begin{bmatrix} A_{[1,1]} & A_{[1,2]} & A_{[1,3]} \\ A_{[2,1]} & A_{[2,2]} & A_{[2,3]} \\ A_{[3,1]} & A_{[3,2]} & A_{[3,3]} \end{bmatrix} \begin{bmatrix} x_1 \\ x_2 \\ x_3 \end{bmatrix} \\ &= \begin{bmatrix} 0.40 & 0.15 & 0.20 \\ 0.15 & 0.30 & 0.25 \\ 0.20 & 0.25 & 0.30 \end{bmatrix} \begin{bmatrix} x_1 \\ x_2 \\ x_3 \end{bmatrix} \\ &= \begin{bmatrix} (0.40)x_1 + (0.15)x_2 + (0.20)x3 \\ (0.15)x_1 + (0.30)x_2 + (0.25)x3 \\ (0.20)x_1 + (0.25)x_2 + (0.30)x_3 \end{bmatrix} . \end{aligned}$$

We have seen that the coordinates of the vector $A\mathbf{x}$ show the amount of output needed from each sector of the economy to produce the combined output of all productive sectors. In general, output from the productive sectors of an economy is also used by a variety of nonproductive sectors, such as purchases by consumers and governments. Let us lump all of the nonproductive sectors into one *open sector*. Let the value d_i denote the aggregate demand by the open sector for the output from productive sector i. Thus, the vector $\mathbf{d} = \begin{bmatrix} d_1 & d_2 & \cdots & d_N \end{bmatrix}^T$ is the demand vector for the open sector. If we have completely accounted for every sector of the economy and there is no waste, then the total output vector $\mathbf{x}$ should exactly account for all of the output $A\mathbf{x}$ needed by the productive sectors as well as the demand $\mathbf{d}$ of the open sector. In other words, we should have

$$\mathbf{x} = A\mathbf{x} + \mathbf{d}\,, \text{ or } (\mathcal{I} - A)\mathbf{x} = \mathbf{d}\,. \tag{4.8}$$

This is a linear system with coefficient matrix $(\mathcal{I} - A)$. The unknowns are the coordinates of the output vector $\mathbf{x}$. Solving the system for $\mathbf{x}$ will tell us how much output is needed from each productive sector of the economy in order to satisfy the needs of all sectors combined with no waste. Assuming the matrix $(\mathcal{I} - A)$ is invertible, we get the solution

$$\mathbf{x} = (\mathcal{I} - A)^{-1}\mathbf{d}\,. \tag{4.9}$$

Example 4.25 Returning to Example 4.24, we now have

$$\mathcal{I} - A = \begin{bmatrix} 1 & 0 & 0 \\ 0 & 1 & 0 \\ 0 & 0 & 1 \end{bmatrix} - \begin{bmatrix} 0.40 & 0.15 & 0.20 \\ 0.15 & 0.30 & 0.25 \\ 0.20 & 0.25 & 0.30 \end{bmatrix}$$

$$= \begin{bmatrix} 0.60 & -0.15 & -0.20 \\ -0.15 & 0.70 & -0.25 \\ -0.20 & -0.25 & 0.70 \end{bmatrix}.$$

The inverse is

$$(\mathcal{I} - A)^{-1} \approx \begin{bmatrix} 2.162 & 0.784 & 0.898 \\ 0.784 & 1.922 & 0.910 \\ 0.898 & 0.910 & 2.010 \end{bmatrix}.$$

Suppose the open sector demand vector is $\mathbf{d} = \begin{bmatrix} 100 & 50 & 95 \end{bmatrix}^T$. Then the desired outputs from the productive sectors are given by

$$\mathbf{x} = (\mathcal{I} - A)^{-1}\mathbf{d} \approx \begin{bmatrix} 340.64 \\ 260.94 \\ 326.23 \end{bmatrix}.$$

(Again, these numbers may not be realistic.)

When we apply Leontief input–output analysis to a national economy, the open demand vector $\mathbf{d}$ represents the gross national product (GNP). An economic analyst could, for instance, play around with different choices for $\mathbf{d}$ to explore how changes in the GNP would affect the productive sectors of the economy. The United States Bureau of Economic Analysis, on its (website [22]), provides data for input–output analysis of the economy of the United States based on 15, 71, or 405 industries.

4.7 Cubic Splines

Many data analysis problems involve *sampling*, in which we have computed a finite set of values of some function or signal whose exact formula we do not know. To fill in the gaps, we must somehow assign values to the function at points in between. This process of filling in values for a sampled function is called *interpolation*. Exactly how the interpolated values are assigned depends on other features we would like the overall function to have, such as being continuous or differentiable. The degree of computational difficulty incurred in assigning the new values is also an important consideration.

One interpolation scheme that has a long history and is still used today is called **cubic splines**. In this method, we start with a sample of known points, say

$$\{(x_0,\ y_0),\ (x_1,\ y_1), \ldots,\ (x_N,\ y_N)\},\ \text{with}\ x_0 < x_1 < \ldots < x_N\ .$$

We then connect each pair of successive points $(x_{k-1},\ y_{k-1})$ and $(x_k,\ y_k)$ by a piece of a cubic curve, say $h_k(x) = a_k x^3 + b_k x^2 + c_k x + d_k$, for $1 \le k \le N$. We select the pieces so that they join together smoothly, producing an overall curve with a continuous second-order derivative. Each cubic curve has four coefficients to be determined, so we need a system of $4N$ equations in order to find them. We get $2N$ of these equations from the requirement that the curves must fit the sample points. That is, for each $k = 1, \ldots,\ N$, we need $h_k(x_{k-1}) = y_{k-1}$ and $h_k(x_k) = y_k$. Next, at each of the intermediate points $x_1, \ldots,\ x_{N-1}$ in the sample, we want the two cubic pieces that meet there to join together smoothly, meaning that the first- and second-order derivatives of the two pieces must agree at the transition. Thus, for each $k = 1, \ldots,\ (N-1)$, we need $h_k'(x_k) = h_{k+1}'(x_k)$ and $h_k''(x_k) = h_{k+1}''(x_k)$. This produces $2(N-1)$ additional equations to the system, giving us $4N-2$ equations so far. The remaining two equations are obtained by prescribing values for the initial slope of the first piece and the final slope of the last piece. That is, we assign values for $h_1'(x_0)$ and $h_N'(x_N)$. Now we have the $4N$ equations we need. We are always evaluating at a specific value of x_k, so these equations are linear in the unknown coefficients of the different cubic pieces.

Formula 4.26 For a cubic $h(x) = ax^3 + bx^2 + cx + d$, the first- and second-order derivatives are

$$h'(x) = 3ax^2 + 2bx + c \text{ and } h''(x) = 6ax + 2b\,.$$

The equations in the system to determine the full spline are as follows.

- For each $k = 1, \ldots,\ N$, the equations $h_k(x_{k-1}) = y_{k-1}$ and $h_k(x_k) = y_k$ become, respectively,

$$\begin{cases} a_k(x_{k-1})^3 + b_k(x_{k-1})^2 + c_k x_{k-1} + d_k = y_{k-1} \text{ and} \\ a_k(x_k)^3 + b_k(x_k)^2 + c_k x_k + d_k = y_k\ . \end{cases}$$

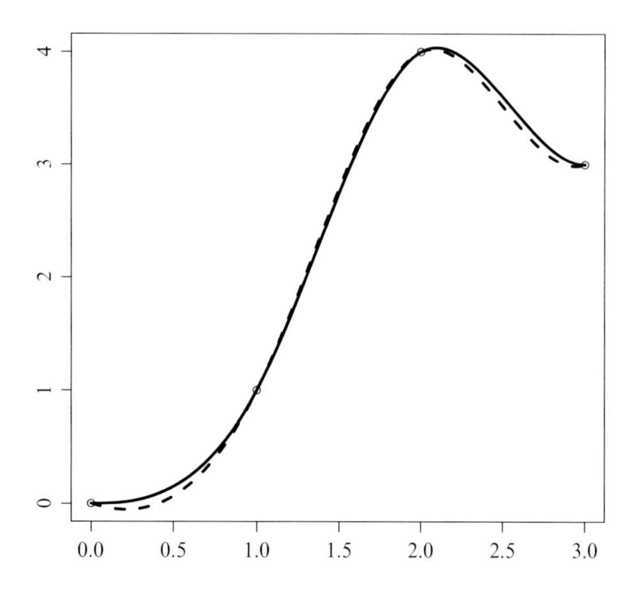

Fig. 4.3 Two cubic splines that pass through the sample points (0, 0), (1, 1), (2, 4), and (3, 3). The solid curve, whose formula is given in Example 4.27, has derivative 0 at both endpoints $x = 0$ and $x = 3$. The dashed curve has slopes -0.5 at $x = 0$ and 0.5 at $x = 3$

- For each $k = 1, \ldots, (N-1)$, the equations $h_k'(x_k) = h_{k+1}'(x_k)$ and $h_k''(x_k) = h_{k+1}''(x_k)$ become, respectively,

$$\begin{cases} 3a_k(x_k)^2 + 2b_k(x_k) + c_k = 3a_{k+1}(x_k)^2 + 2b_{k+1}(x_k) + c_{k+1} \text{ and} \\ \qquad\qquad 6a_k(x_k) + 2b_k = 6a_{k+1}(x_k) + 2b_{k+1} \,. \end{cases}$$

- If desired, we may specify values $h_1'(x_0) = m_0$ and $h_N'(x_N) = m_N$. The corresponding equations in the system are

$$\begin{cases} 3a_1(x_0)^2 + 2b_1(x_0) + c_1 = m_0 \text{ and} \\ 3a_N(x_N)^2 + 2b_N(x_N) + c_N = m_N \,. \end{cases}$$

To form the augmented matrix for the system, we would rewrite the equations to have all of the unknown coefficients a_k, b_k, c_k, and d_k on the left side and always in the same order. As no equation uses more than six of the $4N$ unknowns, the augmented matrix will typically have many entries of 0.

Example 4.27 Suppose we wish to fit a cubic spline to the four sample points (0, 0), (1, 1), (2, 4), and (3, 3). There are three subintervals between the points, so we will need three cubic pieces. Call them h_1, h_2, and h_3. Referring to Formula 4.26, we get this system of equations to be solved.

$$
\begin{aligned}
h_1(x_0) = y_0 &\Rightarrow a_1(0)^3 + b_1(0)^2 + c_1(0) + d_1 = 0\,,\\
h_1(x_1) = y_1 &\Rightarrow a_1(1)^3 + b_1(1)^2 + c_k(1) + d_k = 1\,,\\
h_2(x_1) = y_1 &\Rightarrow a_2(1)^3 + b_2(1)^2 + c_2(1) + d_2 = 1\,,\\
h_2(x_2) = y_2 &\Rightarrow a_2(2)^3 + b_2(2)^2 + c_2(2) + d_2 = 4\,,\\
h_3(x_2) = y_2 &\Rightarrow a_3(2)^3 + b_3(2)^2 + c_3(2) + d_3 = 4\,,\\
h_3(x_3) = y_3 &\Rightarrow a_3(3)^3 + b_3(3)^2 + c_3(3) + d_3 = 3\,.
\end{aligned}
$$

Then,

$$
\begin{aligned}
h_1'(x_1) = h_2'(x_1) &\Rightarrow 3a_1(1)^2 + 2b_1(1) + c_1 = 3a_2(1)^2 + 2b_2(1) + c_2\,,\\
h_2'(x_2) = h_3'(x_2) &\Rightarrow 3a_2(2)^2 + 2b_2(2) + c_2 = 3a_3(2)^2 + 2b_3(2) + c_3\,,\\
h_1''(x_1) = h_2''(x_1) &\Rightarrow 6a_1(1) + 2b_1 = 6a_2(1) + 2b_2\,,\\
h_2''(x_2) = h_3''(x_2) &\Rightarrow 6a_2(2) + 2b_2 = 6a_3(2) + 2b_3\,.
\end{aligned}
$$

Finally, let us specify that $h_1'(0) = 0$ and $h_3'(3) = 0$. That is,

$$3a_1(0)^2 + 2b_1(0) + c_1 = 0 \text{ and } 3a_3(3)^2 + 2b_3(3) + c_3 = 0\,.$$

With the unknown coefficients in the order of the functions h_1, h_2, and h_3, we get the following augmented matrix for this system.

$$
\left[\begin{array}{cccccccccccc|c}
0&0&0&1&0&0&0&0&0&0&0&0&0\\
1&1&1&1&0&0&0&0&0&0&0&0&1\\
0&0&0&0&1&1&1&1&0&0&0&0&1\\
0&0&0&0&8&4&2&1&0&0&0&0&4\\
0&0&0&0&0&0&0&0&8&4&2&1&4\\
0&0&0&0&0&0&0&0&27&9&3&1&3\\
3&2&1&0&-3&-2&-1&0&0&0&0&0&0\\
0&0&0&0&12&4&1&0&-12&-4&-1&0&0\\
6&2&0&0&-6&-2&0&0&0&0&0&0&0\\
0&0&0&0&12&2&0&0&-12&-2&0&0&0\\
0&0&1&0&0&0&0&0&0&0&0&0&0\\
0&0&0&0&0&0&0&0&27&6&1&0&0
\end{array}\right].
$$

Solving this system (with a computer) yields the following solution, shown as the solid curve in Figure 4.3:

$$h(x) = \begin{cases} h_1(x) = 0.8 \cdot x^3 + 0.2 \cdot x^2 & \text{if } 0 \le x \le 1, \\ h_2(x) = -2.4 \cdot x^3 + 9.8 \cdot x^2 - 9.6 \cdot x + 3.2 & \text{if } 1 \le x \le 2, \\ h_3(x) = 2.8 \cdot x^3 - 21.4 \cdot x^2 + 52.8 \cdot x - 38.4 & \text{if } 2 \le x \le 3. \end{cases}$$

The dashed curve in Figure 4.3 shows a spline through the same sample points but with slopes $h_1'(0) = -0.5$ and $h_3'(3) = 0.5$ at the endpoints.

4.8 Solutions to $AX = B$

To solve a linear system $A\mathbf{x} = \mathbf{b}$, we transform the augmented matrix $\left[\, A \middle| \mathbf{b} \,\right]$ into reduced row echelon form. At that stage, any free variables in the system will be revealed. Suppose there are K free variables. Then there are K vectors, $\mathbf{v}_1$ through $\mathbf{v}_K$, say, as well as a specific vector $\mathbf{p}$, determined by the right-most column of the reduced augmented matrix, such that every solution to the original system has the form

$$\mathbf{x} = \mathbf{p} + t_1\mathbf{v}_1 + \cdots + t_K\mathbf{v}_K\,, \tag{4.10}$$

for some choice of the real numbers t_1 through t_K. Here we are using the vector form of the solutions, as shown in Example 4.11.

If the original right-hand side is $\mathbf{b} = \mathbf{0}$, then the fixed solution to the homogeneous system $A\mathbf{x} = \mathbf{0}$ is $\mathbf{p} = \mathbf{0}$. The free variables implied by the system remain the same. Thus, the general solution to $A\mathbf{x} = \mathbf{0}$ has the form

$$\mathbf{x} = t_1\mathbf{v}_1 + \cdots + t_K\mathbf{v}_K\,, \tag{4.11}$$

where t_1 through t_K are real number parameters.

Definition 4.28 Let A be an $M \times N$ matrix, and let $\mathbf{0}$ denote the zero vector in $\mathbb{R}^M$. The set $\left\{\mathbf{x} : \mathbf{x} \in \mathbb{R}^N \text{ and } A\mathbf{x} = \mathbf{0}\right\}$ is called the **nullspace** of the matrix A and is denoted by $Null(A)$.

From (4.11), it follows that each of the individual vectors $\mathbf{v}_i$, for $1 = 1, \ldots, k$, is in $Null(A)$. Also, according to (4.11), $Null(A)$ consists of all possible sums of numerical multiples of the vectors $\{\mathbf{v}_i\}$. We express this by saying that these vectors **span** $Null(A)$ and also that $Null(A)$ is the **span** of these vectors. Notice that $Null(A)$ as a whole is self-contained under vector addition and scalar multiplication. Finally, because each of the vectors $\mathbf{v}_i$ has one of its coordinates equal to 1 in a position where the others all have 0s, it follows that none of the $\mathbf{v}_i$ can be expressed as a sum of numerical multiples of the other ones. The common terminology for this property is to say that the collection $\{\mathbf{v}_i\}$ is **linearly independent**. A linearly independent collection of vectors is called a **basis** for the larger set that it spans, consisting of all sums of numerical multiples of these vectors. The number of vectors in the linearly independent collection is the **dimension** of the space spanned by the vectors. Thus, the dimension of $Null(A)$ is equal to the number of free variables that emerge when A is transformed into reduced row echelon form. This is the same as the number of columns of $RREF(A)$ that do not contain the leading 1 of any row. In particular, if A is invertible, then $Null(A)$ consists of just the zero vector, $\mathbf{0}$.

Remark To prove that the notion of *dimension* just described is uniquely defined requires a somewhat subtle argument that amounts to showing also that $RREF(A)$ is unique, as mentioned in the remark following Algorithm 4.14. We will not attempt to present such an argument here. Suffice it to say that Algorithm 4.14 does indeed produce the correct answers.

As equations (4.10) and (4.11) indicate, every solution to the system $A\mathbf{x} = \mathbf{b}$ has the form $\mathbf{x} = \mathbf{p} + \mathbf{v}$, for some vector $\mathbf{v}$ in the nullspace $Null(A)$.

More generally, given an $M \times N$ matrix A and an $M \times J$ matrix B, we may ask if there is an $N \times J$ matrix X for which $AX = B$. This single matrix equation may be viewed as a collection of J systems of linear equations, all having the same coefficients but different right-hand sides. For each j between 1 and J, the jth system is given by $A\mathbf{x}_j = \mathbf{b}_j$, where $\mathbf{x}_j$ is the jth column vector of the unknown matrix X and $\mathbf{b}_j$ is the jth column vector of the matrix B on the right.

Assuming all of these systems are consistent, then, for $1 \leq j \leq J$, there are fixed vectors $\mathbf{p}^{(j)}$ and vectors $\mathbf{v}^{(j)}$ in $Null(A)$ such that each $\mathbf{x}_j$ has the form

$$\mathbf{x}_j = \mathbf{p}^{(j)} + \mathbf{v}^{(j)},$$

as indicated by (4.10). When we put this all together, we see that the general solution to $AX = B$ has the form $X = P + V$, where $P = \left[\mathbf{p}^{(1)} \cdots \mathbf{p}^{(J)}\right]$ and $V = \left[\mathbf{v}^{(1)} \cdots \mathbf{v}^{(J)}\right]$. The matrix P is a particular solution satisfying $AP = B$. The columns of V are vectors in $Null(A)$, so $AV = [\mathbf{0}]$, the zero matrix.

Example 4.29 Consider the matrix equation $AX = B$, where

$$A = \begin{bmatrix} 2 & 1 & 1 & 5 \\ -1 & 9 & 3 & 0 \\ 0 & 3 & 1 & 1 \\ 2 & -7 & 0 & -1 \end{bmatrix} \text{ and } B = \begin{bmatrix} 4 & 6 & -3 \\ -3 & 3 & 8 \\ 0 & 2 & 1 \\ -1 & -1 & 11 \end{bmatrix}.$$

To find X, we compute

$$\text{RREF}\left(\left[\,A \middle| B\,\right]\right) = \left[\begin{array}{cccc|ccc} 1 & 0 & 0 & 3 & 3 & 3 & -5 \\ 0 & 1 & 0 & 1 & 1 & 1 & -3 \\ 0 & 0 & 1 & -2 & -3 & -1 & 10 \\ 0 & 0 & 0 & 0 & 0 & 0 & 0 \end{array}\right].$$

The bottom row of all 0s indicates that the entire system is consistent. The final three columns of the reduced augmented matrix give us the fixed solution

$$P = \begin{bmatrix} 3 & 3 & -5 \\ 1 & 1 & -3 \\ -3 & -1 & 10 \\ 0 & 0 & 0 \end{bmatrix}.$$

Also, x_4 is a free variable. Every vector in the nullspace $Null(A)$ has the form $t \cdot \begin{bmatrix} -3 & -1 & 2 & 1 \end{bmatrix}^T$, for some real number t. Thus, each column of any solution matrix X may also include such a vector. The value of t can be different for each column. Thus, we may write the general solution to $AX = B$ as

$$X = \begin{bmatrix} 3 & 3 & -5 \\ 1 & 1 & -3 \\ -3 & -1 & 10 \\ 0 & 0 & 0 \end{bmatrix} + \begin{bmatrix} -3 \\ -1 \\ 2 \\ 1 \end{bmatrix} \begin{bmatrix} t_1 & t_2 & t_3 \end{bmatrix},$$

for some numbers t_1, t_2, and t_3.

As indicated in Example 4.21, if the coefficient matrix A is invertible, then $Null(A)$ consists of just $\mathbf{0}$, the zero vector. The equation $AX = B$ has the unique solution $X = A^{-1}B$.

4.9 LU Decomposition

In this section, we show how Gaussian elimination leads to a natural way to factor a matrix as a product of **triangular** matrices.

Definition 4.30 An $M \times N$ matrix A is said to be **upper triangular** if the $[i,\ j]$ matrix entry $a_{i,j} = 0$, whenever $i > j$. Similarly, a matrix B is said to be **lower triangular** if the $[i,\ j]$ matrix entry $b_{i,j} = 0$,whenever $i < j$. A matrix that is both upper triangular and lower triangular is called a **diagonal matrix**.

Remark Notice that the transpose of an upper triangular matrix is a lower triangular matrix, and vice versa. A square diagonal matrix is symmetric. It is also true that the product of two upper triangular matrices is again an upper triangular matrix, while the product of two lower triangular matrices is lower triangular. See Exercise 23 to try your hand at proving this. In some texts, the terms upper triangular, lower triangular, and diagonal are applied only to square matrices.

We also need the concept of an **elementary matrix**.

Definition 4.31 An $M \times M$ matrix E is an **elementary matrix** if it can be produced by performing one elementary row operation on $\mathcal{I}_M$, the $M \times M$ identity matrix.

Each time we perform an elementary row operation on an arbitrary $M \times N$ matrix A, the result is the same as multiplying A on the left by an appropriate $M \times M$ elementary matrix. For example, suppose we add c times the second row of A to the first row. The corresponding elementary matrix E comes from adding c times the second row of $\mathcal{I}_M$ to its first row. Thus,

$$E = \begin{bmatrix} 1 & c & \cdots \\ 0 & 1 & \cdots \\ \vdots & \vdots & \ddots \end{bmatrix} \text{ and } EA = \begin{bmatrix} a_{1,1} + ca_{2,1} & a_{1,2} + ca_{2,2} & \cdots \\ a_{2,1} & a_{2,2} & \cdots \\ \vdots & \vdots & \ddots \end{bmatrix}.$$

Suppose now that A can be transformed using elementary row operations into the matrix U in *row echelon form*. In particular, U is an upper triangular matrix. Using elementary matrices, we have $U = E_k \cdots E_1 A$. For simplicity, assume we never have to switch two rows to get from A to U. Then the elementary matrices we need are of just two types: Either we multiply a row of $\mathcal{I}$ by a constant or we add a multiple of a higher row of $\mathcal{I}$ to a lower row. Either way, each of the elementary matrices $E_1, \ldots, E_k$ is lower triangular. Therefore, the product $E_k \cdots E_1$ is also lower triangular. Next, observe that every elementary matrix has an inverse, since every elementary row operation can be undone. The inverse of $E_k \cdots E_1$ is $L = {E_1}^{-1} \cdots {E_k}^{-1}$. This is also a lower triangular matrix, since the inverse of a lower triangular matrix is lower triangular. Since $U = L^{-1}A$, we see that $A = LU$. Thus, the original matrix A factors as a product of two matrices, one of which is lower triangular and the other upper triangular. This is called an **LU decomposition** of the matrix A. Computationally, the solutions to $A\mathbf{x} = \mathbf{b}$ may be obtained by solving $U\mathbf{x} = L^{-1}\mathbf{b}$ via back substitution. The matrix L provides a record of the elementary row operations needed to transform A into U.

Example 4.32 Compute the LU decomposition of $A = \begin{bmatrix} 1 & 5 & 1 \\ 2 & 1 & -2 \\ 1 & 7 & 2 \end{bmatrix}$. Three elementary row operations are required to transform A into an upper triangular matrix in row echelon form. First, add $-2*$ Row 1 to Row 2. This transforms A into $E_1 A$, with

$$E_1 A = \begin{bmatrix} 1 & 0 & 0 \\ -2 & 1 & 0 \\ 0 & 0 & 1 \end{bmatrix} \cdot \begin{bmatrix} 1 & 5 & 1 \\ 2 & 1 & -2 \\ 1 & 7 & 2 \end{bmatrix} = \begin{bmatrix} 1 & 5 & 1 \\ 0 & -9 & -4 \\ 1 & 7 & 2 \end{bmatrix}. \tag{4.12}$$

Next, add $-1*$ Row 1 to Row 3 to get $E_2 E_1 A$:

$$E_2E_1A = \begin{bmatrix} 1 & 0 & 0 \\ 0 & 1 & 0 \\ -1 & 0 & 1 \end{bmatrix} \cdot \begin{bmatrix} 1 & 5 & 1 \\ 0 & -9 & -4 \\ 1 & 7 & 2 \end{bmatrix} = \begin{bmatrix} 1 & 5 & 1 \\ 0 & -9 & -4 \\ 0 & 2 & 1 \end{bmatrix}. \tag{4.13}$$

Lastly, add (2/9)$*$ Row 2 to Row 3 to get $E_3E_2E_1A$:

$$E_3E_2E_1A = \begin{bmatrix} 1 & 0 & 0 \\ 0 & 1 & 0 \\ 0 & 2/9 & 1 \end{bmatrix} \cdot \begin{bmatrix} 1 & 5 & 1 \\ 0 & -9 & -4 \\ 0 & 2 & 1 \end{bmatrix} = \begin{bmatrix} 1 & 5 & 1 \\ 0 & -9 & -4 \\ 0 & 0 & 1/9 \end{bmatrix}. \tag{4.14}$$

This yields the upper triangular matrix $U = E_3E_2E_1A = \begin{bmatrix} 1 & 5 & 1 \\ 0 & -9 & -4 \\ 0 & 0 & 1/9 \end{bmatrix}$. To get from U back to A, we would carry out the inverse elementary row operations in the reverse order. Thus,

$$\begin{aligned} A &= \begin{bmatrix} 1 & 0 & 0 \\ +2 & 1 & 0 \\ 0 & 0 & 1 \end{bmatrix} \cdot \begin{bmatrix} 1 & 0 & 0 \\ 0 & 1 & 0 \\ +1 & 0 & 1 \end{bmatrix} \cdot \begin{bmatrix} 1 & 0 & 0 \\ 0 & 1 & 0 \\ 0 & -2/9 & 1 \end{bmatrix} \cdot U \\ &= \begin{bmatrix} 1 & 0 & 0 \\ 2 & 1 & 0 \\ 1 & -2/9 & 1 \end{bmatrix} \cdot U \\ &= LU\,, \end{aligned} \tag{4.15}$$

where $L = \begin{bmatrix} 1 & 0 & 0 \\ 2 & 1 & 0 \\ 1 & -2/9 & 1 \end{bmatrix}$ is a lower triangular matrix.

The factorization $A = LU$ is not unique. Row echelon form does not require the leading nonzero entry of each row to be a 1. If we want, we can include the row operations required to force U to have that feature. This changes both U and L.

Example 4.33 Returning to Example 4.32, we can also factor the matrix A as

$$A = \widehat{L}\widehat{U} = \begin{bmatrix} 1 & 0 & 0 \\ 2 & -9 & 0 \\ 1 & 2 & 1/9 \end{bmatrix} \begin{bmatrix} 1 & 5 & 1 \\ 0 & 1 & 4/9 \\ 0 & 0 & 1 \end{bmatrix}. \tag{4.16}$$

Another option is to introduce a third factor in the form of a diagonal matrix that incorporates the steps where we multiply a row by a constant. In this example, let $D = \begin{bmatrix} 1 & 0 & 0 \\ 0 & -9 & 0 \\ 0 & 0 & 1/9 \end{bmatrix}$. We then have $LD = \widehat{L}$, where L is the lower triangular factor from Example 4.32. Thus,

$$A = LD\widehat{U} = \begin{bmatrix} 1 & 0 & 0 \\ 2 & 1 & 0 \\ 1 & -2/9 & 1 \end{bmatrix} \begin{bmatrix} 1 & 0 & 0 \\ 0 & -9 & 0 \\ 0 & 0 & 1/9 \end{bmatrix} \begin{bmatrix} 1 & 5 & 1 \\ 0 & 1 & 4/9 \\ 0 & 0 & 1 \end{bmatrix}. \tag{4.17}$$

This factorization of A is known as an **LDU decomposition**. This is not unique, as we also have

$$A = \begin{bmatrix} 1 & 0 & 0 \\ 2 & -9 & 0 \\ 1 & 2 & 1/9 \end{bmatrix} \begin{bmatrix} 1 & 0 & 0 \\ 0 & -9 & 0 \\ 0 & 0 & 1/9 \end{bmatrix} \begin{bmatrix} 1 & 5 & 1 \\ 0 & -1/9 & -4/81 \\ 0 & 0 & 9 \end{bmatrix}.$$

There is something appealing about the factorization in (4.17), where both L and $\widehat{U}$ have all 1s on their diagonals.

4.10 Affine Projections

Suppose we have a hyperplane in $\mathbb{R}^N$ formed by the terminal points of all vectors $\mathbf{x}$ that satisfy the equation $\mathbf{a}^T\mathbf{x} = b$, for some specified nonzero vector $\mathbf{a}$ and some number b. Next, suppose $\mathbf{u}$ is some arbitrary vector in $\mathbb{R}^N$. Consider the problem of finding the vector $\mathbf{u}_{\text{aff}}$ whose terminal point is the point in the hyperplane that is *closest* to the terminal point of $\mathbf{u}$. That is, we wish to find a vector $\mathbf{u}_{\text{aff}}$ that satisfies $\mathbf{a}^T\mathbf{u}_{\text{aff}} = b$ and for which $||\mathbf{u} - \mathbf{u}_{\text{aff}}||$ is as small as possible among all solutions to the equation of the hyperplane.

From the Pythagorean theorem, we know that, in order to move from the terminal point of $\mathbf{u}$ to the closest point in the hyperplane, we should move parallel to the vector $\mathbf{a}$. That is because the vector $\mathbf{a}$ is orthogonal to the hyperplane that it defines. Thus, the vector $\mathbf{u}_{\text{aff}}$ that we seek must have the form $\mathbf{u}_{\text{aff}} = \mathbf{u} - c \cdot \mathbf{a}$ for some number c.

To find the correct value of c, we substitute $\mathbf{u}_{\text{aff}} = \mathbf{u} - c \cdot \mathbf{a}$ into the equation $\mathbf{a}^T\mathbf{u}_{\text{aff}} = b$ that defines the hyperplane. Solving this for c yields

$$c = \frac{(\mathbf{a}^T\mathbf{u}) - b}{\mathbf{a}^T\mathbf{a}} = \frac{(\mathbf{a}^T\mathbf{u}) - b}{||\mathbf{a}||^2}.$$

Thus,

$$\mathbf{u}_{\text{aff}} = \mathbf{u} - \left(\frac{(\mathbf{a}^T\mathbf{u}) - b}{||\mathbf{a}||^2}\right)\mathbf{a}. \tag{4.18}$$

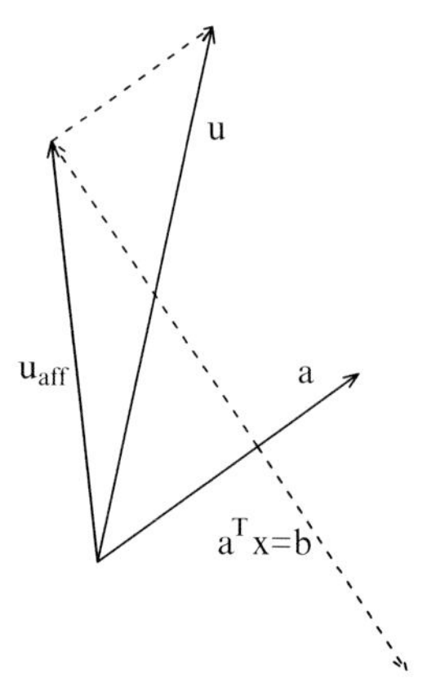

Fig. 4.4 The figure shows a vector $\mathbf{a}$, a hyperplane $\mathbf{a}^T\mathbf{x} = b$ (shown as a dashed line), a vector $\mathbf{u}$, and its affine projection $\mathbf{u}_{\text{aff}}$ into the hyperplane. Notice that $\mathbf{u} - \mathbf{u}_{\text{aff}}$ is a multiple of $\mathbf{a}$

The vector $\mathbf{u}_{\text{aff}}$ just computed is called the **affine projection** of $\mathbf{u}$ into the given hyperplane. Figure 4.4 illustrates the relationship between $\mathbf{u}$ and $\mathbf{u}_{\text{aff}}$.

For an alternate description of the affine projection, note that the projection of $\mathbf{u}$ along $\mathbf{a}$ is given by $\text{proj}_{\mathbf{a}}\mathbf{u} = \left(\mathbf{a}^T\mathbf{u}/||\mathbf{a}||^2\right)\mathbf{a}$. The vector $\left(b/||\mathbf{a}||^2\right)\mathbf{a}$ is the multiple of $\mathbf{a}$ itself that satisfies $\mathbf{a}^T\mathbf{x} = b$. Thus, from (4.18), we may write

$$\mathbf{u}_{\text{aff}} = \left(\mathbf{u} - \text{proj}_{\mathbf{a}}\mathbf{u}\right) + \left(b/||\mathbf{a}||^2\right)\mathbf{a}\,. \tag{4.19}$$

The distance from $\mathbf{u}$ to the hyperplane is given by

$$||\mathbf{u} - \mathbf{u}_{\text{aff}}|| = |c| \cdot ||\mathbf{a}|| = \frac{|\mathbf{a}^T\mathbf{u} - b|}{||\mathbf{a}||}\,. \tag{4.20}$$

Example 4.34 To find the point closest to $(3,\ 1,\ -2)$ in the plane with equation $5x_1 - x_2 + 2x_3 = 4$, we apply (4.18). In this case,

$$\mathbf{a} = \begin{bmatrix} 5 & -1 & 2 \end{bmatrix}^T,\ b = 4,\ \text{and}\ \mathbf{u} = \begin{bmatrix} 3 & 1 & -2 \end{bmatrix}^T.$$

With $c = \left(\mathbf{a}^T\mathbf{u} - b\right)/||\mathbf{a}||^2 = (10-4)/30 = 1/5$, we get

$$\mathbf{u}_{\text{aff}} = \mathbf{u} - c \cdot \mathbf{a} = \begin{bmatrix} 2 & 6/5 & -12/5 \end{bmatrix}^T.$$

Thus, the point closest to $(3,\ 1,\ -2)$ in the given plane is $(2,\ 6/5,\ -12/5)$. The distance from $(3,\ 1,\ -2)$ to the plane is the same as the distance between these two points, which is $\sqrt{1.2} \approx 1.0954$.

4.10.1 *Kaczmarz's Method*

Many of the linear systems that arise in real-world problems are over- or under-determined and have no exact solutions. **Kaczmarz's method** is an iterative algorithm that uses a succession of affine projections to reach an approximate solution to a linear system $A\mathbf{x} = \mathbf{b}$. The method was introduced in 1937 by the Polish mathematician Stefan Kaczmarz (1895–1939), who died, or possibly was killed, soon after Germany invaded Poland at the start of World War II. In 1970, the method was rediscovered as a medical imaging technique in the development of the first CAT scan machines.

To see how the method works, observe that each equation in the $M \times N$ system $A\mathbf{x} = \mathbf{b}$ defines a separate hyperplane in $\mathbb{R}^N$. (We assume that A has no rows of all 0s.) Let $\mathbf{r}_i$ denote the ith row vector of A and b_i the ith coordinate of the vector $\mathbf{b}$. Then a solution $\mathbf{x}$ lies in the hyperplane $\mathbf{r}_i\mathbf{x} = b_i$, for every value of i. Kaczmarz's method proceeds by starting with an initial guess at a solution and then computing the affine projection of this initial guess into the first hyperplane in our list. This projection is then projected into the next hyperplane in the list and so on until we have gone through the entire list of hyperplanes. This constitutes one iteration of the algorithm. The result of this iteration becomes the starting point for the next iteration.

Algorithm 4.35 Kaczmarz's method

(a) Select a starting guess for $\mathbf{x}$; call it $\mathbf{x}^0$.
(b) Next set $\mathbf{x}^{0,0} = \mathbf{x}^0$.
(c) The inductive step is this: Once the vector $\mathbf{x}^{0,i}$ has been determined, define

$$\mathbf{x}^{0,i+1} = \mathbf{x}^{0,i} - \left(\frac{\mathbf{r}_{i+1}\mathbf{x}^{0,i} - b_{i+1}}{||\mathbf{r}_{i+1}||^2}\right)\mathbf{r}_{i+1}^T. \tag{4.21}$$

We have used the affine projection formula (4.18).
(d) Note that if the matrix A has M rows, then the vectors $\mathbf{x}^{0,1}, \mathbf{x}^{0,2}, \ldots, \mathbf{x}^{0,M}$ will be computed.
(e) Once $\mathbf{x}^{0,M}$ has been computed, define $\mathbf{x}^1 = \mathbf{x}^{0,M}$. Begin the process again starting with $\mathbf{x}^1$. That is, now set $\mathbf{x}^{1,0} = \mathbf{x}^1$ and compute the vectors $\mathbf{x}^{1,1}$, $\mathbf{x}^{1,2}, \ldots, \mathbf{x}^{1,M}$, as in (4.21).
(f) Then let $\mathbf{x}^2 = \mathbf{x}^{1,M}$ and repeat the process starting with $\mathbf{x}^{2,0} = \mathbf{x}^2$.
(g) Stop when we have had enough!

In principle, the successive vectors $\mathbf{x}^0, \mathbf{x}^1, \mathbf{x}^2, \ldots$ should get closer to a vector that satisfies the original system $A\mathbf{x} = \mathbf{b}$. However, the end result of each iteration is only guaranteed to satisfy the final equation in the system. If the system has a solution, then Kaczmarz showed that the algorithm will converge to it, though many iterations may be required to get a good approximation. If the system has no solution, then the vectors computed from the algorithm may settle into a specific pattern or might exhibit chaotic behavior. In general, it is computationally feasible to run only a few iterations of the algorithm. The main convergence theorem for Kaczmarz's method is as follows. (For a proof, see [5].)

Theorem 4.36 *If the linear system* $A\mathbf{x} = \mathbf{b}$ *has at least one solution, then Kaczmarz's method converges to a solution of this system. Moreover, if* $\mathbf{x}^0$ *is orthogonal to the nullspace of* A*, then the method converges to the solution with the smallest norm.*

Example 4.37 Apply Kaczmarz's method to the system consisting of just the two lines $x + 2y = 5$ and $x - y = 1$. So $\mathbf{r}_1 = \begin{bmatrix} 1 & 2 \end{bmatrix}$, $\mathbf{r}_2 = \begin{bmatrix} 1 & -1 \end{bmatrix}$, $b_1 = 5$, and $b_2 = 1$. With the initial guess $\mathbf{x}^0 = \begin{bmatrix} 0.5 & 0.5 \end{bmatrix}^T$, the diagram on the left in Figure 4.5 shows that the solution $\begin{bmatrix} 7/3 & 4/3 \end{bmatrix}^T$ is quickly found.

However, when we include the third line $4x + y = 6$, so $\mathbf{r}_3 = \begin{bmatrix} 4 & 1 \end{bmatrix}$ and $b_3 = 6$, then the successive iterations settle into a triangular pattern, shown in the diagram on the right in the figure.

Remark When we implement Kaczmarz's method on a computer, we can choose how much information we want to keep. For instance, we may opt to display the last estimated solution, $\mathbf{x}^{j,\,M}$, from the jth iteration of the algorithm. Then set $\mathbf{x}^{j,\,M} = \mathbf{x}^{j+1,\,0}$ to start the next iteration. Or, we could save only the very last approximate solution, $\mathbf{x}^{J,\,M}$, after all J iterations are complete. At the other extreme, we can display every vector $\mathbf{x}^{j,\,i}$ from every step of every iteration. This approach was used to create Figure 4.5. For a very large system of equations, this last approach may not be practical.

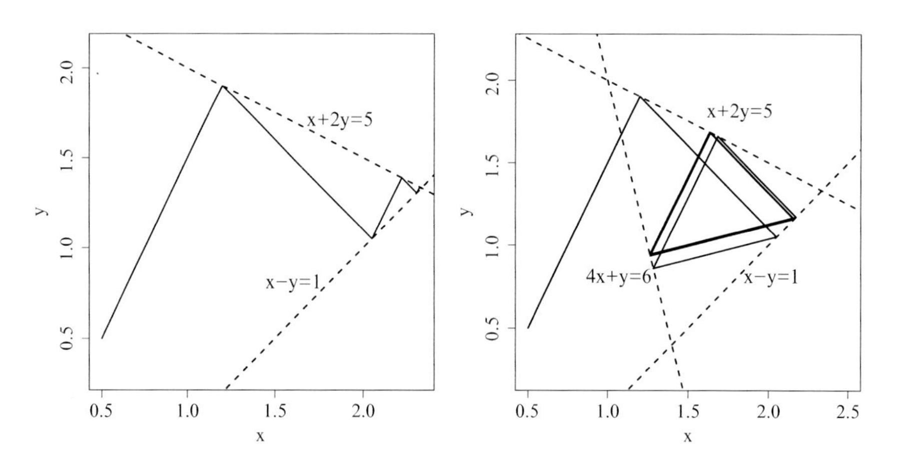

Fig. 4.5 On the left, Kaczmarz's method converges quickly to the point of intersection of two lines. On the right, the successive iterations settle into a triangular pattern for a system of three lines

4.10.2 Fixed Point of an Affine Transformation

An **affine transformation** of $\mathbb{R}^2$ is a mapping of the form

$$\begin{bmatrix} x \\ y \end{bmatrix} \mapsto \begin{bmatrix} a & b \\ c & d \end{bmatrix} \begin{bmatrix} x \\ y \end{bmatrix} + \begin{bmatrix} e \\ f \end{bmatrix} = \begin{bmatrix} ax + by + e \\ cx + dy + f \end{bmatrix}. \tag{4.22}$$

The effect of such a mapping is to apply a matrix multiplication, such as a shear or a rotation, and then shift the output away from the origin by adding a fixed vector. Matrix multiplication alone always maps the **0** vector to itself, so it is interesting to know if an affine transformation will also have a fixed point.

To resolve this issue, imagine that the action of the affine transformation in (4.22) is taking place not in $\mathbb{R}^2$, but in the horizontal plane $z = 1$ in $\mathbb{R}^3$. Then our input vector is not $\begin{bmatrix} x \\ y \end{bmatrix}$ but $\begin{bmatrix} x \\ y \\ 1 \end{bmatrix}$ instead. The output in the plane $z = 1$ is now

$$\begin{bmatrix} ax + by + e \\ cx + dy + f \\ 1 \end{bmatrix} = \begin{bmatrix} a & b & e \\ c & d & f \\ 0 & 0 & 1 \end{bmatrix} \begin{bmatrix} x \\ y \\ 1 \end{bmatrix}. \tag{4.23}$$

Thus, by moving the problem into the plane $z = 1$, the affine transformation is represented as multiplication by the matrix $\begin{bmatrix} a & b & e \\ c & d & f \\ 0 & 0 & 1 \end{bmatrix}$.

A fixed point $\begin{bmatrix} x_* \\ y_* \end{bmatrix}$ for the original affine transformation satisfies

$$\begin{bmatrix} x_* \\ y_* \end{bmatrix} = \begin{bmatrix} a & b \\ c & d \end{bmatrix} \begin{bmatrix} x_* \\ y_* \end{bmatrix} + \begin{bmatrix} e \\ f \end{bmatrix} = \begin{bmatrix} ax_* + by_* + e \\ cx_* + dy_* + f \end{bmatrix}.$$

Moving to the plane $z = 1$, we now solve the system

$$\begin{bmatrix} a & b & e \\ c & d & f \\ 0 & 0 & 1 \end{bmatrix} \begin{bmatrix} x_* \\ y_* \\ 1 \end{bmatrix} = \begin{bmatrix} x_* \\ y_* \\ 1 \end{bmatrix}.$$

This is the same as

$$\begin{bmatrix} a-1 & b & e \\ c & d-1 & f \\ 0 & 0 & 0 \end{bmatrix} \begin{bmatrix} x_* \\ y_* \\ 1 \end{bmatrix} = \begin{bmatrix} 0 \\ 0 \\ 0 \end{bmatrix}.$$

We now carry out row reduction on $\begin{bmatrix} a-1 & b & e \\ c & d-1 & f \\ 0 & 0 & 0 \end{bmatrix}$. Provided the 2×2 matrix $\begin{bmatrix} a-1 & b \\ c & d-1 \end{bmatrix}$ is invertible, meaning that $(a-1)(d-1) - bc \neq 0$, the reduced row echelon form of the full 3×3 matrix has the form $\begin{bmatrix} 1 & 0 & r_1 \\ 0 & 1 & r_2 \\ 0 & 0 & 0 \end{bmatrix}$. Thus, $x = -r_1 z$ and $y = -r_2 z$, where z is a free variable. We insist that $z = 1$, though, so the fixed point is

$$\begin{bmatrix} x_* \\ y_* \\ 1 \end{bmatrix} = \begin{bmatrix} -r_1 \\ -r_2 \\ 1 \end{bmatrix}.$$

Moving back to $\mathbb{R}^2$, the fixed point for the original affine transformation is

$$\begin{bmatrix} x_* \\ y_* \end{bmatrix} = \begin{bmatrix} -r_1 \\ -r_2 \end{bmatrix}.$$

4.11 Exercises

1. Find all solutions to each system of equations. Write down both the augmented matrix for the system and the reduced row echelon form of the augmented matrix.

   (a) $\begin{cases} x + 3y + 2z = 2 \\ 2x + 7y + 7z = -1 \\ 2x + 5y + 2z = 7 \end{cases}$

   (b) $\begin{cases} 3x_1 - 6x_2 + x_3 + 13x_4 = 15 \\ 3x_1 - 6x_2 + 3x_3 + 21x_4 = 21 \\ 2x_1 - 4x_2 + 5x_3 + 26x_4 = 23 \end{cases}$

   (c) $\begin{cases} x_1 + 2x_2 + 3x_3 = 7 \\ 2x_1 + 5x_2 + 5x_3 = -3 \\ 3x_1 + 5x_2 + 8x_3 = 8 \end{cases}$

2. Find all solutions to each system of equations. Write down both the augmented matrix for the system and the reduced row echelon form of the augmented matrix.

(a) $\begin{cases} 3x_1 + x_2 - 3x_3 = 6 \\ 2x_1 + 7x_2 + x_3 = -9 \\ 2x_1 + 5x_2 = -5 \end{cases}$

(b) $\begin{cases} 3x_1 - 6x_2 - 2x_3 = 1 \\ 2x_1 - 4x_2 + x_3 = 17 \\ x_1 - 2x_2 - 2x_3 = -9 \end{cases}$

(c) $\begin{cases} x_1 - 2x_2 + 5x_3 = 3 \\ 3x_1 - 3x_2 + 3x_3 = -2 \\ 4x_1 - 5x_2 + 8x_3 = 1 \end{cases}$

3. Find all solutions to each system of equations. Write down both the augmented matrix for the system and the reduced row echelon form of the augmented matrix.

(a) $\begin{cases} x_1 + \quad + 5x_3 = 3 \\ 2x_2 + 4x_3 = -2 \\ 2x_1 \quad + 10x_3 = 6 \end{cases}$

(b) $\begin{cases} x_1 - 4x_2 + 5x_3 + 5x_4 = -1 \\ 2x_1 + 3x_2 - 5x_3 = 0 \\ 7x_1 + x_2 - x_3 + x_4 = 10 \end{cases}$

(c) $\begin{cases} x_1 - x_2 + 3x_3 + 2x_4 = 7 \\ -2x_1 + x_2 + 5x_3 + x_4 = 5 \\ -3x_1 + 2x_2 + 2x_3 - x_4 = -2 \\ 4x_1 - 3x_2 + x_3 + 3x_4 = 1 \end{cases}$

4. Determine conditions on the right-hand side values b_i such that the given system has solutions.

(a) $\begin{cases} x_1 - 2x_2 + 5x_3 = b_1 \\ 3x_1 - 3x_2 + 3x_3 = b_2 \\ 4x_1 - 5x_2 + 8x_3 = b_3 \end{cases}$

(b) $\begin{cases} x_1 - x_2 + 3x_3 + 2x_4 = b_1 \\ -2x_1 + x_2 + 5x_3 + x_4 = b_2 \\ -3x_1 + 2x_2 + 2x_3 - x_4 = b_3 \\ 4x_1 - 3x_2 + x_3 + 3x_4 = b_4 \end{cases}$

5. For risk assessment in investment and portfolio management, we consider the possibility that a given asset can provide different expected returns depending on the "state of the world." Suppose there are three possible states, which occur with probabilities x_1, x_2, and x_3, respectively. Suppose asset A has expected returns of 10%, 2%, and −5%, depending on the state. Suppose asset B has expected returns of 0%, 30%, and −22%, depending on the state. Meanwhile, suppose there is a risk-free investment that returns 6% regardless of the state of the world. We would like to determine values for the probabilities x_1, x_2, and x_3 such that assets A and B are *risk neutral* in the sense that they will be expected

to return 6% regardless of how the state of the world plays out. To do this, solve the following system of equations.

$$\begin{aligned} x_1 + x_2 + x_3 &= 1 \\ 1.1x_1 + 1.02x_2 + 0.95x_3 &= 1.06 \\ x_1 + 1.3x_2 + 0.78x_3 &= 1.06\,. \end{aligned}$$

6. Consider the homogeneous system of linear equations

$$\begin{aligned} x_1 - 4x_2 - 3x_3 - 7x_4 &= 0 \\ 2x_1 - x_2 + x_3 + 7x_4 &= 0 \\ x_1 + 2x_2 + 3x_3 + 11x_4 &= 0 \end{aligned}$$

   Find two solution vectors $\mathbf{u}$ and $\mathbf{v}$ such that every solution to the system has the form $s\mathbf{u} + t\mathbf{v}$ for some real numbers s and t.
7. For each of the 12 2×2 matrix transformations listed in Section 3.8, either write down the inverse of the matrix or explain why the inverse does not exist. Explain your answers geometrically.
8. For each given matrix, use Formula 4.23 to write down the inverse matrix or explain why the inverse does not exist.

$$\text{(a)} \begin{bmatrix} 3 & -16 \\ 1 & -5 \end{bmatrix} \quad \text{(b)} \begin{bmatrix} 6 & -3 \\ -4 & 2 \end{bmatrix}$$
$$\text{(c)} \begin{bmatrix} 1 & 4 \\ 2 & 7 \end{bmatrix} \quad \text{(d)} \begin{bmatrix} 2 & 5 \\ 3 & 6 \end{bmatrix}$$

9. For each given matrix, use Algorithm 4.18 to compute the inverse matrix. Check your answers using matrix multiplication.

$$\text{(a)} \begin{bmatrix} 1 & 2 & 1 \\ 2 & 5 & 0 \\ 3 & 3 & 8 \end{bmatrix} \quad \text{(b)} \begin{bmatrix} 1 & 0 & 1 \\ 0 & 1 & 1 \\ 1 & 1 & 0 \end{bmatrix}$$
$$\text{(c)} \begin{bmatrix} 3 & -4 & -1 \\ 4 & 1 & 2 \\ 2 & -8 & -4 \end{bmatrix} \quad \text{(d)} \begin{bmatrix} \sqrt{2} & 2\sqrt{2} & 0 \\ -\sqrt{2} & \sqrt{2} & 0 \\ 0 & 0 & 1 \end{bmatrix}$$

10. (a) Write down the inverse of the matrix $\begin{bmatrix} 7 & 6 \\ 8 & 7 \end{bmatrix}$.

    (b) Find all matrices $\mathbf{X}$ for which

$$\begin{bmatrix} 7 & 6 \\ 8 & 7 \end{bmatrix} \mathbf{X} = \begin{bmatrix} 2 & 0 & 4 \\ 0 & 5 & -3 \end{bmatrix}.$$

Check your answer using matrix multiplication.

11. Assuming the relevant values of a, b, c, and d are all nonzero, find the inverses of the following matrices. Check your answers using matrix multiplication.

(a) $\begin{bmatrix} a & 0 \\ 0 & d \end{bmatrix}$ (b) $\begin{bmatrix} 0 & b \\ c & 0 \end{bmatrix}$

(c) $\begin{bmatrix} a & 1 \\ 0 & 1 \end{bmatrix}$ (d) $\begin{bmatrix} a & 1 \\ 0 & b \end{bmatrix}$

(e) $\begin{bmatrix} a & 1 & 0 \\ 0 & a & 1 \\ 0 & 0 & a \end{bmatrix}$ (f) $\begin{bmatrix} a & 1 & 0 \\ 0 & b & 1 \\ 0 & 0 & c \end{bmatrix}$

12. In each case, find all values of x for which the given matrix *does not* have an inverse. Explain your work.

(a)

$$\begin{bmatrix} x-1 & x^2 & x^3 \\ 0 & x+2 & x^4 \\ 0 & 0 & x-4 \end{bmatrix}$$

(b)

$$\begin{bmatrix} 1-x & 0 & 0 \\ 0 & 1 & 2-x \\ 0 & 1+x & 0 \end{bmatrix}$$

13. Find all possible 2×2 matrices X such that

$$\begin{bmatrix} 1 & 2 \\ 2 & 4 \end{bmatrix} X = \begin{bmatrix} 0 & 0 \\ 0 & 0 \end{bmatrix}.$$

14. Let $A = \begin{bmatrix} 1 & 0 & -3 \\ 3 & 2 & 4 \\ 2 & -3 & 5 \end{bmatrix}$.

(a) Apply Algorithm 4.18 to verify that

$$A^{-1} = \frac{1}{61} \begin{bmatrix} 22 & 9 & 6 \\ -7 & 11 & -13 \\ -13 & 3 & 2 \end{bmatrix}.$$

(b) Compute the vector $\mathbf{x}$ such that $A\mathbf{x} = \begin{bmatrix} 7 \\ -4 \\ 3 \end{bmatrix}$.

(c) Find the 3×3 matrix X such that

$$AX = \begin{bmatrix} 1 & 5 & 1 \\ 2 & 1 & -2 \\ 1 & 7 & 2 \end{bmatrix}.$$

15. Let $B = \begin{bmatrix} 7 & -4 & 3 \\ 1 & 5 & -2 \\ 0 & 3 & 9 \end{bmatrix}$.

(a) Apply Algorithm 4.18 to verify that

$$B^{-1} = \frac{1}{402} \begin{bmatrix} 51 & 45 & -7 \\ -9 & 63 & 17 \\ 3 & -21 & 39 \end{bmatrix}.$$

(b) Compute the vector $\mathbf{x}$ such that $B\mathbf{x} = \begin{bmatrix} -13 \\ 3 \\ 2 \end{bmatrix}$.

(c) Find the 3×3 matrix X such that

$$BX = \begin{bmatrix} 1 & 5 & 1 \\ 2 & 1 & -2 \\ 1 & 7 & 2 \end{bmatrix}.$$

16. (a) Write down the inverse of the matrix $T = \begin{bmatrix} 1 & 0 \\ 1 & 1 \end{bmatrix}$.

(b) Find the inverse of the matrix $T = \begin{bmatrix} 1 & 0 & 0 \\ 1 & 1 & 0 \\ 1 & 0 & 1 \end{bmatrix}$.

(c) Let T be the $N \times N$ matrix that has entries of all 1 s in the first column, all 1 s along the main diagonal, and all 0 s everywhere else. That is, let

$$T = \left[\begin{array}{c|c} 1 & \mathbf{0}_{N-1}^T \\ \hline \mathbf{1}_{N-1} & \mathcal{I}_{N-1} \end{array}\right].$$

Find the inverse matrix T^{-1}. Prove that your answer is correct.

17. For the hypothetical economy of Examples 4.24 and 4.25, the Leontief input–output matrix is

$A =$

$\mapsto$	Agriculture	Manufacturing	Power
Agriculture	0.40	0.15	0.20
Manufacturing	0.15	0.30	0.25
Power	0.20	0.25	0.30

(a) How many dollars' worth of manufacturing output are needed to produce $500 worth of agricultural output? Briefly explain.
(b) Compute the inverse matrix $(\mathcal{I} - A)^{-1}$.
(c) For each open sector demand vector $\mathbf{d}$, compute the optimal output vector

$$\mathbf{x} = (\mathcal{I} - A)^{-1}\mathbf{d}.$$

$$(i)\ \mathbf{d} = \begin{bmatrix} 100 \\ 30 \\ 80 \end{bmatrix};\ (ii)\ \mathbf{d} = \begin{bmatrix} 100 \\ 30 \\ 105 \end{bmatrix};\ (iii)\ \mathbf{d} = \begin{bmatrix} 100 \\ 60 \\ 90 \end{bmatrix}.$$

18. Consider a hypothetical economy with three productive sectors—agriculture, manufacturing, and electrical power—and Leontief input–output matrix given by

$A =$

$\mapsto$	Agriculture	Manufacturing	Power
Agriculture	0.30	0.10	0.10
Manufacturing	0.25	0.30	0.30
Power	0.15	0.20	0.20

(a) How many dollars' worth of electrical power are needed to produce $500 worth of manufacturing output? Briefly explain.
(b) Compute the inverse matrix $(\mathcal{I} - A)^{-1}$.
(c) For each open sector demand vector $\mathbf{d}$, compute the optimal output vector

$$\mathbf{x} = (\mathcal{I} - A)^{-1}\mathbf{d}.$$

$$(i)\ \mathbf{d} = \begin{bmatrix} 100 \\ 30 \\ 80 \end{bmatrix};\ (ii)\ \mathbf{d} = \begin{bmatrix} 100 \\ 30 \\ 105 \end{bmatrix};\ (iii)\ \mathbf{d} = \begin{bmatrix} 100 \\ 60 \\ 90 \end{bmatrix}.$$

19. From Example 4.32, an LU decomposition of $A = \begin{bmatrix} 1 & 5 & 1 \\ 2 & 1 & -2 \\ 1 & 7 & 2 \end{bmatrix}$ is given by

$$A = LU = \begin{bmatrix} 1 & 0 & 0 \\ 2 & 1 & 0 \\ 1 & -2/9 & 1 \end{bmatrix} \cdot \begin{bmatrix} 1 & 5 & 1 \\ 0 & -9 & -4 \\ 0 & 0 & 1/9 \end{bmatrix}.$$

Now solve the system $A \begin{bmatrix} x_1 \\ x_2 \\ x_3 \end{bmatrix} = \begin{bmatrix} -2 \\ 5 \\ 1 \end{bmatrix}$ as follows.

(a) Set $\mathbf{v} = \begin{bmatrix} v_1 \\ v_2 \\ v_3 \end{bmatrix} = U \begin{bmatrix} x_1 \\ x_2 \\ x_3 \end{bmatrix}$. Solve $L\mathbf{v} = \begin{bmatrix} -2 \\ 5 \\ 1 \end{bmatrix}$ for $\mathbf{v}$ using *forward substitution.*

(b) Set $\mathbf{x} = \begin{bmatrix} x_1 \\ x_2 \\ x_3 \end{bmatrix}$. Solve $U\mathbf{x} = \mathbf{v}$ for $\mathbf{x}$ using *backward substitution.*

(c) Check the solution by computing $A\mathbf{x}$.

20. Let $A = \begin{bmatrix} 1 & 4 & 1 \\ 2 & 7 & 4 \\ 2 & 8 & 3 \end{bmatrix}$.

(a) Compute the LU decomposition of A, using the method from Example 4.32.

(b) Using $A = LU$, solve the system $A \begin{bmatrix} x_1 \\ x_2 \\ x_3 \end{bmatrix} = \begin{bmatrix} 7 \\ 6 \\ 4 \end{bmatrix}$, like so.

(i) Use forward substitution to solve $L\mathbf{v} = \begin{bmatrix} 7 \\ 6 \\ 4 \end{bmatrix}$ for $\mathbf{v}$.

(ii) Use backward substitution to solve $U\mathbf{x} = \mathbf{v}$ for $\mathbf{x}$.

(iii) Check the solution by computing $A\mathbf{x}$.

21. In each part, find the point in the given plane that is closest to the given point. See Example 4.34.

(a) $5x_1 - x_2 + 2x_3 = 4$; $(4,\ 2,\ 3)$.

(b) $3x_1 + x_2 - 3x_3 = -4$; $(6,\ 1,\ -5)$.
(c) $x_1 + 3x_2 + 3x_3 = 13$; $(2,\ 2,\ 2)$.
(d) $2x_1 + 5x_2 + 4x_3 = 23$; $(4,\ 2,\ 5)$.
(e) $x_1 + x_2 + x_3 = 1$; $(1,\ -2,\ 1)$.
(a) $3x_1 - 6x_2 + 3x_3 = 21$; $(2,\ 2,\ 3)$.

22. In each part, find the fixed point for the given affine transformation by first shifting the problem to the plane $z = 1$, as shown in equations (4.22) and (4.23) from Section 4.10.2.

(a) $\begin{bmatrix} x \\ y \end{bmatrix} \mapsto \begin{bmatrix} 4 & 2 \\ -1 & 3 \end{bmatrix} \begin{bmatrix} x \\ y \end{bmatrix} + \begin{bmatrix} 1 \\ 1 \end{bmatrix}$.

(b) $\begin{bmatrix} x \\ y \end{bmatrix} \mapsto \begin{bmatrix} 5 & -1 \\ 3 & 4 \end{bmatrix} \begin{bmatrix} x \\ y \end{bmatrix} + \begin{bmatrix} 1 \\ -1 \end{bmatrix}$.

(c) $\begin{bmatrix} x \\ y \end{bmatrix} \mapsto \begin{bmatrix} 6 & 7 \\ 2 & 4 \end{bmatrix} \begin{bmatrix} x \\ y \end{bmatrix} + \begin{bmatrix} 3 \\ 1 \end{bmatrix}$.

23. *(Proof Problem)* Show that the product of two upper triangular matrices is again an upper triangular matrix. (Upper triangular matrices are defined in Definition 4.30.) Use transposes to show that the product of two lower triangular matrices is also lower triangular.
24. *(Proof Problem)* Prove that the inverse of an invertible upper triangular matrix is also upper triangular. (*Hint:* Think about the elementary row operations needed to transform an upper triangular matrix into $\mathcal{I}$.) Use transposes to prove that the inverse of an invertible lower triangular matrix is also lower triangular.

4.12 Projects

Project 4.1 (Leontief Input–Output Analysis)
Access the industry input–output data from the United States Bureau of Economic Analysis. (See the websites [22].) It is not realistic for us to compute the optimal output vector here. Instead, try to understand and explain what some of the data represent. For instance, are there sectors that require no or very little contribution from certain other sectors? Are there some sectors that are heavily dependent on certain other sectors? Which sectors rely most heavily on the sector or sectors in which you are, or wish to be, employed? Make any such observations that seem interesting to you.

Project 4.2 (Cubic Splines)
Refer for guidance to the cubic spline problem from Formula 4.26, Example 4.27, and Figure 4.3.

1. Define the cubic spline h by

$$h(x) = \begin{cases} h_1(x) = 0.8 \cdot x^3 + 0.2 \cdot x^2 & \text{if } 0 \leq x \leq 1, \\ h_2(x) = -2.4 \cdot x^3 + 9.8 \cdot x^2 - 9.6 \cdot x + 3.2 & \text{if } 1 \leq x \leq 2, \\ h_3(x) = 2.8 \cdot x^3 - 21.4 \cdot x^2 + 52.8 \cdot x - 38.4 & \text{if } 2 \leq x \leq 3. \end{cases}$$

   (a) Plot the graph of h on the interval $0 \leq x \leq 3$.
   (b) Verify that the graph of h passes through the sample points $(0,\ 0)$, $(1,\ 1)$, $(2,\ 4)$, and $(3,\ 3)$.
   (c) Verify that h is continuous at the intermediate points $(1,\ 1)$ and $(2,\ 4)$.
   (d) Verify that the first derivative $h'(x)$ is continuous at the intermediate points $(1,\ 1)$ and $(2,\ 4)$.
   (e) Verify that the second-order derivative $h''(x)$ is continuous at the intermediate points $(1,\ 1)$ and $(2,\ 4)$.
   (f) Verify that $h_1'(0) = 0$ and $h_3'(3) = 0$.

2. A cubic spline through the points $(0,\ 0)$, $(1,\ 1)$, $(2,\ 4)$, and $(3,\ 3)$ has three cubic pieces; call these $h_1(x)$, $h_2(x)$, and $h_3(x)$. Perform the following steps:

   (a) Express each of $h_1(x)$, $h_2(x)$, and $h_3(x)$ as a generic cubic with four coefficients to be determined.
   (b) Write down the six (6) linear equations that the coefficients of $h_1(x)$, $h_2(x)$, and $h_3(x)$ must satisfy to ensure that the splines go through the prescribed points.
   (c) Write down the four (4) linear equations involving the first- and second-order derivatives of $h_1(x)$, $h_2(x)$, and $h_3(x)$ needed to ensure that the pieces match up smoothly at the intermediate points.
   (d) Write down the two (2) linear equations needed to ensure that the slopes at the endpoints satisfy $h_1'(0) = -0.5$ and $h_3'(3) = 0.5$.
   (e) Use a computer to solve the resulting system of twelve (12) linear equations. This gives the coefficients of $h_1(x)$, $h_2(x)$, and $h_3(x)$.
   (f) Plot the resulting spline. (*Hint:* The plot should be the same as the dashed curve in Figure 4.3.)

3. Repeat the steps of the previous problem to compute and plot a cubic spline through the points $(0,\ 0)$, $(1,\ 3)$, $(2,\ 1)$, and $(3,\ 2)$, with slopes at the endpoints given by $h_1'(0) = 0$ and $h_3'(3) = 0$.

Project 4.3 (Kaczmarz's Method)
Using a computer, apply Kaczmarz's method, described in Algorithm 4.35, to find approximate solutions to the following linear systems. In each case, select a starting guess and choose the number of iterations of the algorithm to be applied.

1. $\begin{cases} x_1 - 2x_2 = -2 \\ x_1 + \ x_2 = \ \ 4 \end{cases}$ Sketch your results, as in Figure 4.5.

2. $\begin{cases} x_1 - 2x_2 = -2 \\ x_1 + x_2 = 4 \\ 3x_1 - x_2 = 1 \end{cases}$ Sketch your results, as in Figure 4.5.

3. $\begin{cases} 2x_1 + 8x_2 + 3x_3 = 2 \\ x_1 + 3x_2 + 2x_3 = 5 \\ 2x_1 + 7x_2 + 4x_3 = 8 \end{cases}$

4. $\begin{cases} x_1 + 3x_2 + 3x_3 = 13 \\ 2x_1 + 5x_2 + 4x_3 = 23 \\ 2x_1 + 7x_2 + 8x_3 = 29 \end{cases}$

5. $\begin{cases} 3x_1 + x_2 - 3x_3 = -4 \\ x_1 + x_2 + x_3 = 1 \\ 5x_1 + 6x_2 + 8x_3 = 8 \end{cases}$

6. $\begin{cases} 3x_1 - 6x_2 + x_3 + 13x_4 = 15 \\ 3x_1 - 6x_2 + 3x_3 + 21x_4 = 21 \\ 2x_1 - 4x_2 + 5x_3 + 26x_4 = 23 \end{cases}$

Chapter 5
Least Squares and Matrix Geometry

5.1 The Column Space of a Matrix

A system of M linear equations with N unknowns has the form $A\mathbf{x} = \mathbf{b}$, where $A = \left[a_{i,j}\right]$ is the $M \times N$ matrix of coefficients, $\mathbf{x} = \left[x_1 \cdots x_N\right]^T$ is an unknown vector in $\mathbb{R}^N$, and $\mathbf{b} = \left[b_1 \cdots b_M\right]^T$ is a given vector in $\mathbb{R}^M$. The left-hand side, $A\mathbf{x}$, is equivalent to a sum of numerical multiples of the column vectors of A. Namely,

$$A\mathbf{x} = x_1 \cdot \begin{bmatrix} a_{1,1} \\ \vdots \\ a_{M,1} \end{bmatrix} + x_2 \cdot \begin{bmatrix} a_{1,2} \\ \vdots \\ a_{M,2} \end{bmatrix} + \cdots + x_N \cdot \begin{bmatrix} a_{1,N} \\ \vdots \\ a_{M,N} \end{bmatrix}. \tag{5.1}$$

A solution to $A\mathbf{x} = \mathbf{b}$ consists of specific values for the coordinates of $\mathbf{x}$ that result in a true equation. Thus, the system $A\mathbf{x} = \mathbf{b}$ has a solution if, and only if, the vector $\mathbf{b}$ on the right-hand side is a linear combination of the column vectors of A. In Definition 2.11, we defined the column space of A, or $Col(A)$, to be the span of the collection of column vectors of A. The computation in (5.1) shows, equivalently, that $Col(A)$ is the set of all vectors $\mathbf{b}$ for which the system $A\mathbf{x} = \mathbf{b}$ has a solution.

Example 5.1 The column space of $A = \begin{bmatrix} 5 & 1 & 5 \\ 2 & 5 & -3 \\ -2 & -3 & 4 \end{bmatrix}$ is all of $\mathbb{R}^3$. To see this, take a generic vector $\mathbf{b} = \begin{bmatrix} b_1 \\ b_2 \\ b_3 \end{bmatrix}$, and observe that

---

Supplementary Information The online version contains supplementary material available at https://doi.org/10.1007/978-3-031-39562-8_5.

T. G. Feeman, *Applied Linear Algebra and Matrix Methods*,
Springer Undergraduate Texts in Mathematics and Technology,
https://doi.org/10.1007/978-3-031-39562-8_5

$$RREF(\left[\,A\middle|\mathbf{b}\,\right]) = \left[\begin{array}{ccc|c} 1 & 0 & 0 & (11b_1 - 19b_2 - 28b_3)/73 \\ 0 & 1 & 0 & (-2b_1 + 30b_2 + 25b_3)/73 \\ 0 & 0 & 1 & (4b_1 + 13b_2 + 23b_3)/73 \end{array}\right].$$

Thus, given $\mathbf{b}$ in $\mathbb{R}^3$, the vector $\mathbf{x} = \frac{1}{73} \cdot \begin{bmatrix} 11b_1 - 19b_2 - 28b_3 \\ -2b_1 + 30b_2 + 25b_3 \\ 4b_1 + 13b_2 + 23b_3 \end{bmatrix}$ is a solution to $A\mathbf{x} = \mathbf{b}$.

Example 5.2 To find $Col(A)$, where $A = \begin{bmatrix} 3 & -6 & -2 \\ 2 & -4 & 1 \\ 1 & -2 & -2 \end{bmatrix}$, take a generic vector $\mathbf{b} = \begin{bmatrix} b_1 & b_2 & b_3 \end{bmatrix}^T$, and observe that

$$RREF(\left[\,A\middle|\mathbf{b}\,\right]) = \left[\begin{array}{ccc|c} 1 & -2 & 0 & (2b_2 + b_3)/5 \\ 0 & 0 & 1 & (b_2 - 2b_3)/5 \\ 0 & 0 & 0 & (5b_1 - 4b_2 - 7b_3)/5 \end{array}\right]. \tag{5.2}$$

This shows that the system $A\mathbf{x} = \mathbf{b}$ has a solution if, and only if, the coordinates of $\mathbf{b}$ satisfy the equation $5b_1 - 4b_2 - 7b_3 = 0$. This is the equation of a plane in $\mathbb{R}^3$.

In Section 5.4, we look at column spaces in more detail.

5.2 Least Squares: Projection into $Col(A)$

Suppose A is an $M \times N$ matrix and $\mathbf{b}$ is a vector in $\mathbb{R}^M$ that is not in $Col(A)$. Thus, the system $A\mathbf{x} = \mathbf{b}$ does not have a solution. To *approximately* solve the system, we look for a vector $\widehat{\mathbf{x}}$ such that $A\widehat{\mathbf{x}}$ is *closest* to $\mathbf{b}$ among all vectors in $Col(A)$. That is, we wish to find a vector $\widehat{\mathbf{x}}$ in $\mathbb{R}^N$ such that

$$||A\widehat{\mathbf{x}} - \mathbf{b}|| = \min\left\{||A\mathbf{x} - \mathbf{b}|| \,:\, \mathbf{x} \in \mathbb{R}^N\right\}. \tag{5.3}$$

This is called a **least squares problem** because the norm of a vector can be expressed as a sum of squares and we are trying to minimize such a norm. A vector $\widehat{\mathbf{x}}$ that satisfies (5.3) is called a **least squares solution** to the problem. The problem of simple linear regression, discussed in Section 3.5, is an example of a least squares problem.

Geometrically, for $A\widehat{\mathbf{x}}$ to be the closest vector to $\mathbf{b}$ in $Col(A)$, the difference $(A\widehat{\mathbf{x}} - \mathbf{b})$ must be *orthogonal* to $Col(A)$. See Figure 5.1. That way, we have a

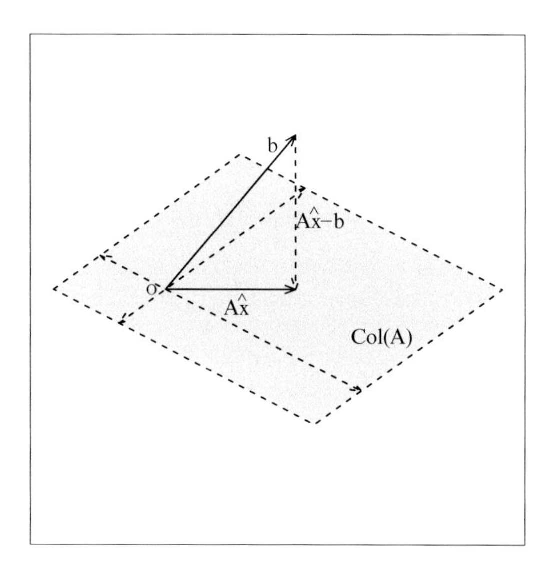

Fig. 5.1 For $\mathbf{b}$ not in $Col(A)$, a least squares solution $\widehat{\mathbf{x}}$ has the property that $A\widehat{\mathbf{x}} - \mathbf{b}$ is orthogonal to $Col(A)$

right triangle with $\mathbf{b}$ as its hypotenuse and the vectors $A\widehat{\mathbf{x}}$ and $(A\widehat{\mathbf{x}} - \mathbf{b})$ as the adjacent sides. A vector that is orthogonal to $Col(A)$ must have inner product equal to 0 with every column vector of A. For the least squares problem, this means that $\mathbf{a}_j^T\,(A\widehat{\mathbf{x}} - \mathbf{b}) = 0$, for every column vector $\mathbf{a}_j$ of A. It follows that $\widehat{\mathbf{x}}$ satisfies (5.3) if, and only if,

$$A^T\,(A\widehat{\mathbf{x}} - \mathbf{b}) = 0\,.$$

This is the same as

$$A^T A\widehat{\mathbf{x}} = A^T\mathbf{b}\,.$$

This last equation is called the **normal equation**.

Definition 5.3 A **least squares solution** to the equation $A\mathbf{x} = \mathbf{b}$ is a vector $\widehat{\mathbf{x}}$ that satisfies the **normal equation**

$$A^T A\widehat{\mathbf{x}} = A^T\mathbf{b}\,. \tag{5.4}$$

In the event that $A^T A$ is invertible, the normal equation (5.4) has the unique solution $\widehat{\mathbf{x}} = \left(A^T A\right)^{-1} A^T\mathbf{b}$. In the event that $A^T A$ is not invertible, the solution $\widehat{\mathbf{x}}$ is not unique. Nonetheless, the vector $A\widehat{\mathbf{x}}$ that is closest to $\mathbf{b}$ in $Col(A)$ *is* unique. (See Exercise #17 in section 5.7.) This unique vector $A\widehat{\mathbf{x}}$ is called the **projection** of $\mathbf{b}$ into $Col(A)$, denoted by $\mathrm{proj}_{Col(A)}\mathbf{b}$.

Formula 5.4 Suppose A is an $M \times N$ matrix for which $A^T A$ is invertible. Then the following statements hold.

(i) For every vector $\mathbf{b}$ in $\mathbb{R}^M$, the normal equation $A^T A\mathbf{x} = A^T\mathbf{b}$ has the *unique* solution

$$\widehat{\mathbf{x}} = \left(A^T A\right)^{-1} A^T\mathbf{b}. \tag{5.5}$$

(ii) For every vector $\mathbf{b}$ in $\mathbb{R}^M$, the **projection** of $\mathbf{b}$ into $Col(A)$ is given by

$$\text{proj}_{Col(A)}\mathbf{b} = A\widehat{\mathbf{x}} = A\left(A^T A\right)^{-1} A^T\mathbf{b}. \tag{5.6}$$

(iii) The matrix

$$A\left(A^T A\right)^{-1} A^T \tag{5.7}$$

implements projection into the column space of A.

Remark Inspired by equation (5.5), the matrix $\left(A^T A\right)^{-1} A^T$ is called the **pseudoinverse** of A, denoted by A^+. The least squares solution to $A\mathbf{x} = \mathbf{b}$ is then given by $\widehat{\mathbf{x}} = A^+\mathbf{b}$. This assumes that $A^T A$ is invertible. When A itself is invertible, then A^T is also invertible and

$$A^+ = \left(A^T A\right)^{-1} A^T = A^{-1}\left(A^T\right)^{-1} A^T = A^{-1}.$$

Thus, the pseudoinverse is the same as the inverse when that exists. In Chapter 10, we define the pseudoinverse of any matrix A, even when $A^T A$ is not invertible.

Remark Least squares approximation can be viewed as a calculus problem. Given an $M \times N$ matrix A and a vector $\mathbf{b}$ in $\mathbb{R}^M$, define the real-valued function $F : \mathbb{R}^N \to \mathbb{R}$ by

$$F(\mathbf{x}) = ||A\mathbf{x} - \mathbf{b}||^2 = (A\mathbf{x} - \mathbf{b})^T (A\mathbf{x} - \mathbf{b}), \text{ for all } \mathbf{x} \text{ in } \mathbb{R}^N.$$

One can show—this is the calculus part—that the *gradient vector* of F satisfies

$$\nabla F(\mathbf{x}) = 2\left(A^T A\mathbf{x} - A^T\mathbf{b}\right), \text{ for all } \mathbf{x} \text{ in } \mathbb{R}^N. \tag{5.8}$$

A least squares solution $\widehat{\mathbf{x}}$ produces a minimum value for F and, hence, $\nabla F(\widehat{\mathbf{x}}) = \mathbf{0}$. According to (5.8), then, $\widehat{\mathbf{x}}$ satisfies the normal equation $A^T A\widehat{\mathbf{x}} = A^T\mathbf{b}$.

5.3 Least Squares: Two Applications

5.3.1 Multiple Linear Regression

In Section 3.5, we used the *normal equation* to find a *regression line* that best fit a data set of paired observations of two related random variables. More generally, the value of a particular output variable might depend on several input variables rather than on just one. Finding the combination of the inputs that best explains the output is a general least squares problem.

Mathematically, suppose we have input variables $X_1, X_2, \ldots, X_N$, and an output variable Y, and we suspect that there may be a linear relationship between them of the form

$$\beta_0 + \beta_1 X_1 + \beta_2 X_2 + \cdots \beta_N X_N \approx Y \,,$$

where the parameters β_i are constants to be determined. To test this, we collect a sample of M observations, where each observation includes measurements of all the input variables along with the output variable. For the ith observation, we denote the measurement of the variable X_j by $x_{i,j}$ and the measurement of Y by y_i. If the linear relationship holds, then, for each i with $1 \leq i \leq M$,

$$\beta_0 + \beta_1 x_{i,1} + \beta_2 x_{i,2} + \cdots \beta_N x_{i,N} \approx y_i \,.$$

To express this in matrix form, set $\mathbf{y} = \begin{bmatrix} y_1 & \cdots & y_M \end{bmatrix}^T$, and take $\mathbf{x}_j$ to be the column vector of measured values of the variable X_j from the different observations in the sample. Now set

$$A = \begin{bmatrix} \mathbf{1}_M \middle| \mathbf{x}_1 \middle| \cdots \middle| \mathbf{x}_N \end{bmatrix} \text{ and } \beta = \begin{bmatrix} \beta_0 \\ \beta_1 \\ \vdots \\ \beta_N \end{bmatrix} .$$

Our conjecture is that $A\beta \approx \mathbf{y}$. The least squares solution is a vector $\widehat{\beta}$ that satisfies the *normal equation* (5.4):

$$A^T A \widehat{\beta} = A^T \mathbf{y} \,.$$

By equation (5.5), and assuming the matrix $A^T A$ is invertible,

$$\widehat{\beta} = \left(A^T A\right)^{-1} A^T \mathbf{y} \,. \tag{5.9}$$

Once we have a model, we should assess how well it explains the data from the sample. To do this, we compare the error $||A\widehat{\beta} - \mathbf{y}||$ that arises from estimating $\mathbf{y}$ as $A\widehat{\beta}$ to the overall variability in the measurements of Y just on their own, without taking the inputs X_j into account. As in Section 3.5, we quantify this with the value R^2, defined as

$$R^2 = 1 - \frac{||A\widehat{\beta} - \mathbf{y}||^2}{||\mathbf{y} - \bar{y}\mathbf{1}||^2}\,. \tag{5.10}$$

Here, $\bar{y} = (y_1 + \cdots + y_M)/M$ is the average of the observed values of Y. If the linear model is perfect, then $R^2 = 1$. If the regression errors are large, then R^2 will be close to 0.

Example 5.5 Examples 1.16 and 3.15 refer to data from 1980 on child mortality and female literacy rates along with per capita gross national product for 57 developing countries. The data are given in Table 1.2. Figure 1.8 shows scatter plots for different pairs of measurements. Equation (1.28) shows the corresponding correlation coefficients. In a least squares simple linear regression model for child mortality as a function of female literacy, we found $R^2 \approx 0.647$, indicating that the female literacy rate alone only partly explains the variability in child mortality rates. Figure 3.6 shows the regression line along with the data.

We are now equipped to analyze how well female literacy rates and per capita GNP *together* explain the child mortality rates. We denote by $\mathbf{y}$ the column vector of CM data. The 57×3 matrix A has all 1s in the first column, the FLR data in the second column, and the pcGNP data in the third column. The least squares solution to $A\beta = \mathbf{y}$ is given by

$$\widehat{\beta} = \left(A^T A\right)^{-1} A^T\mathbf{y}\,, \text{ as in (5.9).}$$

Using a computer, we find

$$A^T A = \begin{bmatrix} 57 & 2736 & 41520 \\ 2736 & 167214 & 2428620 \\ 41520 & 2428620 & 46842200 \end{bmatrix} \text{ and } A^T\mathbf{y} = \begin{bmatrix} 8681 \\ 335001 \\ 4888110 \end{bmatrix}.$$

Then

$$\widehat{\beta} \approx \begin{bmatrix} 267.28 \\ -1.80 \\ -0.04 \end{bmatrix}.$$

The corresponding linear model is

$$\text{CM} \approx 267.28 - 1.80 \cdot (\text{FLR}) - 0.04 \cdot (\text{pcGNP})\,.$$

The coefficient -1.80 indicates that an increase of 1% in the female literacy rate would explain a reduction of about 1.8 in the number of child deaths per 1000 live births. An increase in pcGNP of \$25 is associated with a reduction in the number of child deaths by 1 per 1000 births.

To complete the analysis, we compute

$$||A\widehat{\beta} - \mathbf{y}||^2 \approx 83042.54 \text{ and } ||\mathbf{y} - \bar{y}\mathbf{1}||^2 \approx 287377.9\,.$$

Thus,

$$R^2 = 1 - \frac{||A\widehat{\beta} - \mathbf{y}||^2}{||\mathbf{y} - \bar{y}\mathbf{1}||^2} \approx 0.7076\,.$$

This suggests that using both FLR and pcGNP is better at explaining CM than just FLR by itself. But the model is still less than perfect. The data do not take into account a variety of factors affecting women's lives such as income distribution, access to education and health care, and participation in the work force outside the home.

Example 5.6 One application of multiple regression is to a *time series*, where we look at how a given economic statistic changes over time. This example is simplified from Example 8.1, in the 3rd edition of the book [9]. The output variable is personal consumption expenditure (PCE $= Y$) as a function of the input variables personal disposable income (PDI $= X_1$) and time ($= X_2$). We expect both PDI and PCE to go up over time. The question is whether the rise in PCE is explained only by time or also by the rise in PDI. For the 15 years from 1956 (year 1) to 1970 (year 15), we record PDI and PCE, measured in hundreds of dollars.

PCE vs. PDI and time; 1956 to 1970					
Year	1 (1956)	2	3	4	5
PDI (× \$100)	309.3	316.1	318.8	333.0	340.3
PCE (× \$100)	281.4	288.1	290.0	307.3	316.1
Year	6	7	8	9	10
PDI (× \$100)	350.5	367.2	381.2	408.1	434.8
PCE (× \$100)	322.5	338.4	353.3	373.7	397.7
Year	11	12	13	14	15 (1970)
PDI (× \$100)	458.9	477.5	499.0	513.5	533.2
PCE (× \$100)	418.1	430.1	452.7	469.1	476.9

Create the 15×3 matrix A that has all 1s in the first column, the PDI data in the second column, and the year (1–15) in the third column. Form the vector $\mathbf{y}$ containing the PCE data. We use the normal equation $A^T A\widehat{\beta} = A^T\mathbf{y}$ to find the least squares solution $\widehat{\beta}$ that minimizes $||A\beta - \mathbf{y}||^2$. Using a computer, we get

$$A^T A = \begin{bmatrix} 15 & 6041.4 & 120 \\ 6041.4 & 2518089.4 & 53127.2 \\ 120 & 53127.2 & 1240 \end{bmatrix}.$$

Also,

$$A^T \mathbf{y} = \begin{bmatrix} 5515.4 \\ 2296160.9 \\ 48374.1 \end{bmatrix}.$$

Finally,

$$\widehat{\beta} = (A^T A)^{-1} A^T \mathbf{y} = \begin{bmatrix} 53.16 \\ 0.727 \\ 2.736 \end{bmatrix}.$$

The corresponding linear model is

$$PCE \approx 53.16 + 0.727 \cdot (PDI) + 2.736 \cdot (\text{year}).$$

To compute the R^2 value, we have

$$||A\widehat{\beta} - \mathbf{y}||^2 = 77.17 \text{ and } ||\mathbf{y} - \bar{y}\mathbf{1}||^2 = 66042.3.$$

Thus,

$$R^2 = 1 - \frac{||A\widehat{\beta} - \mathbf{y}||^2}{||\mathbf{y} - \bar{y}\mathbf{1}||^2} = 0.9988.$$

The model is an excellent fit that explains the output well in terms of the input variables.

In the model, the coefficient of PDI is 0.727. This indicates that about 73 cents of each additional dollar of disposable income is spent as part of consumption expenditure. Meanwhile, consumption expenditure increases by about \$273 each year because of other factors such as price increases. This is for the years 1956 to 1970. We should not assume that the same model will work today.

5.3.2 Curve Fitting with Least Squares

Simple linear regression uses a least squares problem to determine a straight line model that is a good fit for the available data for two related random variables. However, some other curve may be more appropriate.

Example 5.7 Suppose we have two variables X and Y with observed values

$$\mathbf{x} = \begin{bmatrix} -3 \\ -2 \\ -1 \\ 0 \\ 1 \\ 2 \\ 3 \end{bmatrix} \text{ and } \mathbf{y} = \begin{bmatrix} 1 \\ 5 \\ 2 \\ 2 \\ 0 \\ -1 \\ 0 \end{bmatrix}.$$

For simple linear regression, we set $A = \left[\mathbf{1} \middle| \mathbf{x}\right]$ and solve the normal equation $A^T A\beta = A^T\mathbf{y}$. This yields the regression line

$$Y \approx 1.286 - (0.607)X.$$

With $R^2 = 0.44$, this line is not a very good fit. A plot of the data points suggests that a cubic curve might provide a better fit. Thus, we would like coefficients β_0, β_1, β_2, and β_3 such that

$$Y \approx \beta_0 + \beta_1 X + \beta_2 X^2 + \beta_3 X^3.$$

In this model, the output Y depends not only on the value of X but on X^2 and X^3 as well. To frame our least squares problem, we need the matrix

$$A = \left[\mathbf{1} \middle| X \middle| X^2 \middle| X^3\right] = \begin{bmatrix} 1 & -3 & 9 & -27 \\ 1 & -2 & 4 & -8 \\ 1 & -1 & 1 & -1 \\ 1 & 0 & 0 & 0 \\ 1 & 1 & 1 & 1 \\ 1 & 2 & 4 & 8 \\ 1 & 3 & 9 & 27 \end{bmatrix}.$$

Solving the normal equation

$$A^T A \begin{bmatrix} \beta_0 \\ \beta_1 \\ \beta_2 \\ \beta_3 \end{bmatrix} = A^T\mathbf{y}$$

yields the curve

$$Y \approx 1.714 - 1.968X - 0.107X^2 + 0.194X^3.$$

Now we have $R^2 = 0.83$, much higher than before. Figure 5.2 shows the data along with this cubic and the linear regression solution.

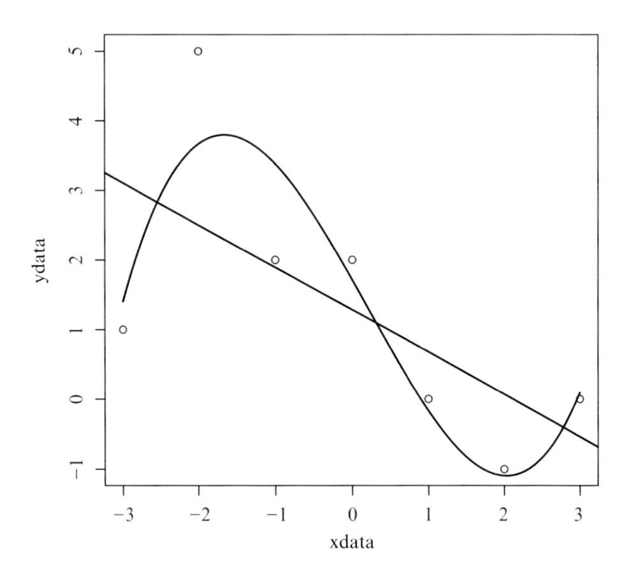

Fig. 5.2 A cubic curve fits the data better than the regression line

5.4 Four Fundamental Subspaces

According to the normal equation (5.4), when A is an $M \times N$ matrix, every vector $\mathbf{b}$ in $\mathbb{R}^M$ can be written as $\mathbf{b} = \mathbf{y} + \mathbf{u}$, where $\mathbf{y}$ is in the column space $Col(A)$ and $\mathbf{u}$ is in the nullspace $Null(A^T)$. Indeed, we may take $\mathbf{y} = A\widehat{\mathbf{x}}$, where $\widehat{\mathbf{x}}$ is any solution to (5.4), and $\mathbf{u} = \mathbf{b} - A\widehat{\mathbf{x}}$. The summands $\mathbf{y}$ and $\mathbf{u}$ are orthogonal to each other by construction. Moreover, this decomposition of $\mathbf{b}$ is unique because $A\widehat{\mathbf{x}}$ is the same for every least squares solution $\widehat{\mathbf{x}}$.

Looking at this decomposition from a different angle, suppose a vector $\mathbf{u}$ is orthogonal to every vector in $Col(A)$. Then, for every $\mathbf{x}$ in $\mathbb{R}^N$, we get

$$\left(A^T\mathbf{u}\right)^T \mathbf{x} = \left(\mathbf{u}^T A\right)\mathbf{x} = \mathbf{u}^T\,(A\mathbf{x}) = 0\,.$$

This means that $A^T\mathbf{u}$ is orthogonal to every vector in $\mathbb{R}^N$, including, necessarily, itself. Hence, $A^T\mathbf{u} = \mathbf{0}$. In other words, $\mathbf{u}$ belongs to $Null(A^T)$.

Conversely, suppose $\mathbf{u}$ belongs to $Null(A^T)$, and $\mathbf{y} = A\mathbf{x}$ is a generic element of $Col(A)$. Then we have

$$\mathbf{u}^T\mathbf{y} = \mathbf{u}^T\,(A\mathbf{x}) = \left(\mathbf{u}^T A\right)\mathbf{x} = \left(A^T\mathbf{u}\right)^T \mathbf{x} = \mathbf{0}^T\mathbf{x} = [0]\,.$$

Thus, $\mathbf{u}$ is orthogonal to every vector in $Col(A)$. This shows that the set $Null(A^T)$ is exactly the set of all vectors in $\mathbb{R}^M$ orthogonal to every vector in $Col(A)$.

That last sentence is a bit of a mouthful, so we conveniently have some terminology and notation to use as shortcuts.

Definition 5.8 For any nonempty collection $\mathcal{S}$ of vectors in $\mathbb{R}^N$, the set of all vectors in $\mathbb{R}^N$ that are *orthogonal to every vector in* $\mathcal{S}$ is called the **orthogonal complement** of $\mathcal{S}$, denoted by the symbol $\mathcal{S}^\perp$. That is,

$$\mathcal{S}^\perp = \{\mathbf{x} \in \mathbb{R}^N \ : \ \mathbf{y}^T\mathbf{x} = 0\,, \ \text{for all } \mathbf{y} \text{ in } \mathcal{S}\}. \tag{5.11}$$

The set $\mathcal{S}^\perp$ is often nicknamed "S perp" (short for "perpendicular").

Thus, for every $M \times N$ matrix A, we have shown that $Null(A^T)$ is the orthogonal complement of $Col(A)$ in $\mathbb{R}^M$, and we write

$$Null(A^T) = Col(A)^\perp\,. \tag{5.12}$$

Reversing the roles of A and A^T in (5.12), we see that $Null(A)$ is the orthogonal complement of $Col(A^T)$ in $\mathbb{R}^N$, written as

$$Null(A) = Col(A^T)^\perp\,. \tag{5.13}$$

It follows that every vector $\mathbf{v}$ in $\mathbb{R}^N$ can be written uniquely as $\mathbf{v} = \mathbf{w} + \mathbf{x}$, with $\mathbf{w}$ in $Col(A^T)$, $\mathbf{x}$ in $Null(A)$, and $\mathbf{x}^T\mathbf{w} = 0$. Figure 5.3 summarizes our discussion.

We have identified four spaces associated with an $M \times N$ matrix A:

- The *column space* of A, denoted $Col(A)$;
- The *nullspace* of A, denoted $Null(A)$;
- The *column space* of A^T, denoted $Col(A^T)$; and
- The *nullspace* of A^T, denoted $Null(A^T)$.

These spaces are popularly known as **the four fundamental subspaces** thanks to the efforts of the renowned mathematician, teacher, and author, Gilbert Strang. Some examples will help us to develop a more detailed understanding of these spaces. In particular, we will see how to find sets of vectors that generate these spaces.

Example 5.9 Let $A = \begin{bmatrix} 1 & -4 & -3 & -3 \\ 2 & -6 & -5 & -5 \\ 3 & -1 & -4 & -5 \end{bmatrix}$. Then A has the reduced row echelon form

$$RREF(A) = \begin{bmatrix} 1 & 0 & 0 & 2 \\ 0 & 1 & 0 & -1 \\ 0 & 0 & 1 & 3 \end{bmatrix}.$$

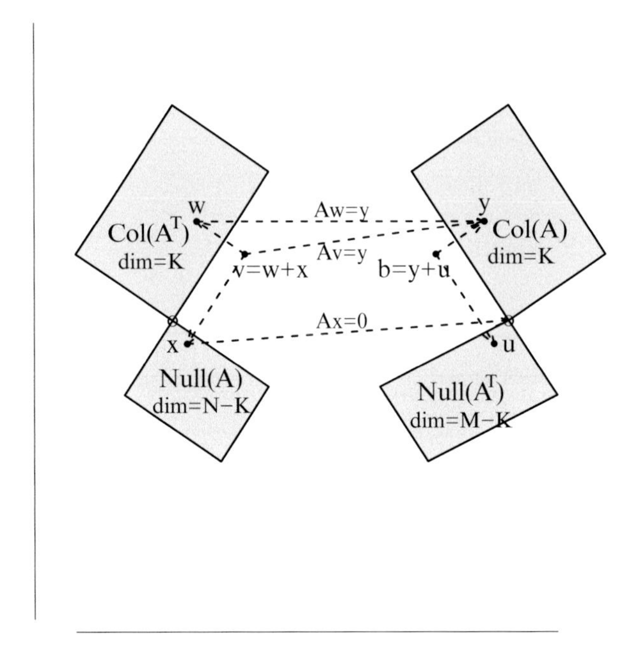

Fig. 5.3 The four fundamental subspaces associated with an $M \times N$ matrix A. On the left side, every vector $\mathbf{v}$ in $\mathbb{R}^N$ is a sum of orthogonal vectors from $Col(A^T)$ and $Null(A)$. On the right, every vector $\mathbf{b}$ in $\mathbb{R}^M$ is a sum of orthogonal vectors from $Col(A)$ and $Null(A^T)$. The arrows joining the left and right sides show the action of A on the different components of $\mathbf{v}$

The first three columns of $RREF(A)$ contain leading 1s, while the fourth column does not. It follows that the fourth column of A is itself a linear combination of the first three columns. Specifically,

$$\begin{bmatrix} -3 \\ -5 \\ -5 \end{bmatrix} = 2 \cdot \begin{bmatrix} 1 \\ 2 \\ 3 \end{bmatrix} - 1 \cdot \begin{bmatrix} -4 \\ -6 \\ -1 \end{bmatrix} + 3 \cdot \begin{bmatrix} -3 \\ -5 \\ -4 \end{bmatrix}.$$

None of the first three column vectors of A can be expressed as a linear combination of the other two, however. We conclude that the set

$$\left\{ \begin{bmatrix} 1 \\ 2 \\ 3 \end{bmatrix} \begin{bmatrix} -4 \\ -6 \\ -1 \end{bmatrix} \begin{bmatrix} -3 \\ -5 \\ -4 \end{bmatrix} \right\}$$

spans $Col(A)$. No two of these columns vectors will span $Col(A)$. Thus, we say that $Col(A)$ has *dimension* equal to 3, which also means that $Col(A)$ is all of $\mathbb{R}^3$ in this example. The first three column vectors of A are a *basis* of $Col(A)$.

The subspace $Col(A^T)$ also has dimension 3. To see why, note that the *rows* of A are the *transposed columns* of A^T. When we perform elementary row operations on a matrix, we are computing linear combinations of its rows. In this example, every row of $RREF(A)$ contains a leading 1. That means that no row of A can be eliminated by writing it as a linear combination of the other rows. For A^T, this means that none of the *columns* of A^T can be expressed as a linear combination of the other columns. We can use either the three original columns of A^T or the transposed rows of $RREF(A)$ as a spanning set for $Col(A^T)$. That is, each of the sets

$$\left\{ \begin{bmatrix} 1 \\ -4 \\ -3 \\ -3 \end{bmatrix} \begin{bmatrix} 2 \\ -6 \\ -5 \\ -5 \end{bmatrix} \begin{bmatrix} 3 \\ -1 \\ -4 \\ -5 \end{bmatrix} \right\} \text{ and } \left\{ \begin{bmatrix} 1 \\ 0 \\ 0 \\ 2 \end{bmatrix} \begin{bmatrix} 0 \\ 1 \\ 0 \\ -1 \end{bmatrix} \begin{bmatrix} 0 \\ 0 \\ 1 \\ 3 \end{bmatrix} \right\}$$

forms a *basis* for $Col(A^T)$. Either way, it takes three vectors to span $Col(A^T)$, so this is a 3-dimensional subspace of $\mathbb{R}^4$. Notice that $Col(A)$ and $Col(A^T)$ have the *same dimension*, equal to the number of rows or columns of $RREF(A)$ that contain leading 1s.

The nullspace $Null(A)$ is the set of all vectors $\mathbf{x}$ in $\mathbb{R}^4$ such that $A\mathbf{x} = \mathbf{0}$. Looking again at $RREF(A)$, we see that x_4 is a free variable. Every solution to $A\mathbf{x} = \mathbf{0}$ has the form $\mathbf{x} = t \cdot \begin{bmatrix} -2 \\ 1 \\ -3 \\ 1 \end{bmatrix}$. Thus, $Null(A)$ has dimension 1. Notice that the dimensions of $Col(A)$ and $Null(A)$ must add up to $N - 4$, since every column of $RREF(A)$ either contains a leading 1 or corresponds to a free variable in the system $A\mathbf{x} = \mathbf{0}$.

To wrap up this example, we look at $Null(A^T)$, the nullspace of A^T. One way to find a description of $Null(A^T)$ is to transform A^T into reduced row echelon form, identify any free variables, and form the corresponding vectors in the nullspace, as we did with $Null(A)$. Another approach is to recognize that $Null(A^T)$ is the orthogonal complement to $Col(A)$ in $\mathbb{R}^3$. Since $Col(A)$ is already using up all of $\mathbb{R}^3$ in this example, the nullspace of A^T must be just the zero vector $\mathbf{0}$. The dimension of $Null(A^T)$ in this example is equal to 0.

Example 5.10 Take $A = \begin{bmatrix} 3 & 1 & 1 & 6 \\ 1 & -2 & 5 & -5 \\ 4 & 1 & 2 & 7 \end{bmatrix}$. In this case, we find that

$$RREF(A) = \begin{bmatrix} 1 & 0 & 1 & 1 \\ 0 & 1 & -2 & 3 \\ 0 & 0 & 0 & 0 \end{bmatrix}.$$

The leading 1s of $RREF(A)$ are in the first two columns, so $Col(A)$ has dimension 2 and is generated by $\begin{bmatrix} 3 \\ 1 \\ 4 \end{bmatrix}$ and $\begin{bmatrix} 1 \\ -2 \\ 1 \end{bmatrix}$, the first two column vectors of A itself.

Turning to $Null(A)$, we see from $RREF(A)$ that there are two free variables when solving $A\mathbf{x} = \mathbf{0}$. So the nullspace of A is the 2-dimensional subspace of $\mathbb{R}^4$ spanned by the two vectors $\begin{bmatrix} -1 \\ 2 \\ 1 \\ 0 \end{bmatrix}$ and $\begin{bmatrix} -1 \\ -3 \\ 0 \\ 1 \end{bmatrix}$. Again, note that the dimensions of $Col(A)$ and $Null(A)$ add up to $N = 4$, the total number of columns of A.

Next, the columns of A^T are the transposed rows of A. Since the third row of $RREF(A)$ is all 0s, that means that the third column of A^T can be obtained as a linear combination of the first two columns. Thus, $Col(A^T)$ has dimension 2. For a basis of $Col(A^T)$, we can use either the nonzero rows of $RREF(A)$, transposed into columns, or the first two columns of A^T itself. That is, each of the sets

$$\left\{ \begin{bmatrix} 1 \\ 0 \\ 1 \\ 1 \end{bmatrix} \begin{bmatrix} 0 \\ 1 \\ -2 \\ 3 \end{bmatrix} \right\} \text{ or } \left\{ \begin{bmatrix} 3 \\ 1 \\ 1 \\ 6 \end{bmatrix} \begin{bmatrix} 1 \\ -2 \\ 5 \\ -5 \end{bmatrix} \right\}$$

is a basis for $Col(A^T)$. Notice that the basis vectors for $Col(A^T)$ are orthogonal to those for $Null(A)$.

Finally, the space $Null(A^T)$ has dimension 1, equal to the number of rows of all 0s in $RREF(A)$, and also equal to $M - \dim(Col(A)) = 3 - 2$. One approach to finding a basis vector for $Null(A^T)$ uses the fact that $Null(A^T) = Col(A)^{\perp}$. Augment A with a generic right-hand side before row reduction. In this example, we get

$$\left[\begin{array}{cccc|c} 3 & 1 & 1 & 6 & b_1 \\ 1 & -2 & 5 & -5 & b_2 \\ 4 & 1 & 2 & 7 & b_3 \end{array}\right] \xrightarrow{RREF} \left[\begin{array}{cccc|c} 1 & 0 & 1 & 1 & (2b_1 + b_2)/7 \\ 0 & 1 & -2 & 3 & (b_1 - 3b_2)/7 \\ 0 & 0 & 0 & 0 & (9b_1 + b_2 - 7b_3)/63 \end{array}\right].$$

Thus, $Col(A)$ consists of all $\mathbf{b}$ in $\mathbb{R}^3$ for which $9b_1 + b_2 - 7b_3 = 0$. This means that every $\mathbf{b}$ in $Col(A)$ is orthogonal to the vector $\begin{bmatrix} 9 \\ 1 \\ -7 \end{bmatrix}$. Therefore, this vector spans $Null(A^T)$. Notice that this vector is indeed orthogonal to the column vectors of A. We can write down a new basis for $Col(A)$ from this, too, since two linearly independent vectors that satisfy $9b_1 + b_2 - 7b_3 = 0$ are $\begin{bmatrix} 7 \\ 0 \\ 9 \end{bmatrix}$ and $\begin{bmatrix} 0 \\ 7 \\ 1 \end{bmatrix}$.

Formula 5.11 The following facts pertain to the four fundamental subspaces of an arbitrary $M \times N$ matrix A.

1. $Col(A)^{\perp} = Null(A^T)$, in $\mathbb{R}^M$, and $Col(A^T)^{\perp} = Null(A)$, in $\mathbb{R}^N$.
2. The dimension of $Col(A)$ is the number of columns of $RREF(A)$ that contain leading 1s. The original column vectors of A in the columns that contain the leading 1s of $RREF(A)$ form a basis of $Col(A)$.
3. The dimension of $Col(A^T)$ is the number of rows of $RREF(A)$ that contain leading 1s. A basis for $Col(A^T)$ consists of the nonzero rows of $RREF(A)$ transposed into column vectors.
4. The number of columns of $RREF(A)$ that contain leading 1s is the same as the number of rows of $RREF(A)$ that contain leading 1s. Therefore, $Col(A)$ and $Col(A^T)$ have the same dimension. We write

$$\dim\left(Col(A)\right) = \dim\left(Col(A^T)\right). \tag{5.14}$$

5. The dimension of the nullspace $Null(A)$ is equal to the number of columns of $RREF(A)$ that do not contain leading 1s. Since every column either contains a leading 1 or does not, the dimensions of $Col(A)$ and $Null(A)$ must add up to N, the total number of columns of A. This observation is recorded as a famous theorem.

   Theorem 5.12 (Rank–Plus–Nullity) *For every matrix A, the sum of the dimension of the column space, $Col(A)$, and the dimension of the nullspace, $Null(A)$, is equal to the total number of columns of A. That is, for an $M \times N$ matrix A,*

$$\dim\left(Col(A)\right) + \dim\left(Null(A)\right) = N. \tag{5.15}$$

6. By the Rank–plus–Nullity Theorem 5.12, if $Col(A)$ and $Col(A^T)$ both have dimension K, then $Null(A)$ has dimension $N-K$, and $Null(A^T)$ has dimension $M-K$.

5.4.1 Column–Row Factorization

As we have seen, the original column vectors of A in the columns that contain the leading 1s of $RREF(A)$ form a basis of $Col(A)$. At the same time, the entries in the columns of $RREF(A)$ that do not contain leading 1s show how to express the other columns of A as linear combinations of the leading columns. These two facts together offer a way to factor any matrix A.

Algorithm 5.13 (Column–Row Factorization)
Let A be an $M \times N$ matrix.

1. Take K to be the number of columns of $RREF(A)$ with leading 1s. This is the same as the dimension of $Col(A)$.
2. Form the $M \times K$ matrix A_{col} whose columns are the original column vectors of A in the columns that contain the leading 1s of $RREF(A)$. Thus, $Col(A_{col}) = Col(A)$.
3. Form the $K \times N$ matrix A_{row} equal to $RREF(A)$ but with any rows of all 0s removed. The rows of A_{row} form a basis of $Col(A^T)$ when transposed into columns.
4. Then, $A = A_{col} \cdot A_{row}$.

Example 5.14 The matrix $A = \begin{bmatrix} 1 & 2 & -1 & 7 \\ 2 & 4 & 3 & 4 \\ 3 & 6 & 1 & 13 \end{bmatrix}$ has reduced row echelon form given by $RREF(A) = \begin{bmatrix} 1 & 2 & 0 & 5 \\ 0 & 0 & 1 & -2 \\ 0 & 0 & 0 & 0 \end{bmatrix}$. Thus, set

$$A_{col} = \begin{bmatrix} 1 & -1 \\ 2 & 3 \\ 3 & 1 \end{bmatrix} \text{ and } A_{row} = \begin{bmatrix} 1 & 2 & 0 & 5 \\ 0 & 0 & 1 & -2 \end{bmatrix}.$$

Matrix multiplication confirms that

$$A_{col} \cdot A_{row} = \begin{bmatrix} 1 & -1 \\ 2 & 3 \\ 3 & 1 \end{bmatrix} \cdot \begin{bmatrix} 1 & 2 & 0 & 5 \\ 0 & 0 & 1 & -2 \end{bmatrix} = \begin{bmatrix} 1 & 2 & -1 & 7 \\ 2 & 4 & 3 & 4 \\ 3 & 6 & 1 & 13 \end{bmatrix} = A\,.$$

Remark When the columns of the matrix A are linearly independent, then every column of $RREF(A)$ has a leading 1. This yields $A_{row} = \mathcal{I}$ and $A_{col} = A$. The column–row factorization becomes $A = A \cdot \mathcal{I}$.

5.5 Geometry of Transformations

As described in Section 3.8, each $M \times N$ matrix A defines a function with domain space $\mathbb{R}^N$ and target space $\mathbb{R}^M$. Specifically, we map each vector $\mathbf{x}$ in $\mathbb{R}^N$ to the vector $A\mathbf{x}$ in $\mathbb{R}^M$. The function defined by a matrix A in this way has the property that vector arithmetic in the domain space is carried over into the target space. Indeed, suppose $\mathbf{u}$ and $\mathbf{v}$ are any vectors in $\mathbb{R}^N$ and s and t are any real numbers. Then

$$A(s \cdot \mathbf{u} + t \cdot \mathbf{v}) = s \cdot A\mathbf{u} + t \cdot A\mathbf{v}\,. \tag{5.16}$$

That is,

$$s \cdot \mathbf{u} + t \cdot \mathbf{v} \quad \overset{\text{is mapped to}}{\longmapsto} \quad s \cdot A\mathbf{u} + t \cdot A\mathbf{v}\,.$$

Geometrically, according to the property in (5.16), the line generated by $\mathbf{u}$ gets mapped to the line generated by $A\mathbf{u}$, and the line generated by $\mathbf{v}$ gets mapped to the line generated by $A\mathbf{v}$. Then, points in the plane generated by $\mathbf{u}$ and $\mathbf{v}$ get mapped to corresponding points in the plane generated by $A\mathbf{u}$ and $A\mathbf{v}$. (*Caution:* $A\mathbf{u}$ and $A\mathbf{v}$ can lie on the same line even if $\mathbf{u}$ and $\mathbf{v}$ do not. Also, either or both of $A\mathbf{u}$ and $A\mathbf{v}$ might be $\mathbf{0}$.)

Due to property (5.16), we say that the mapping $\mathbf{x} \mapsto A\mathbf{x}$, defined by multiplication by A, is a **linear transformation.** In this context, matrix arithmetic corresponds to combining functions. Namely, the matrix $A + B$ implements the sum of two functions, while the matrix BA corresponds to the composite mapping *first multiply by* A*, then multiply by* B. Schematically,

$$\mathbf{x} \mapsto A\mathbf{x} \mapsto B(A\mathbf{x}) = (BA)\mathbf{x}\,.$$

The two main spaces associated with a linear transformation are its **range** and its **nullspace**.

Definition 5.15 Consider an $M \times N$ matrix A and the corresponding linear transformation defined by the mapping $\mathbf{x} \mapsto A\mathbf{x}$, for all $\mathbf{x}$ in $\mathbb{R}^N$.

- The **range** of this transformation is the set $\{A\mathbf{x} : \mathbf{x} \in \mathbb{R}^N\}$ of all outputs under multiplication by A. Equivalently, the range is the set of all vectors $\mathbf{b}$ in $\mathbb{R}^M$ for which the system $A\mathbf{x} = \mathbf{b}$ has a solution. Thus, the range of the transformation coincides with $Col(A)$, the column space of the matrix A.
- The **nullspace** of this transformation is the set $\{\mathbf{x} \in \mathbb{R}^N : A\mathbf{x} = \mathbf{0}\}$. This coincides with $Null(A)$, the nullspace of the matrix A.

Notice that each column vector of a given $M \times N$ matrix A is in the range of the transformation defined by the A. Indeed, for each $1 \leq j \leq N$, denote by $\mathbf{e}_j$ the jth column vector of $\mathcal{I}_N$. Then $A\mathbf{e}_j = \mathbf{a}_j$, the jth column vector of A. The vectors $\{\mathbf{e}_j\}$ are called the **standard basis vectors** for $\mathbb{R}^N$. Therefore, the columns of A are the images of the standard basis vectors under multiplication by A.

For an arbitrary vector $\mathbf{x} = \begin{bmatrix} x_1 & \cdots & x_N \end{bmatrix}^T$ in $\mathbb{R}^N$, observe that

$$\mathbf{x} = x_1\mathbf{e}_1 + x_2\mathbf{e}_2 + \cdots + x_N\mathbf{e}_N\,.$$

Thus,

$$\begin{aligned} A\mathbf{x} &= A(x_1\mathbf{e}_1 + x_2\mathbf{e}_2 + \cdots + x_N\mathbf{e}_N) \\ &= x_1 A\mathbf{e}_1 + x_2 A\mathbf{e}_2 + \cdots + x_N A\mathbf{e}_N \\ &= x_1\mathbf{a}_1 + x_2\mathbf{a}_2 + \cdots + x_N\mathbf{a}_N\,. \end{aligned} \tag{5.17}$$

This confirms that the range of the transformation coincides with $Col(A)$.

The transformation defined by an $M \times N$ matrix A is said to be **surjective**, or **onto**, when the range is the full target space $\mathbb{R}^M$. That is, the transformation is surjective if $Col(A)$ has dimension M. According to the Rank–plus–Nullity Theorem 5.12, the dimensions of the range and nullspace must add up to N. Hence, if $M > N$, then the transformation cannot be surjective. If $M \leq N$, then the transformation is surjective when $RREF(A)$ has M columns containing leading 1s.

The transformation is said to be **injective**, or **one-to-one**, if $Null(A)$ contains just the zero vector $\mathbf{0}$ in $\mathbb{R}^N$. To explain the terminology, suppose there are two vectors $\mathbf{u} \neq \mathbf{v}$ in $\mathbb{R}^N$ such that $A\mathbf{u} = A\mathbf{v}$. Let $\mathbf{w} = \mathbf{u} - \mathbf{v}$. Then $A\mathbf{w} = A\mathbf{u} - A\mathbf{v} = \mathbf{0}$. So $\mathbf{w}$ is in $Null(A)$ but $\mathbf{w} \neq \mathbf{0}$. It follows that, if $\mathbf{0}$ is the *only vector* in $Null(A)$, then different inputs $\mathbf{u} \neq \mathbf{v}$ must go to different outputs $A\mathbf{u} \neq A\mathbf{v}$. In that sense, the transformation is *one-to-one*.

Appealing again to the Rank–plus–Nullity Theorem 5.12, if $M < N$, then the transformation defined by A cannot possibly be injective. Indeed, when A has more columns than rows, $RREF(A)$ must have columns that do not contain leading 1s. Each such column produces nonzero vectors in $Null(A)$.

For a transformation to be both surjective and injective, we must have $M = N$, meaning that A is a square matrix. Moreover, every column of $RREF(A)$ must contain a leading 1. In other words, $RREF(A) = \mathcal{I}_N$, so that A is invertible. A transformation that is both surjective and injective is said to be **bijective**. In this context, then, a linear transformation is bijective if, and only if, the matrix A that defines it is invertible.

5.6 Matrix Norms

The norm of a vector $\mathbf{x}$, in $\mathbb{R}^N$, is $||\mathbf{x}|| = \left(\mathbf{x}^T\mathbf{x}\right)^{1/2}$. Geometrically, $||\mathbf{x}||$ is the length of the long diagonal of a particular N-dimensional box, as indicated in Figure 1.1. We now ask, "What is the norm of a matrix?" It turns out there is more than one useful answer.

Every $M \times N$ matrix A can be viewed as a vector in $\mathbb{R}^{MN}$ whose coordinates have been rearranged to form N columns side-by side instead of being stored in one long column. In this view, we compute the norm of the matrix by taking the square root of the sum of the squares of all the entries of the matrix. This is called the **Frobenius norm** of the matrix.

Alternatively, we may view an $M \times N$ matrix A as representing the linear transformation $\mathbf{x} \mapsto A\mathbf{x}$. From this point of view, we compare the size of each output vector $A\mathbf{x}$ to the norm of the input vector $\mathbf{x}$ that produced it. That is, for each $\mathbf{x}$ in $\mathbb{R}^N$, consider the ratio $||A\mathbf{x}||/||\mathbf{x}||$. Of course, this ratio does not make sense if $\mathbf{x} = \mathbf{0}$. It is a result of advanced calculus that this ratio will achieve its maximum at some nonzero vector $\mathbf{x}$. This maximum ratio is the largest factor by which the length of any input vector is rescaled when we multiply by A. This notion of the size of a matrix is called the **operator norm**.

Definition 5.16 Let A be an $M \times N$ matrix.

(i) The **Frobenius norm** of A, denoted $||A||_F$, is given by

$$||A||_F = \left(\sum_{j=1}^{N} \sum_{i=1}^{M} (a_{i,\,j})^2 \right)^{1/2} . \tag{5.18}$$

This is the square root of the sum of the squares of all entries of A.

(ii) The **operator norm** of A, denoted $||A||_{op}$, is given by

$$||A||_{op} = \max \left\{ ||A\mathbf{x}||/||\mathbf{x}|| \; : \; \mathbf{x} \neq \mathbf{0} \text{ in } \mathbb{R}^N \right\} . \tag{5.19}$$

Both the Frobenius norm and the operator norm satisfy the properties of the vector norm discussed in the Remark following Formula 1.6 as well as the triangle inequality (1.9). In addition, both of these norms satisfy the following properties.

Proposition 5.17 *Both the Frobenius norm and the operator norm satisfy the following:*

(i) $||A^T|| = ||A||$, *for all matrices* A.

(ii) For all suitable matrices A *and* B,

$$||AB|| \leq ||A|| \cdot ||B|| . \tag{5.20}$$

Proof The statement (i) is immediate for the Frobenius norm, since A and A^T have exactly the same entries, just with a different arrangement. For the operator norm, we have a bit more work to do. Assuming that A is an $M \times N$ matrix, then A^T is $N \times M$. Let $\mathbf{y}$ be a nonzero vector in $\mathbb{R}^M$. We will show that $||A^T\mathbf{y}|| \leq ||A||_{op} \cdot ||\mathbf{y}||$. It then follows from the definition of the operator norm that $||A^T||_{op} \leq ||A||_{op}$. Then, because $A = (A^T)^T$, it will follow that $||A||_{op} \leq ||A^T||_{op}$ and, so, $||A^T||_{op} = ||A||_{op}$, as claimed.

We need to show that $||A^T\mathbf{y}|| \leq ||A||_{op} \cdot ||\mathbf{y}||$ for $\mathbf{y} \neq \mathbf{0}$ in $\mathbb{R}^M$. This inequality is immediately true if $A^T\mathbf{y} = \mathbf{0}$; so we'll assume that $A^T\mathbf{y} \neq \mathbf{0}$. Then we get

$$\begin{aligned} ||A^T\mathbf{y}||^2 &= (A^T\mathbf{y})^T(A^T\mathbf{y}) \\ &= (AA^T\mathbf{y})^T\mathbf{y} \\ &\leq ||AA^T\mathbf{y}|| \cdot ||\mathbf{y}|| \text{ (by the Cauchy–Schwarz inequality (1.20))} \\ &\leq ||A||_{op} \cdot ||A^T\mathbf{y}|| \cdot ||\mathbf{y}|| \text{ (by the definition of } ||A||_{op}) . \end{aligned}$$

Dividing both sides by $||A^T\mathbf{y}|| \neq 0$ yields the inequality

$$||A^T\mathbf{y}|| \leq ||A||_{op} \cdot ||\mathbf{y}|| \, .$$

This holds for all nonzero $\mathbf{y}$, so $||A^T||_{op} \leq ||A||_{op}$, as claimed.

To prove inequality (ii) for the Frobenius norm, take $A \in \mathbb{M}_{M\times N}(\mathbb{R})$ and $B \in \mathbb{M}_{N\times K}(\mathbb{R})$. The entry in row i and column j of the matrix product AB is given by $(AB)_{i,\,j} = \sum_{l=1}^{N} a_{i,\,l} b_{l,\,j}$. This is the inner product of the ith row vector of A and the jth column vector of B. By the Cauchy–Schwarz inequality,

$$(AB)_{ij}^2 \leq \left(\sum_{l=1}^{N} (a_{il})^2\right) \cdot \left(\sum_{l=1}^{N} (b_{lj})^2\right) .$$

For each fixed value of i, summing up both sides of this last inequality over all j gives

$$\sum_{j=1}^{K} (AB)_{i,\,j}^2 \leq \left(\sum_{l=1}^{N} (a_{,})^2\right) \cdot \sum_{j=1}^{K} \left(\sum_{l=1}^{N} (b_{l,\,j})^2\right) = \left(\sum_{l=1}^{N} (a_{i,\,l})^2\right) \cdot ||B||_F^2 \, .$$

Summing up these inequalities over all i yields the result

$$||AB||_F^2 \leq \left(\sum_{i=1}^{M} \left(\sum_{l=1}^{N} (a_{i,\,l})^2\right)\right) \cdot ||B||_F^2 = ||A||_F^2 \cdot ||B||_F^2 \, ,$$

as in (5.20).

For the operator norm, it follows from formula (5.19) that, for all $\mathbf{x}$,

$$||(AB)\mathbf{x}|| \leq ||A||_{op} \cdot ||B\mathbf{x}|| \leq ||A||_{op} \cdot ||B||_{op} \cdot ||\mathbf{x}|| \, .$$

Thus,

$$||AB||_{op} = \max\left\{||(AB)\mathbf{x}||/||\mathbf{x}|| \,:\, \mathbf{x} \neq \mathbf{0}\right\} \leq ||A||_{op} \cdot ||B||_{op} \, ,$$

as claimed in (5.20). □

Remark Unlike with the Frobenius norm, there is no simple formula for computing the operator norm of a matrix directly from the matrix entries. For this reason, it is often easier to use the Frobenius norm in computational applications.

Remark As with the vector norm, there are other reasonable notions of the norm of a matrix, for example, summing up the absolute values of all the entries. We will not pursue this further here.

Remark There is another way to compute the Frobenius norm. Suppose A is an $M \times N$ matrix and look at the $N \times N$ symmetric matrix $A^T A$. The diagonal entries of $A^T A$ are $\sum_{i=1}^{M} (a_{i,j})^2$ for $j = 1, \ldots, N$. It follows that the sum of the diagonal entries of $A^T A$ coincides with the square of the Frobenius norm of A. The sum of the diagonal entries of a square matrix is called the *trace* of the matrix. Thus, the Frobenius norm of a matrix A is equal to the square root of the **trace** of $A^T A$:

$$||A||_F = \sqrt{trace(A^T A)}. \tag{5.21}$$

5.7 Exercises

1. Express the vector $\begin{bmatrix} 5 \\ 30 \\ -21 \end{bmatrix}$ as a sum of numerical multiples of the column vectors of the matrix $A = \begin{bmatrix} 5 & 1 & 5 \\ 2 & 5 & -3 \\ -2 & -3 & 4 \end{bmatrix}$.

2. Let $A = \begin{bmatrix} 3 & -6 & -2 \\ 2 & -4 & 1 \\ 1 & -2 & -2 \end{bmatrix}$. Determine whether or not each given vector is in $Col(A)$. If it is, write it as a linear combination of the column vectors of A.

$$(a) \begin{bmatrix} 1 \\ 17 \\ -9 \end{bmatrix}; \ (b) \begin{bmatrix} 1 \\ 1 \\ 1 \end{bmatrix}; \ (c) \begin{bmatrix} 7 \\ 7 \\ 1 \end{bmatrix}; \ (d) \begin{bmatrix} 5 \\ 1 \\ 3 \end{bmatrix}; \ (e) \begin{bmatrix} 5 \\ 3 \\ 1 \end{bmatrix}.$$

3. Let $A = \begin{bmatrix} 3 & 1 & -3 \\ 1 & 1 & 1 \\ 5 & 6 & 8 \end{bmatrix}$. Determine whether or not each given vector is in the column space $Col(A)$. If it is, write it as a linear combination of the column vectors of A.

$$(a) \begin{bmatrix} 1 \\ 1 \\ 6 \end{bmatrix}; \ (b) \begin{bmatrix} 5 \\ 1 \\ 4 \end{bmatrix}; \ (c) \begin{bmatrix} 2 \\ 1 \\ 5 \end{bmatrix}; \ (d) \begin{bmatrix} 8 \\ 2 \\ 9 \end{bmatrix}; \ (e) \begin{bmatrix} 5 \\ 1 \\ 3 \end{bmatrix}.$$

4. For each given matrix A, find an equation, or equations, that the coordinates of the vector **b** must satisfy so that **b** belongs to the column space $Col(A)$.

(a) $A = \begin{bmatrix} 1 & 3 & 3 \\ 2 & 5 & 4 \\ 2 & 7 & 8 \end{bmatrix}$;

(b) $A = \begin{bmatrix} 2 & 5 & 12 \\ 3 & 1 & 5 \\ 5 & 8 & 21 \end{bmatrix}$;

(c) $A = \begin{bmatrix} 4 & -2 & -3 & 1 \\ 2 & -2 & -5 & 5 \\ 4 & 1 & 2 & -7 \\ 3 & 0 & 1 & -4 \end{bmatrix}$.

5. For each given inconsistent system $A\mathbf{x} = \mathbf{b}$, do the following:

   (i) Compute $A^T A$ and $A^T \mathbf{b}$, as in (5.4).
   (ii) Compute the inverse $\left(A^T A\right)^{-1}$.
   (iii) Compute the least squares solution $\widehat{\mathbf{x}} = \left(A^T A\right)^{-1} A^T \mathbf{b}$, as in (5.5).
   (iv) Compute the projection $\text{proj}_{Col(A)}\mathbf{b} = A\widehat{\mathbf{x}}$, as in (5.6).
   (v) Compute $||A\widehat{\mathbf{x}} - \mathbf{b}||^2$, the sum of square errors.

(a) $\begin{bmatrix} -1 & 2 \\ 2 & -3 \\ -1 & 3 \end{bmatrix} \begin{bmatrix} x_1 \\ x_2 \end{bmatrix} = \begin{bmatrix} 4 \\ 1 \\ 2 \end{bmatrix}$;

(b) $\begin{bmatrix} 2 & 4 \\ 1 & 3 \\ 5 & 8 \end{bmatrix} \begin{bmatrix} x_1 \\ x_2 \end{bmatrix} = \begin{bmatrix} 3 \\ -1 \\ 1 \end{bmatrix}$;

(c) $\begin{bmatrix} 1 & 2 \\ -2 & 3 \\ 7 & 4 \end{bmatrix} \begin{bmatrix} x_1 \\ x_2 \end{bmatrix} = \begin{bmatrix} 3 \\ -5 \\ 1 \end{bmatrix}$;

(d) $\begin{bmatrix} 2 & 1 \\ 8 & 3 \\ 3 & 2 \end{bmatrix} \begin{bmatrix} x_1 \\ x_2 \end{bmatrix} = \begin{bmatrix} 4 \\ -2 \\ 1 \end{bmatrix}$.

6. For each given inconsistent system $A\mathbf{x} = \mathbf{b}$, do the following:

   (i) Compute $A^T A$ and $A^T \mathbf{b}$, as in (5.4).
   (ii) Compute the inverse $\left(A^T A\right)^{-1}$.
   (iii) Compute the least squares solution $\widehat{\mathbf{x}} = \left(A^T A\right)^{-1} A^T \mathbf{b}$, as in (5.5).
   (iv) Compute the projection $\text{proj}_{Col(A)}\mathbf{b} = A\widehat{\mathbf{x}}$, as in (5.6).
   (v) Compute $||A\widehat{\mathbf{x}} - \mathbf{b}||^2$, the sum of square errors.

(a) $\begin{bmatrix} 1 & 2 & 3 \\ -4 & -6 & -1 \\ -3 & -5 & -4 \\ -3 & -5 & -5 \end{bmatrix} \begin{bmatrix} x_1 \\ x_2 \\ x_3 \end{bmatrix} = \begin{bmatrix} 3 \\ 2 \\ 1 \\ 2 \end{bmatrix}$;

(b) $\begin{bmatrix} 2 & 3 & 4 \\ -2 & 0 & -2 \\ -5 & 1 & -3 \\ 0 & 1 & 1 \end{bmatrix} \begin{bmatrix} x_1 \\ x_2 \\ x_3 \end{bmatrix} = \begin{bmatrix} 3 \\ 2 \\ 1 \\ 2 \end{bmatrix}$;

(c) $\begin{bmatrix} 2 & 4 & 3 \\ -2 & 1 & 0 \\ -5 & 2 & 1 \\ 1 & 1 & 1 \end{bmatrix} \begin{bmatrix} x_1 \\ x_2 \\ x_3 \end{bmatrix} = \begin{bmatrix} 2 \\ -1 \\ 3 \\ 1 \end{bmatrix}$;

(d) $\begin{bmatrix} 4 & 5 & -2 \\ 1 & 0 & 2 \\ 3 & 2 & 2 \\ 5 & 3 & 3 \end{bmatrix} \begin{bmatrix} x_1 \\ x_2 \\ x_3 \end{bmatrix} = \begin{bmatrix} 4 \\ 1 \\ 2 \\ 1 \end{bmatrix}$;

(e) $\begin{bmatrix} 2 & 1 & 5 \\ 4 & 3 & 8 \\ -1 & 2 & -7 \\ -2 & -7 & 6 \\ 2 & 3 & 1 \end{bmatrix} \begin{bmatrix} x_1 \\ x_2 \\ x_3 \end{bmatrix} = \begin{bmatrix} 1 \\ 3 \\ -1 \\ 2 \\ 2 \end{bmatrix}$.

7. Take $A = \begin{bmatrix} 1 & 1 & 0 \\ 1 & 1 & 0 \\ 1 & 0 & 1 \\ 1 & 0 & 1 \end{bmatrix}$ and $\mathbf{b} = \begin{bmatrix} 1 \\ 3 \\ 8 \\ 2 \end{bmatrix}$.

(a) Show that the system $A\mathbf{x} = \mathbf{b}$ does not have a solution.

(b) Show that the vector $\mathbf{u} = \begin{bmatrix} -1 \\ 1 \\ 1 \end{bmatrix}$ satisfies $A\mathbf{u} = \mathbf{0}$.

(c) Compute $A^T A$ and $\widehat{\mathbf{b}} = A^T \mathbf{b}$.

(d) Show that, for every real number t, the vector

$$\mathbf{x} = \begin{bmatrix} 5 \\ -3 \\ 0 \end{bmatrix} + t \cdot \mathbf{u}$$

satisfies the normal equation $A^T A\mathbf{x} = A^T \mathbf{b}$.

(e) Show that, for all $\mathbf{x}$ of the form in part (d), we have

$$A\mathbf{x} = \begin{bmatrix} 2 \\ 2 \\ 5 \\ 5 \end{bmatrix}.$$

This is the unique vector in $Col(A)$ that is closest to $\mathbf{b}$.

(f) Compute $||A\mathbf{x} - \mathbf{b}||^2$, where $\mathbf{x}$ has the form in part (d).

(g) Determine the value of t for which the corresponding vector in part (d) has the minimum norm. Compute that minimum norm.

8. Take $A = \begin{bmatrix} 1 & 2 & 0 \\ 0 & -2 & 2 \\ 5 & 6 & 4 \\ 3 & 8 & -2 \end{bmatrix}$ and $\mathbf{y} = \begin{bmatrix} 1 \\ 1 \\ 3 \\ 3 \end{bmatrix}$.

(a) Show that the system $A\mathbf{x} = \mathbf{y}$ does not have a solution.

(b) Show that the vector $\mathbf{u} = \begin{bmatrix} -2 \\ 1 \\ 1 \end{bmatrix}$ satisfies $A\mathbf{u} = \mathbf{0}$.

(c) Compute $A^T A$ and $\widehat{\mathbf{y}} = A^T \mathbf{y}$.

(d) Show that, for every real number t, the vector

$$\mathbf{x} = \begin{bmatrix} 87/161 \\ 5/46 \\ 0 \end{bmatrix} + t \cdot \mathbf{u}$$

satisfies the *normal equation* $A^T A\mathbf{x} = A^T \mathbf{y}$.

(e) Compute $||A\mathbf{x} - \mathbf{y}||^2$, where $\mathbf{x}$ has the form in part (d).

(f) Determine the value of t for which the corresponding vector in part (d) has the minimum norm. Compute that minimum norm.

9. For each set of paired measurements, do the following:

(i) Create a scatter plot of the data.

(ii) Does the scatter plot suggest a possible linear relationship between the variables? Explain.

(iii) Find the equation of the least squares regression line. Graph the line and the data in the same plot picture.

(a)

i	1	2	3	4	5	6	7	8	9	10	11
x_i	7.5	9	15	10	6	7	13	5	8	14.5	10.5
y_i	36	16	10	22	12	14	8	4	24	15	40

(b)

i	1	2	3	4	5	6	7	8
x_i	30	20	25	15	35	40	37	27
y_i	3	1.5	1.5	1	4	6	5.5	3.5

(c)

i	1	2	3	4	5	6	7	8	9	10
x_i	77	95	30	45	85	50	65	60	63	82
y_i	1.5	4.0	0.5	1.4	2.0	0.8	2.5	2.0	1.7	2.8

10. The table shows a set of paired measurements (x_i, y_i).

i	x_i	y_i
1	3	4
2	4	7
3	8	7
4	12	11
5	17	10

(a) For the line L_1 with equation $y = 3 + 0.75x$, compute the sum of square errors $\sum_{i=1}^{5} (y_i - (3 + 0.75x_i))^2$.
(b) For the line L_2 with equation $y = 3 + 0.5x$, compute the sum of square errors $\sum_{i=1}^{5} (y_i - (3 + 0.5x_i))^2$.
(c) Which of the lines L_1 and L_2 provides a better fit of the data?
(d) Find the equation of the line that minimizes the sum of square errors.

11. The share prices of nine growth stocks were measured in January 1981 (x_i) and January 1982 (y_i), as shown in the table. Figure 5.4 shows a plot of the data along with two lines L_1 and L_2 that seem to fit the data reasonably well.

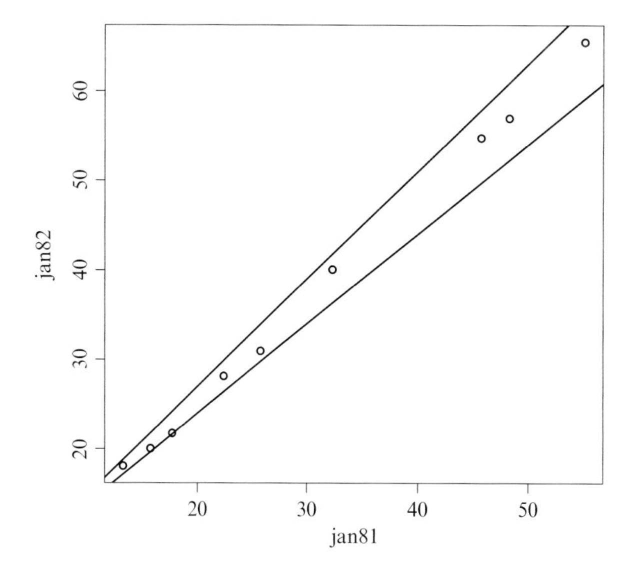

Fig. 5.4 A scatter plot of share prices of nine stocks from January 1981 and January 1982, along with two candidates for the regression line

i	x_i	y_i
	share price \$	share price \$
stock	Jan. 1981	Jan. 1982
1	22.40	28.15
2	45.70	54.75
3	15.65	20.05
4	48.25	56.95
5	32.31	40.06
6	55.14	65.45
7	17.65	21.75
8	25.78	30.95
9	13.17	18.10

(a) For the line L_1 with equation $y = 4 + x$, compute the sum of square errors $\sum_{i=1}^{9} (y_i - (4 + x_i))^2$.
(b) For the line L_2 with equation $y = 3 + 1.2x$, compute the sum of square errors $\sum_{i=1}^{9} (y_i - (3 + 1.2x_i))^2$.
(c) Which of the lines L_1 and L_2 provides a better fit of the data?
(d) Find the equation of the line that minimizes the sum of square errors.

12. In this exercise, we use multiple regression to examine the effect of team home runs (HR) and team earned run average (ERA) on team wins in Major League Baseball. The data used here are from the 2021 regular season for six teams in the National League: Washington Nationals (WAS), Philadelphia Phillies (PHL), Milwaukee Brewers (MIL), Cincinnati Reds (CIN), Los Angeles Dodgers (LAD), and Colorado Rockies (COL). The data are shown in the table.

Team:	WAS	PHL	MIL	CIN	LAD	COL
HR:	182	198	194	222	237	182
ERA:	4.80	4.39	3.50	4.40	3.01	4.82
Wins:	65	82	95	83	106	74

(a) Create the 6×3 matrix A that has all 1s in the first column, the HR data in the second column, and the ERA data in the third column.
(b) Referring to the normal equation (5.4), compute the matrix $A^T A$ and the vector $A^T \mathbf{w}$ where $\mathbf{w}$ is the vector of team wins data.
(c) Solve the normal equation, as in (5.9), to find the regression parameters

$$\widehat{\beta} = \begin{bmatrix} \beta_0 \\ \beta_1 \\ \beta_2 \end{bmatrix} = \left(A^T A\right)^{-1} A^T \mathbf{w}\,.$$

(Use a computer!)

(d) Explain the meaning of the slope parameters β_1 and β_2 in the context of this example.
(e) The vector $A\widehat{\beta}$ shows the number of wins predicted by the model for each team. Compute this vector, rounded off to the nearest whole number. Then compute the difference $A\widehat{\beta} - \mathbf{w}$ to see how close the predictions are to the actual team performances.
(f) Compute the sum of square errors $SSE = ||A\widehat{\beta} - \mathbf{w}||^2$.
(g) Compute the value R^2, as in (5.10). Remember that the closer this is to 1, the better the model explains the data.

13. In #12, we computed a least squares model for predicting team wins based on team home runs and team ERA. In this exercise, we apply the model to six other teams to see how well our predictions turn out. Here are the relevant data from the 2021 regular season for six other teams in the National League: Atlanta Braves (ATL), Miami Marlins (MIA), St. Louis Cardinals (STL), Pittsburgh Pirates (PIT), San Francisco Giants (SFG), and San Diego Padres (SDP).

Team:	ATL	MIA	STL	PIT	SFG	SDP
HR:	239	158	198	124	241	180
ERA:	3.88	3.96	3.98	5.08	3.24	4.10
Wins:	88	67	90	61	107	79

(a) Create the 6×3 matrix B that has all 1s in the first column, the HR data in the second column, and the ERA data in the third column.
(b) Compute the vector $B\widehat{\beta}$ rounded off to the nearest whole number. This shows the number of wins predicted by the model for each team.

$$\text{Note: } \widehat{\beta} = \begin{bmatrix} \beta_0 \\ \beta_1 \\ \beta_2 \end{bmatrix} \approx \begin{bmatrix} 129.177 \\ 0.119 \\ -16.657 \end{bmatrix} \text{ from the previous exercise.}$$

(c) Finally, compute the difference between $B\widehat{\beta}$ and the actual wins to see how close the predictions are to the actual team performances. Which team underachieved the most when compared to the model?

14. Use least squares to find the quadratic curve $y = \beta_0 + \beta_1 x + \beta_2 x^2$ that best fits the given data. In each case, compute the value of R^2, and create a plot that shows the data and the quadratic in the same picture. Refer to Example 5.7 for guidance.

(a)

i	1	2	3	4	5
x_i	0	1	2	3	4
y_i	3	3.1	3.7	4.3	4.7

(b)

i	1	2	3	4	5
x_i	0	1	2	3	4
y_i	2.3	4.8	6.1	5.1	1.7

(c)

i	1	2	3	4	5	6
x_i	0	1	2	3	4	5
y_i	31	8	10	29	71	125

15. Use least squares to find the cubic curve that best fits the given data. In each case, compute the value of R^2 and create a plot that shows the data and the cubic curve in the same picture. Refer to Example 5.7 for guidance.

(a)

i	1	2	3	4	5
x_i	−1	0	1	2	3
y_i	3	5	5	3	7

(b)

i	1	2	3	4	5
x_i	0	1	2	3	4
y_i	2.5	1.8	5.9	7.8	2.3

(c)

i	1	2	3	4	5
x_i	0	1	2	3	4
y_i	31	53	65	60	4

16. For each of the four fundamental subspaces $Col(A)$, $Col(A^T)$, $Null(A)$, and $Null(A^T)$, find its dimension and a set of vectors that generates it. Also, compute the *Column–Row* factorization of A, as in Algorithm 5.13.

(a) $A = \begin{bmatrix} 3 & -6 & -2 \\ 2 & -4 & 1 \\ 1 & -2 & -2 \end{bmatrix}$;

(b) $A = \begin{bmatrix} 1 & 3 & 2 \\ 2 & 7 & 7 \\ 2 & 5 & 2 \end{bmatrix}$;

(c) $A = \begin{bmatrix} 3 & 1 & 1 & 6 \\ 1 & -2 & 5 & -5 \\ 4 & 1 & 2 & 7 \end{bmatrix}$.

(d) $A = \begin{bmatrix} 1 & 0 & 5 & 3 \\ 0 & 2 & 4 & -2 \\ 2 & 0 & 10 & 6 \end{bmatrix}$;

(e) $A = \begin{bmatrix} 1 & -3 & -7 & 7 & -4 \\ 3 & 4 & 5 & 1 & 5 \\ 2 & 5 & 8 & 0 & 3 \end{bmatrix}$;

(f) $A = \begin{bmatrix} 1 & 2 & 0 \\ 0 & -2 & 2 \\ 5 & 6 & 4 \\ 3 & 8 & -2 \end{bmatrix}$;

(g) $A = \begin{bmatrix} 4 & 5 & -2 & 7 \\ 1 & 0 & 2 & 3 \\ 3 & 2 & 2 & 7 \\ 5 & 3 & 4 & 12 \\ 5 & 8 & -6 & 7 \end{bmatrix}$.

17. *(Proof Problem)* Suppose $\widehat{\mathbf{x}}$ and $\widetilde{\mathbf{x}}$ are two solutions to the normal equation (5.4). That is, suppose

$$A^T A\widehat{\mathbf{x}} = A^T A\widetilde{\mathbf{x}} = A^T \mathbf{b}\,.$$

Prove that

$$A\widehat{\mathbf{x}} = A\widetilde{\mathbf{x}}$$

even if $A^T A$ is not invertible. (*Hint:* Show that $A\left(\widehat{\mathbf{x}} - \widetilde{\mathbf{x}}\right)$ is both in $Col(A)$ and orthogonal to $Col(A)$.)

18. *(Proof Problem)* Let $\mathcal{S}$ be a nonempty collection of vectors in $\mathbb{R}^N$. The orthogonal complement of $\mathcal{S}$, denoted by $\mathcal{S}^\perp$, is defined in Definition 5.8.

(a) Show that $\mathcal{S}^\perp$ is a subspace of $\mathbb{R}^N$.
(b) Show that $\mathcal{S}^\perp = (span(\mathcal{S}))^\perp$.

19. *(Proof Problem)* Let A be an $M \times N$ matrix.

(a) **Prove:** For every $\mathbf{x}$ in $\mathbb{R}^N$ and every $\mathbf{y}$ in $\mathbb{R}^M$,

$$(A\mathbf{x})^T\, \mathbf{y} = \mathbf{x}^T \left(A^T \mathbf{y}\right).$$

(b) Illustrate the proposition from part (a) using $A = \begin{bmatrix} 4 & 0 \\ -3 & 2 \\ 1 & 5 \end{bmatrix}$, $\mathbf{x} = \begin{bmatrix} x_1 \\ x_2 \end{bmatrix}$ in $\mathbb{R}^2$, and $\mathbf{y} = \begin{bmatrix} y_1 \\ y_2 \\ y_3 \end{bmatrix}$ in $\mathbb{R}^3$.

(c) Use the proposition from part (a) to **prove** that $Col(A)^\perp = Null(A^T)$ for every matrix A.

5.8 Project

Project 5.1 (Multiple Regression)
Part I This project explores multiple regression, along the lines of Example 5.6. This simplified version of [9, Example 8.5] examines the effect on the quantity of roses sold of the prices of both roses and carnations, a competing product. The data are quarterly for the years 1972–1975 in the Detroit metropolitan area. For each 3-month period, X_1 denotes the wholesale price of roses, in $ per dozen; X_2 is the wholesale price of carnations, in $ per dozen; and Y is the quantity of roses sold, in dozens.

Roses sold vs. price of roses and of carnations								
X_1	2.26	2.54	3.07	2.91	2.73	2.77	3.59	3.23
X_2	3.49	2.85	4.06	3.64	3.21	3.66	3.76	3.49
Y	11484	9348	8429	10079	9240	8862	6216	8253
X_1	2.6	2.89	3.77	3.64	2.82	2.96	4.24	3.69
X_2	3.13	3.2	3.65	3.6	2.94	3.12	3.58	3.53
Y	8038	7476	5911	7950	6134	5868	3160	5872

1. Create the matrix A that has all 1s in the first column, the X_1 data in the second column, and the X_2 values in the third column.
2. Form the vector $\mathbf{y}$ that contains the Y data.
3. Using the normal equation, find the vector $\widehat{\beta}$ that minimizes $||A\beta - \mathbf{y}||^2$, like so.

   (a) Compute $A^T A$ and $A^T \mathbf{y}$.
   (b) Compute $\widehat{\beta} = (A^T A)^{-1} A^T \mathbf{y}$.

4. Write down the model: $Y \approx \beta_1 + \beta_2 \cdot X_1 + \beta_3 \cdot X_2$.
5. Compute R^2, like so.

   (a) Compute $||A\widehat{\beta} - \mathbf{y}||^2$.
   (b) Compute $||\mathbf{y} - \bar{y}\mathbf{1}||^2$.
   (c) Then $R^2 = 1 - \left(||A\widehat{\beta} - \mathbf{y}||^2 / ||\mathbf{y} - \bar{y}\mathbf{1}||^2\right)$.

6. Interpret the model.

   (a) What do the coefficients of X_1 and X_2 signify?
   (b) What does a negative coefficient indicate in the context of the example?
   (c) How well does the model explain the output in terms of the input variables? Discuss/explain your answer.

Part II. Repeat Part I above for a different example of a multiple regression model. Your example must examine the relationship between a single output variable and at least two input variables.

Project 5.2 (Modeling a Logistic Population) Suppose we wish to model the spread of an infection through a population of people. Or, to lighten the tone, we can model the spread of a rumor. For simplicity, we assume that each person is either *infected*, meaning that they have heard the rumor, or *susceptible*, meaning that they have not yet heard the rumor. We further assume that everyone who is infected is actively interested in spreading the rumor by telling it to others and that everyone who is susceptible is available to have the rumor told to them. (There is no quarantining in either group.) Our final assumption is that the total population, of infected and susceptible combined, is constant.

The rate at which the rumor spreads is given by the number of susceptible people who become infected on any given day. For a person to become newly infected, they must interact with a person who is already infected. If only a small number of people are infected, then they will only be able to spread the rumor slowly. Similarly, if only a small number of people remain susceptible, then only a small number of people can become newly infected. A model that combines these features is to say that *the rate at which the rumor can spread is proportional to the product of two numbers: the number of infected people and the number of susceptible people.* If either of those numbers is small, then the product will be relatively small. This model is called the *logistic differential equation.* Symbolically, take P to be the total fixed population, and set $y = y(t)$ to be the number of people who are infected at time t. Therefore, the number of susceptible people at time t is $(P - y(t))$. The logistic differential equation looks like so

$$dy/dt = k \cdot y(t) \cdot (P - y(t))\,. \tag{5.22}$$

The value of the constant k is crucial to the model. A larger value of k means the rumor will spread faster. Our main interest here is in estimating the value of k in a given scenario based on observed data.

Without going into the details of how to integrate this differential equation and arrive at a formula for $y(t)$, suffice it to say that y must satisfy an equation of the form

$$\log\left(\frac{y}{P-y}\right) = \alpha + (kP)t\,. \tag{5.23}$$

Suppose we have observed values of y_i at various time points t_i. Formula (5.23) suggests the following matrix formulation of our problem:

$$\begin{bmatrix} 1 & t_1 \\ 1 & t_2 \\ \vdots & \vdots \\ 1 & t_N \end{bmatrix} \begin{bmatrix} \alpha \\ kP \end{bmatrix} = \begin{bmatrix} \log\left(y_1/(P-y_1)\right) \\ \log\left(y_2/(P-y_2)\right) \\ \vdots \\ \log\left(y_N/(P-y_N)\right) \end{bmatrix}. \tag{5.24}$$

This is exactly the setup for applying the least squares method for determining the values of α and kP that best fit the data.

1. A rumor spreads through a population of $P = 300$ people. Estimates are taken of the number of people who have heard the rumor on successive days. Here are the data.

day t_i	1	2	3	4	5	6	7	8	9
infected pop. y_i	10	20	40	50	105	165	240	255	270

(a) Plug this data in to model (5.24). Note that the right-hand side vector requires us to compute $\log(y_i/(P - y_i))$ for each i.
(b) Solve the normal equation. This will give us the preferred values of α and kP.
(c) In the same picture, plot the data points $(t_i,\ y_i)$ and the graph of the function

$$y(t) = \frac{P}{1 + e^{-(\alpha + kPt)}}.$$

This is the *sigmoid curve* that best fits the data in the least squares sense.

Chapter 6
Orthogonal Systems

6.1 Projections Revisited

A linear system $A\mathbf{x} = \mathbf{b}$ has a solution for $\mathbf{x}$ precisely when $\mathbf{b}$ is in the column space $Col(A)$. Otherwise, solving the *normal equation* $A^T A\widehat{\mathbf{x}} = A^T\mathbf{b}$ yields a vector $\widehat{\mathbf{x}}$ for which $A\widehat{\mathbf{x}}$ is the closest vector to $\mathbf{b}$ in $Col(A)$. Thus, $A\widehat{\mathbf{x}}$ is the *projection* of $\mathbf{b}$ into $Col(A)$. Here, we explore an alternative approach to finding this projection. The new approach efficiently solves the normal equation, one dimension at a time.

To get started, let $A = \left[\mathbf{a}_1 \middle| \mathbf{a}_2\right]$ be an $M \times 2$ matrix whose columns are nonzero, noncollinear vectors in $\mathbb{R}^M$. The column space of A is the plane in $\mathbb{R}^M$ generated by these two column vectors. Replacing $\mathbf{a}_2$ with $\widetilde{\mathbf{a}}_2 = \mathbf{a}_2 - \text{proj}_{\mathbf{a}_1}\mathbf{a}_2$ yields the same plane. The vector $\widetilde{\mathbf{a}}_2$ is orthogonal to $\mathbf{a}_1$. Now form the unit vectors $\mathbf{q}_1 = (1/||\mathbf{a}_1||)\mathbf{a}_1$ and $\mathbf{q}_2 = (1/||\widetilde{\mathbf{a}}_2||)\widetilde{\mathbf{a}}_2$. These unit vectors are orthogonal to each other and generate the same plane as $\mathbf{a}_1$ and $\mathbf{a}_2$. Thus, the $M \times 2$ matrix $Q = \left[\mathbf{q}_1 \middle| \mathbf{q}_2\right]$ has *the same column space* as A.

Example 6.1 Let $A = \begin{bmatrix} 1 & 2 \\ -2 & -3 \\ 1 & 3 \end{bmatrix}$. The column vectors are

$$\mathbf{a}_1 = \begin{bmatrix} 1 \\ -2 \\ 1 \end{bmatrix} \text{ and } \mathbf{a}_2 = \begin{bmatrix} 2 \\ -3 \\ 3 \end{bmatrix}.$$

Supplementary Information The online version contains supplementary material available at https://doi.org/10.1007/978-3-031-39562-8_6.

T. G. Feeman, *Applied Linear Algebra and Matrix Methods*,
Springer Undergraduate Texts in Mathematics and Technology,
https://doi.org/10.1007/978-3-031-39562-8_6

We have $||\mathbf{a}_1|| = \sqrt{6}$, and

$$\text{proj}_{\mathbf{a}_1}\mathbf{a}_2 = \frac{\mathbf{a}_1^T\mathbf{a}_2}{||\mathbf{a}_1||^2}\cdot\mathbf{a}_1 = \frac{11}{6}\cdot\begin{bmatrix}1\\-2\\1\end{bmatrix}.$$

This gives us

$$\widetilde{\mathbf{a}}_2 = \mathbf{a}_2 - \text{proj}_{\mathbf{a}_1}\mathbf{a}_2 = \begin{bmatrix}2\\-3\\3\end{bmatrix} - \frac{11}{6}\cdot\begin{bmatrix}1\\-2\\1\end{bmatrix} = \frac{1}{6}\cdot\begin{bmatrix}1\\4\\7\end{bmatrix}.$$

Finally, we compute the two orthogonal unit vectors

$$\mathbf{q}_1 = \frac{\mathbf{a}_1}{||\mathbf{a}_1||} = \frac{1}{\sqrt{6}}\cdot\begin{bmatrix}1\\-2\\1\end{bmatrix} \text{ and } \mathbf{q}_2 = \frac{\widetilde{\mathbf{a}}_2}{||\widetilde{\mathbf{a}}_2||} = \frac{1}{\sqrt{66}}\cdot\begin{bmatrix}1\\4\\7\end{bmatrix}.$$

The matrix

$$Q = \left[\,\mathbf{q}_1\middle|\mathbf{q}_2\,\right] = \begin{bmatrix}1/\sqrt{6} & 1/\sqrt{66}\\-2/\sqrt{6} & 4/\sqrt{66}\\1/\sqrt{6} & 7/\sqrt{66}\end{bmatrix}$$

has the same column space as A.

With $Col(A) = Col(Q)$, we compute the projection of an arbitrary vector $\mathbf{a}_3$ into $Col(Q)$. This has the form $Q\widehat{\mathbf{x}}$ where $\widehat{\mathbf{x}}$ satisfies the normal equation $Q^TQ\widehat{\mathbf{x}} = Q^T\mathbf{a}_3$, as in (5.4). By the construction of Q,

$$Q^TQ = \begin{bmatrix}\mathbf{q}_1^T\\\hline\mathbf{q}_2^T\end{bmatrix}\left[\,\mathbf{q}_1\middle|\mathbf{q}_2\,\right] = \begin{bmatrix}1&0\\0&1\end{bmatrix} = \mathcal{I}_2\,, \tag{6.1}$$

since $\mathbf{q}_1$ and $\mathbf{q}_2$ are orthogonal unit vectors. The normal equation now becomes simply $\widehat{\mathbf{x}} = Q^T\mathbf{a}_3$. The projection of $\mathbf{a}_3$ into the column space $Col(Q) = Col(A)$ is, therefore,

$$\text{proj}_{Col(Q)}\mathbf{a}_3 = Q\widehat{\mathbf{x}} = QQ^T\mathbf{a}_3\,. \tag{6.2}$$

In particular, **the matrix QQ^T implements projection into** $Col(A)$. Computationally,

$$QQ^T = \left[\,\mathbf{q}_1\middle|\mathbf{q}_2\,\right]\begin{bmatrix}\mathbf{q}_1^T\\\hline\mathbf{q}_2^T\end{bmatrix} = \mathbf{q}_1\mathbf{q}_1^T + \mathbf{q}_2\mathbf{q}_2^T\,, \tag{6.3}$$

and

$$QQ^T\mathbf{a}_3 = (\mathbf{q}_1^T\mathbf{a}_3)\mathbf{q}_1 + (\mathbf{q}_2^T\mathbf{a}_3)\mathbf{q}_2\,. \tag{6.4}$$

This extends formula (3.31) and shows that the projection of $\mathbf{a}_3$ into the plane generated by the column vectors of Q is equal to the sum of the projections of $\mathbf{a}_3$ along those vectors separately.

Example 6.2 Let $A = \begin{bmatrix} 1 & 2 \\ -2 & -3 \\ 1 & 3 \end{bmatrix}$, as in Example 6.1. We computed $Q = \begin{bmatrix} 1/\sqrt{6} & 1/\sqrt{66} \\ -2/\sqrt{6} & 4/\sqrt{66} \\ 1/\sqrt{6} & 7/\sqrt{66} \end{bmatrix}$ with the same column space as A. The columns of Q are orthogonal unit vectors, so $Q^TQ = \mathcal{I}_2$, as one can verify. According to (6.2), the matrix QQ^T implements projection into $Col(Q) = Col(A)$. In this example,

$$\begin{aligned} QQ^T &= \begin{bmatrix} 1/\sqrt{6} & 1/\sqrt{66} \\ -2/\sqrt{6} & 4/\sqrt{66} \\ 1/\sqrt{6} & 7/\sqrt{66} \end{bmatrix} \begin{bmatrix} 1/\sqrt{6} & -2/\sqrt{6} & 1/\sqrt{6} \\ 1/\sqrt{66} & 4/\sqrt{66} & 7/\sqrt{66} \end{bmatrix} \\ &= \frac{1}{11} \cdot \begin{bmatrix} 2 & -3 & 3 \\ -3 & 10 & 1 \\ 3 & 1 & 10 \end{bmatrix}. \end{aligned}$$

Now take $\mathbf{a}_3 = \begin{bmatrix} 4 \\ 1 \\ 2 \end{bmatrix}$, for instance. We compute

$$\text{proj}_{Col(Q)}\mathbf{a}_3 = QQ^T\mathbf{a}_3 = \frac{1}{11} \cdot \begin{bmatrix} 2 & -3 & 3 \\ -3 & 10 & 1 \\ 3 & 1 & 10 \end{bmatrix} \begin{bmatrix} 4 \\ 1 \\ 2 \end{bmatrix} = \begin{bmatrix} 1 \\ 0 \\ 3 \end{bmatrix}.$$

If we need more convincing that this is correct, observe that

$$\mathbf{a}_3 - QQ^T\mathbf{a}_3 = \begin{bmatrix} 3 \\ 1 \\ -1 \end{bmatrix}.$$

This is orthogonal to both column vectors of A. Thus, $QQ^T\mathbf{a}_3$ is the closest vector to $\mathbf{a}_3$ in $Col(Q) = Col(A)$.

Example 6.3 (Volume of a Parallelepiped) Consider the vectors

$$\mathbf{a}_1 = \begin{bmatrix} 1 \\ -2 \\ 1 \end{bmatrix}, \quad \mathbf{a}_2 = \begin{bmatrix} 2 \\ -3 \\ 3 \end{bmatrix}, \quad \text{and } \mathbf{a}_3 = \begin{bmatrix} 4 \\ 1 \\ 2 \end{bmatrix},$$

from Examples 6.1 and 6.2. These vectors generate a parallelepiped in $\mathbb{R}^3$. To compute the volume of this parallelepiped, use the parallelogram generated by $\mathbf{a}_1$ and $\mathbf{a}_2$ as the *base*. From equation (1.34), the area of the base is

$$\left(||\mathbf{a}_1||^2 \cdot ||\mathbf{a}_2||^2 - (\mathbf{a}_1^T \mathbf{a}_2)^2\right)^{1/2} = \left(6 \cdot 22 - (11)^2\right)^{1/2} = \sqrt{11}\,.$$

To get the altitude of the parallelepiped, we "drop a perpendicular" from the tip of $\mathbf{a}_3$ to the plane of the base. This is the vector $(\mathbf{a}_3 - \text{proj}_{Col(A)}\mathbf{a}_3)$, where A is the matrix with columns $\mathbf{a}_1$ and $\mathbf{a}_2$. Forming the matrix Q, as in Example 6.1, we have $\text{proj}_{Col(A)}\mathbf{a}_3 = QQ^T\mathbf{a}_3$. Thus, the altitude of the parallelepiped is

$$||\mathbf{a}_3 - QQ^T\mathbf{a}_3|| = \left\| \begin{bmatrix} 4 \\ 1 \\ 2 \end{bmatrix} - \begin{bmatrix} 1 \\ 0 \\ 3 \end{bmatrix} \right\| = \left\| \begin{bmatrix} 3 \\ 1 \\ -1 \end{bmatrix} \right\| = \sqrt{11}\,.$$

The volume of the parallelepiped is

$$\text{Volume} = (\sqrt{11}) \cdot (\sqrt{11}) = 11\,.$$

In general, three vectors $\mathbf{a}_1$, $\mathbf{a}_2$, and $\mathbf{a}_3$ in $\mathbb{R}^3$ that do not all lie in the same plane will determine a parallelepiped. Taking the parallelogram generated by $\mathbf{a}_1$ and $\mathbf{a}_2$ as the base, then

$$\text{the area of base} = \left(||\mathbf{a}_1||^2 \cdot ||\mathbf{a}_2||^2 - (\mathbf{a}_1^T \mathbf{a}_2)^2\right)^{1/2}.$$

(See (1.34).) The altitude of the parallelepiped is equal to

$$||\mathbf{a}_3 - \text{proj}_{Col(Q)}\mathbf{a}_3|| = ||\mathbf{a}_3 - QQ^T\mathbf{a}_3||\,,$$

where Q is a matrix whose columns are orthogonal unit vectors that generate the same plane as $\mathbf{a}_1$ and $\mathbf{a}_2$. The volume of the parallelepiped generated by $\mathbf{a}_1$, $\mathbf{a}_2$, and $\mathbf{a}_3$ is given by

$$\text{Volume} = \left\|\mathbf{a}_3 - QQ^T\mathbf{a}_3\right\| \cdot \left(||\mathbf{a}_1||^2 \cdot ||\mathbf{a}_2||^2 - (\mathbf{a}_1^T \mathbf{a}_2)^2\right)^{1/2}. \tag{6.5}$$

Do not memorize this formula. The main point right now is that $QQ^T\mathbf{a}_3$ represents the projection of $\mathbf{a}_3$ into the plane generated by $\mathbf{a}_1$ and $\mathbf{a}_2$. Then the norm of $\mathbf{a}_3 - QQ^T\mathbf{a}_3$ is the altitude of the parallelepiped. An alternative formulation for this volume is obtained in Project 6.1.

Remark In general, any three vectors in $\mathbb{R}^M$ that do not lie in the same plane will generate a parallelepiped in $\mathbb{R}^M$. This will have a volume as three-dimensional figure, but, when $M > 3$, technically the M-dimensional volume of this figure is 0. Just as a line has no area in $\mathbb{R}^2$ and a planar region has no volume in $\mathbb{R}^3$, so a three-dimensional figure does not take up any "space" in a higher-dimensional environment.

The matrix $Q = \left[\, \mathbf{q}_1 \middle| \mathbf{q}_2 \,\right]$ constructed above has two special properties that we discovered.

(i) $Q^TQ = \mathcal{I}$; and
(ii) $QQ^T = \mathbf{q}_1\mathbf{q}_1^T + \mathbf{q}_2\mathbf{q}_2^T$ is the matrix for *projection into* $Col(Q)$.

In fact, these properties extend to any $M \times N$ matrix

$$Q = \left[\, \mathbf{q}_1 \middle| \cdots \middle| \mathbf{q}_N \,\right]$$

whose columns are mutually orthogonal unit vectors in $\mathbb{R}^M$. (*Fine print:* We require $M \geq N$ for this to be possible, as we can't have more mutually orthogonal vectors than the dimension of the space where they live.)

For property (i), we note that Q^TQ is the $N \times N$ matrix whose $[i, j]$ entry is equal to

$$Q^TQ_{[i,\,j]} = \mathbf{q}_i^T\mathbf{q}_j = \begin{cases} 1, & \text{if } i = j, \text{ and} \\ 0, & \text{if } i \neq j. \end{cases} \tag{6.6}$$

Thus, $Q^TQ = \mathcal{I}_N$ for such a matrix Q.

For property (ii), matrix multiplication gives us

$$\begin{aligned} QQ^T &= \left[\, \mathbf{q}_1 \middle| \mathbf{q}_2 \middle| \cdots \middle| \mathbf{q}_N \,\right] \begin{bmatrix} \mathbf{q}_1^T \\ \hline \mathbf{q}_2^T \\ \hline \vdots \\ \hline \mathbf{q}_N^T \end{bmatrix} \\ &= \mathbf{q}_1\mathbf{q}_1^T + \mathbf{q}_2\mathbf{q}_2^T + \cdots + \mathbf{q}_n\mathbf{q}_N^T\,. \end{aligned} \tag{6.7}$$

For any vector $\mathbf{b}$ in $\mathbb{R}^M$, its projection into the column space $Col(Q)$ has the form $Q\widehat{\mathbf{x}}$, where $\widehat{\mathbf{x}}$ satisfies the normal equation $Q^TQ\widehat{\mathbf{x}} = Q^T\mathbf{b}$. This reduces to

$\widehat{\mathbf{x}} = Q^T \mathbf{b}$, since $Q^T Q = \mathcal{I}$. Therefore,

$$\begin{aligned}
\mathrm{proj}_{Col(Q)}\mathbf{b} &= Q\widehat{\mathbf{x}} \\
&= QQ^T\mathbf{b} && (6.8) \\
&= \left(\mathbf{q}_1\mathbf{q}_1^T + \mathbf{q}_2\mathbf{q}_2^T \cdots + \mathbf{q}_N\mathbf{q}_N^T\right)\mathbf{b} \\
&= (\mathbf{q}_1^T\mathbf{b})\mathbf{q}_1 + (\mathbf{q}_2^T\mathbf{b})\mathbf{q}_2 + \cdots + (\mathbf{q}_N^T\mathbf{b})\mathbf{q}_N && (6.9) \\
&= \mathrm{proj}_{\mathbf{q}_1}\mathbf{b} + \mathrm{proj}_{\mathbf{q}_2}\mathbf{b} + \cdots + \mathrm{proj}_{\mathbf{q}_N}\mathbf{b}\,.
\end{aligned}$$

In other words, the projection of $\mathbf{b}$ into $Col(Q)$ is equal to the sum of its projections along the individual column vectors $\mathbf{q}_i$.

Example 6.4 Set $Q = \left[\,\mathbf{q}_1 \middle| \mathbf{q}_2 \middle| \mathbf{q}_3\,\right]$, where

$$\mathbf{q}_1 = \begin{bmatrix} 1/2 \\ 1/2 \\ 1/2 \\ 1/2 \end{bmatrix},\ \mathbf{q}_2 = \begin{bmatrix} 1/2 \\ 1/2 \\ -1/2 \\ -1/2 \end{bmatrix},\ \text{and } \mathbf{q}_3 = \begin{bmatrix} 1/2 \\ -1/2 \\ 1/2 \\ -1/2 \end{bmatrix}.$$

These are mutually orthogonal unit vectors in $\mathbb{R}^4$. The matrix for projection into the column space $Col(Q)$ is

$$\begin{aligned}
QQ^T &= \begin{bmatrix} 1/2 & 1/2 & 1/2 \\ 1/2 & 1/2 & -1/2 \\ 1/2 & -1/2 & 1/2 \\ 1/2 & -1/2 & -1/2 \end{bmatrix} \cdot \begin{bmatrix} 1/2 & 1/2 & 1/ & 1/2 \\ 1/2 & 1/2 & -1/2 & -1/2 \\ 1/2 & -1/2 & 1/2 & -1/2 \end{bmatrix} \\
&= \frac{1}{4} \cdot \begin{bmatrix} 3 & 1 & 1 & -1 \\ 1 & 3 & -1 & 1 \\ 1 & -1 & 3 & 1 \\ -1 & 1 & 1 & 3 \end{bmatrix}.
\end{aligned}$$

Accordingly, the projection of a vector $\mathbf{y} = \left[\,y_1\ y_2\ y_3\ y_4\,\right]^T$ in $\mathbb{R}^4$ into $Col(Q)$ is

$$\mathrm{proj}_{Col(Q)}\mathbf{y} = QQ^T\mathbf{y} = \frac{1}{4} \cdot \begin{bmatrix} 3y_1 + y_2 + y_3 - y_4 \\ y_1 + 3y_2 - y_3 + y_4 \\ y_1 - y_2 + 3y_3 + y_4 \\ -y_1 + y_2 + y_3 + 3y_4 \end{bmatrix}.$$

Notice that

$$\mathbf{y} - QQ^T\mathbf{y} = \frac{1}{4} \cdot \begin{bmatrix} y_1 - y_2 - y_3 + y_4 \\ -y_1 + y_2 + y_3 - y_4 \\ -y_1 + y_2 + y_3 - y_4 \\ y_1 - y_2 - y_3 + y_4 \end{bmatrix} = (\mathbf{q}_4^T\mathbf{y})\mathbf{q}_4 \,,$$

where $\mathbf{q}_4 = \begin{bmatrix} 1/2 & -1/2 & -1/2 & 1/2 \end{bmatrix}^T$ is a unit vector orthogonal to the column vectors of Q. Hence, $\mathbf{y} - QQ^T\mathbf{y}$ is orthogonal to $Col(Q)$, as expected.

6.2 Building Orthogonal Sets

Now suppose we have any $M \times N$ matrix A and we wish to compute projections into the column space $Col(A)$. As in the $N = 2$ case above, we identify a set of *mutually orthogonal unit vectors* $\mathbf{q}_1, \mathbf{q}_2, \ldots, \mathbf{q}_N$ such that the column space of the matrix

$$Q = \left[\, \mathbf{q}_1 \middle| \mathbf{q}_2 \middle| \cdots \middle| \mathbf{q}_N \,\right]$$

is *the same as* that of A. Then, for every $\mathbf{y}$ in $\mathbb{R}^M$,

$$\text{proj}_{Col(A)}\mathbf{y} = \text{proj}_{Col(Q)}\mathbf{y} = QQ^T\mathbf{y} \,,$$

as in (6.8).

Before going further, we introduce some convenient terminology.

Definition 6.5

(i) A collection $\{\mathbf{v}_1, \mathbf{v}_2, \ldots, \mathbf{v}_N\}$ of nonzero vectors in $\mathbb{R}^M$ with the property that ${\mathbf{v}_i}^T\mathbf{v}_j = 0$ when $i \neq j$ is called an **orthogonal set** or an **orthogonal family**.

(ii) An orthogonal set of *unit vectors* in $\mathbb{R}^M$ is called an **orthonormal set** or an **orthonormal family**.

(iii) An orthonormal set of M vectors in $\mathbb{R}^M$ is called an **orthonormal basis** of $\mathbb{R}^M$.

(iv) An $M \times M$ matrix whose columns form an orthonormal basis of $\mathbb{R}^M$ is called an **orthogonal matrix**. Note that $V^T = V^{-1}$ for an orthogonal matrix. (See Exercise #16.)

At the start of Section 6.1, we converted a pair of vectors $\mathbf{a}_1$ and $\mathbf{a}_2$ into two orthogonal unit vectors, $\mathbf{q}_1$ and $\mathbf{q}_2$, that generate the same plane. The matrices $A_2 = \left[\, \mathbf{a}_1 \middle| \mathbf{a}_2 \,\right]$ and $Q_2 = \left[\, \mathbf{q}_1 \middle| \mathbf{q}_2 \,\right]$ have the same column space. The matrix $Q_2Q_2^T$ implements projection into this column space.

Now bring in a third vector $\mathbf{a}_3$. As in Example 6.3, the vector

$$\widetilde{\mathbf{a}}_3 = \mathbf{a}_3 - (\mathbf{q}_1^T\mathbf{a}_3)\mathbf{q}_1 - (\mathbf{q}_2^T\mathbf{a}_3)\mathbf{q}_2$$

gives the altitude of the parallelepiped generated by $\mathbf{a}_1$, $\mathbf{a}_2$, and $\mathbf{a}_3$. The corresponding unit vector

$$\mathbf{q}_3 = \widetilde{\mathbf{a}}_3/||\widetilde{\mathbf{a}}_3||$$

is orthogonal to $\mathbf{q}_1$ and $\mathbf{q}_2$. Every step in the construction of $\mathbf{q}_1$, $\mathbf{q}_2$, and $\mathbf{q}_3$ is reversible. Moreover, each successive $\mathbf{q}_j$ is in the span of $\mathbf{a}_1$ through $\mathbf{a}_j$. Hence, we can also recover $\mathbf{a}_1$, $\mathbf{a}_2$, and $\mathbf{a}_3$ from $\mathbf{q}_1$, $\mathbf{q}_2$, and $\mathbf{q}_3$ in such a way that the span remains the same at every step. This means that the matrices

$$A_3 = \left[\,\mathbf{a}_1\middle|\mathbf{a}_2\middle|\mathbf{a}_3\,\right] \text{ and } Q_3 = \left[\,\mathbf{q}_1\middle|\mathbf{q}_2\middle|\mathbf{q}_3\,\right]$$

have the same column space. The matrix $Q_3Q_3^T$ implements projection into this column space, as in (6.8).

Continuing like this, we can convert any set of vectors into an orthonormal set, preserving the column space at each step along the way. To describe the basic algorithm, suppose that the set of vectors we start with is linearly independent. This is not something we would actually check before proceeding, so we will also look at how the algorithm is affected when some of the vectors in the starting set are linear combinations of others.

Algorithm 6.6 Gram–Schmidt algorithm

- **Step 0.** Suppose $\{\mathbf{a}_1,\,\mathbf{a}_2,\,\ldots,\,\mathbf{a}_N\}$ is a set of nonzero vectors in $\mathbb{R}^M$, with $N \le M$.
- **Step 1.** Let $\mathbf{q}_1 = \mathbf{a}_1/||\mathbf{a}_1||$; thus, $\mathbf{q}_1$ is a unit vector.
- **Step 2.** Let $\widetilde{\mathbf{a}}_2 = \mathbf{a}_2 - \left(\mathbf{q}_1^T\mathbf{a}_2\right)\mathbf{q}_1$. Then create the unit vector

$$\mathbf{q}_2 = \widetilde{\mathbf{a}}_2/||\widetilde{\mathbf{a}}_2||\,.$$

  Fine print: If $\widetilde{\mathbf{a}}_2 = \mathbf{0}$, then $\mathbf{a}_2$ is a scalar multiple of $\mathbf{a}_1$. In this event, we drop $\mathbf{a}_2$ from consideration and treat $\mathbf{a}_3$ as the new $\mathbf{a}_2$.
- **Step** j. Having computed the orthonormal set $\{\,\mathbf{q}_1,\,\ldots,\,\mathbf{q}_{j-1}\,\}$, we compute the next vector $\mathbf{q}_j$ like so. Let

$$\widetilde{\mathbf{a}}_j = \mathbf{a}_j - \left(\left({\mathbf{q}_1}^T\mathbf{a}_j\right)\mathbf{q}_1 + \cdots + \left({\mathbf{q}_{j-1}}^T\mathbf{a}_j\right)\mathbf{q}_{j-1}\right)\,. \tag{6.10}$$

  Then define

$$\mathbf{q}_j = \widetilde{\mathbf{a}}_j/||\widetilde{\mathbf{a}}_j||\,. \tag{6.11}$$

Fine print: If $\widetilde{\mathbf{a}}_j = \mathbf{0}$, then $\mathbf{a}_j$ is a linear combination of $\mathbf{a}_1$ through $\mathbf{a}_{j-1}$. In this event, we drop $\mathbf{a}_j$ from consideration and treat $\mathbf{a}_{j+1}$ as the new $\mathbf{a}_j$.

- **Step N. Stop when we have computed $\mathbf{q}_N$.**

The computations (6.10) and (6.11) in Step j of the Gram–Schmidt algorithm are reversible. Also, each successive $\mathbf{q}_j$ is in the span of $\mathbf{a}_1$ through $\mathbf{a}_j$. This guarantees that the matrix

$$Q_j = \left[\mathbf{q}_1 \middle| \cdots \middle| \mathbf{q}_j\right]$$

formed at the end of step j has the same column space as the matrix $A_j = \left[\mathbf{a}_1 \middle| \cdots \middle| \mathbf{a}_j\right]$. The matrix $Q_j Q_j^T$ implements projection into that column space.

Remark The Gram–Schmidt algorithm works by adjusting one vector at a time, projecting each successive vector along the vectors that have already been adjusted. If, at any step along the way, one of the vectors $\mathbf{a}_j$ is a linear combination of the preceding vectors, then $\mathbf{a}_j$ will not contribute anything new to the column space. In place of the next unit vector $\mathbf{q}_j$, we will get $\mathbf{0}$. We can either include $\mathbf{0}$ as the jth column of Q, in which case $Q^T Q$ will not equal the identity matrix, or we can delete that column from Q, in which case Q and A will still have the same column space but will not have the same number of columns.

Example 6.7 The set

$$\left\{\mathbf{a}_1 = \left[1\ 1\ 0\right]^T,\ \mathbf{a}_2 = \left[0\ 1\ 1\right]^T,\ \mathbf{a}_3 = \left[1\ 0\ 1\right]^T\right\}$$

is a basis of $\mathbb{R}^3$. Apply the Gram–Schmidt algorithm to generate an orthonormal basis of $\mathbb{R}^3$.

- Compute $||\mathbf{a}_1|| = \sqrt{2}$; thus,

$$\mathbf{q}_1 = \mathbf{a}_1/||\mathbf{a}_1|| = \left[\tfrac{1}{\sqrt{2}}\ \tfrac{1}{\sqrt{2}}\ 0\right]^T.$$

- Next, compute $\mathbf{q}_1{}^T\mathbf{a}_2 = 1/\sqrt{2}$. Thus,

$$\begin{aligned}\widetilde{\mathbf{a}}_2 &= \mathbf{a}_2 - \left(\mathbf{q}_1{}^T\mathbf{a}_2\right)\mathbf{q}_1 \\ &= \left[0\ 1\ 1\right]^T - \frac{1}{\sqrt{2}}\left[\tfrac{1}{\sqrt{2}}\ \tfrac{1}{\sqrt{2}}\ 0\right]^T \\ &= \left[-\tfrac{1}{2}\ \tfrac{1}{2}\ 1\right]^T.\end{aligned}$$

Since $||\widetilde{\mathbf{a}}_2|| = \sqrt{6}/2$, it follows that

$$\mathbf{q}_2 = \widetilde{\mathbf{a}}_2/||\widetilde{\mathbf{a}}_2|| = \frac{2}{\sqrt{6}}\left[-\tfrac{1}{2}\ \tfrac{1}{2}\ 1\right]^T = \left[-\tfrac{1}{\sqrt{6}}\ \tfrac{1}{\sqrt{6}}\ \tfrac{2}{\sqrt{6}}\right]^T.$$

- Finally, compute ${\mathbf{q}_1}^T \mathbf{a}_3 = 1/\sqrt{2}$ and $\mathbf{q}_2^T \mathbf{a}_3 = 1/\sqrt{6}$. Thus,

$$\begin{aligned}\widetilde{\mathbf{a}}_3 &= \mathbf{a}_3 - \left({\mathbf{q}_1}^T \mathbf{a}_3\right) \mathbf{q}_1 - \left({\mathbf{q}_2}^T \mathbf{a}_3\right) \mathbf{q}_2 \\ &= \begin{bmatrix} 1 & 0 & 1 \end{bmatrix}^T - \begin{bmatrix} \frac{1}{2} & \frac{1}{2} & 0 \end{bmatrix}^T - \begin{bmatrix} -\frac{1}{6} & \frac{1}{6} & \frac{2}{6} \end{bmatrix}^T \\ &= \begin{bmatrix} \frac{2}{3} & -\frac{2}{3} & \frac{2}{3} \end{bmatrix}^T .\end{aligned}$$

Now $||\widetilde{\mathbf{a}}_3|| = 2/\sqrt{3}$. Thus,

$$\mathbf{q}_3 = \widetilde{\mathbf{a}}_3/||\widetilde{\mathbf{a}}_3|| = \frac{\sqrt{3}}{2}\begin{bmatrix} \frac{2}{3} & -\frac{2}{3} & \frac{2}{3} \end{bmatrix}^T = \begin{bmatrix} \frac{1}{\sqrt{3}} & -\frac{1}{\sqrt{3}} & \frac{1}{\sqrt{3}} \end{bmatrix}^T .$$

- The resulting orthonormal basis of $\mathbb{R}^3$ is

$$\left\{ \mathbf{q}_1 = \begin{bmatrix} 1/\sqrt{2} \\ 1/\sqrt{2} \\ 0 \end{bmatrix}, \mathbf{q}_2 = \begin{bmatrix} -1/\sqrt{6} \\ 1/\sqrt{6} \\ 2/\sqrt{6} \end{bmatrix}, \mathbf{q}_3 = \begin{bmatrix} 1/\sqrt{3} \\ -1/\sqrt{3} \\ 1/\sqrt{3} \end{bmatrix} \right\}.$$

It might seem as if this was a lot of work. So the good news is that a computer can be programmed to implement the Gram–Schmidt algorithm quickly and efficiently. Also, less computational effort is required to obtain the matrix QQ^T for projection into $Col(A)$ using this step-by-step approach than to compute $A\left(A^T A\right)^{-1} A^T$ as in formula (5.6), and we get more information along the way as a bonus.

6.3 QR Factorization

The Gram–Schmidt algorithm takes a matrix $A = \left[\mathbf{a}_1 \middle| \cdots \middle| \mathbf{a}_N\right]$ and converts it, one column at a time, into a matrix $Q = \left[\mathbf{q}_1 \middle| \cdots \middle| \mathbf{q}_N\right]$ that has the same column space as A and whose columns form an orthonormal set. Since the computations in each step of the algorithm are reversible, it is possible to recover A from Q. To see how, we revisit the Gram–Schmidt algorithm. But this time we do our bookkeeping.

In Step j of the algorithm, given in formulas (6.10) and (6.11), we compute the inner products ${\mathbf{q}_i}^T \mathbf{a}_j$, for $1 \le i \le (j-1)$, to form $\widetilde{\mathbf{a}}_j$. Then, we compute the norm $||\widetilde{\mathbf{a}}_j||$ to form the unit vector $\mathbf{q}_j$. Let us store these values as entries in column j of a new matrix. Specifically, form the $N \times N$ upper triangular matrix R whose $[i,\ j]$ entry is given by

$$R_{[i,\,j]} = \begin{cases} {\mathbf{q}_i}^T \mathbf{a}_j\,, & \text{if } i < j; \\ \|\widetilde{\mathbf{a}}_j\|\,, & \text{if } i = j; \text{ and} \\ 0\,, & \text{if } i > j. \end{cases} \tag{6.12}$$

It follows from (6.10) and (6.11) that the jth column of the product QR is precisely the vector $\mathbf{a}_j$ from our original collection. Indeed, multiplying Q by the jth column of R, we get

$$\begin{aligned} (QR)_{[*,\,j]} &= \left[\, \mathbf{q}_1 \middle| \mathbf{q}_2 \middle| \cdots \middle| \mathbf{q}_N \,\right] \begin{bmatrix} {\mathbf{q}_1}^T \mathbf{a}_j \\ \vdots \\ {\mathbf{q}_{j-1}}^T \mathbf{a}_j \\ \|\widetilde{\mathbf{a}}_j\| \\ 0 \\ \vdots \\ 0 \end{bmatrix} \\ &= \left(\mathbf{q}^T \mathbf{a}_j\right) \mathbf{q}_1 + \cdots + \left({\mathbf{q}_{j-1}}^T \mathbf{a}_j\right) \mathbf{q}_{j-1} + \|\widetilde{\mathbf{a}}_j\| \mathbf{q}_j \\ &= \mathbf{a}_j\,. \end{aligned} \tag{6.13}$$

In other words, $QR = \left[\, \mathbf{a}_1 \middle| \cdots \middle| \mathbf{a}_N \,\right] = A$. This is called the QR **factorization** of the matrix A. The hidden assumption is that the columns of A form a linearly independent set. Calculating the QR factorization of a matrix is equal in efficiency to applying the Gram–Schmidt algorithm, since the matrices Q and R are formed step-by-step as the algorithm progresses.

Example 6.8 Let $A = \left[\, \mathbf{a}_1 \middle| \mathbf{a}_2 \,\right]$ be an $M \times 2$ matrix whose columns are nonzero and noncollinear. In Section 6.1, we constructed the matrix $Q = \left[\, \mathbf{q}_1 \middle| \mathbf{q}_2 \,\right]$, with orthonormal column vectors, such that $Col(Q) = Col(A)$. The corresponding 2×2 matrix R is

$$R = \begin{bmatrix} \|\mathbf{a}_1\| & \mathbf{a}_1^T \mathbf{a}_2 / \|\mathbf{a}_1\| \\ 0 & \|\widetilde{\mathbf{a}}_2\| \end{bmatrix},$$

where $\widetilde{\mathbf{a}}_2 = \mathbf{a}_2 - \mathrm{proj}_{\mathbf{a}_1} \mathbf{a}_2$. Thus,

$$\begin{aligned} QR &= \left[\, \mathbf{a}_1 / \|\mathbf{a}_1\| \middle| \widetilde{\mathbf{a}}_2 / \|\widetilde{\mathbf{a}}_2\| \,\right] \begin{bmatrix} \|\mathbf{a}_1\| & \mathbf{a}_1^T \mathbf{a}_2 / \|\mathbf{a}_1\| \\ 0 & \|\widetilde{\mathbf{a}}_2\| \end{bmatrix} \\ &= \left[\, \mathbf{a}_1 + \mathbf{0} \middle| \mathrm{proj}_{\mathbf{a}_1} \mathbf{a}_2 + \widetilde{\mathbf{a}}_2 \,\right] \\ &= \left[\, \mathbf{a}_1 \middle| \mathbf{a}_2 \,\right] \\ &= A\,. \end{aligned} \tag{6.14}$$

Example 6.9 For $A = \begin{bmatrix} 1 & 0 & 1 \\ 1 & 1 & 0 \\ 0 & 1 & 1 \end{bmatrix}$, as in Example 6.7, we get

$$Q = \begin{bmatrix} 1/\sqrt{2} & -1/\sqrt{6} & 1/\sqrt{3} \\ 1/\sqrt{2} & 1/\sqrt{6} & -1/\sqrt{3} \\ 0 & 2/\sqrt{6} & 1/\sqrt{3} \end{bmatrix} \text{ and } R = \begin{bmatrix} \sqrt{2} & 1/\sqrt{2} & 1/\sqrt{2} \\ 0 & \sqrt{6}/2 & 1/\sqrt{6} \\ 0 & 0 & 2/\sqrt{3} \end{bmatrix}.$$

The columns of Q form an orthonormal basis of $\mathbb{R}^3$, R is upper triangular, and $A = QR$.

6.4 Least Squares with QR

Consider again the normal equation

$$A^T A\widehat{\mathbf{x}} = A^T \mathbf{b}\,, \tag{6.15}$$

associated with a linear system $A\mathbf{x} = \mathbf{b}$ that cannot be solved exactly. Using the QR factorization $A = QR$, we have

$$A^T A = (QR)^T (QR) = R^T (Q^T Q)R = R^T R\,,$$

since $Q^T Q = \mathcal{I}$. The normal equation now becomes

$$R^T R\widehat{\mathbf{x}} = A^T \mathbf{b}\,. \tag{6.16}$$

To solve this for $\widehat{\mathbf{x}}$, temporarily set $\mathbf{w} = R\widehat{\mathbf{x}}$. The matrix R^T is *lower triangular*, so the system $R^T \mathbf{w} = A^T \mathbf{b}$ can be solved for $\mathbf{w}$ using *forward substitution*. Then, because R is *upper triangular*, we use *backward substitution* to solve the system $R\widehat{\mathbf{x}} = \mathbf{w}$ for $\widehat{\mathbf{x}}$.

Notice that, while we must compute Q in order to know R, once we are finished, we may discard Q to save storage space on our computer.

Example 6.10 We can apply equation (6.16) to tackle simple linear regression, introduced in Section 3.5. The framework is that we have two random variables, X and Y, and a set of paired observations of them, (x_i, y_i), for $1 \le i \le M$. We wish to find values for two unknowns, α and β, in order to minimize the sum of squared errors

$$SSE = \sum_{i=1}^{M} (\alpha + \beta x_i - y_i)^2\,, \tag{6.17}$$

as in (3.19). The corresponding line $y = \alpha + \beta x$ is called the regression line, or line of best fit, for the data. Setting

$$A = \begin{bmatrix} 1 & x_1 \\ 1 & x_2 \\ \vdots & \vdots \\ 1 & x_M \end{bmatrix} \text{ and } \mathbf{y} = \begin{bmatrix} y_1 \\ y_2 \\ \vdots \\ y_M \end{bmatrix},$$

the normal equation is

$$A^T A \begin{bmatrix} \alpha \\ \beta \end{bmatrix} = A^T \mathbf{y}\,. \tag{6.18}$$

The first column of A is the all 1s vector $\mathbf{1}_M$ in $\mathbb{R}^M$. Thus, $R_{1,1} = ||\mathbf{1}|| = \sqrt{M}$ and $\mathbf{q}_1 = (1/\sqrt{M}) \cdot \mathbf{1}_M$. Next, let $\mathbf{x}$ denote the second column of A. The $[1, 2]$ entry of R is

$$\mathbf{q}_1^T \mathbf{x} = (1/\sqrt{M}) \cdot (x_1 + \cdots + x_M) = \sqrt{M} \cdot \bar{x}\,,$$

where $\bar{x} = (x_1 + \cdots + x_M)/M$ is the average value of the observations of X. Finally, for the $[2, 2]$ entry of R, compute the norm

$$\begin{aligned} ||\mathbf{x} - (\mathbf{q}_1^T \mathbf{x})\mathbf{q}_1|| &= ||\mathbf{x} - \bar{x} \cdot \mathbf{1}_M|| \\ &= \sqrt{(x_1 - \bar{x})^2 + \cdots + (x_M - \bar{x})^2}\,. \end{aligned}$$

Denote this value by σ. That is,

$$\sigma = \sqrt{(x_1 - \bar{x})^2 + \cdots + (x_M - \bar{x})^2} = \sqrt{\sum (x_i - \bar{x})^2}\,. \tag{6.19}$$

The matrix R is

$$R = \begin{bmatrix} \sqrt{M} & \sqrt{M} \cdot \bar{x} \\ 0 & \sigma \end{bmatrix}. \tag{6.20}$$

The normal equation becomes $R^T R \begin{bmatrix} \alpha \\ \beta \end{bmatrix} = A^T \mathbf{y}$, as in (6.16). To solve this, first compute

$$A^T \mathbf{y} = \begin{bmatrix} y_1 + y_2 + \cdots + y_M \\ x_1 y_1 + x_2 y_2 + \cdots + x_M y_M \end{bmatrix} = \begin{bmatrix} \sum y_i \\ \sum x_i y_i \end{bmatrix}.$$

For convenience, write $A^T\mathbf{y} = \widehat{\mathbf{y}} = \begin{bmatrix} \widehat{y}_1 \\ \widehat{y}_2 \end{bmatrix}$. Next, with $\mathbf{w} = \begin{bmatrix} w_1 \\ w_2 \end{bmatrix} = R\begin{bmatrix} \alpha \\ \beta \end{bmatrix}$, we solve $R^T\mathbf{w} = \widehat{\mathbf{y}}$ by forward substitution. This gives $w_1 = \widehat{y}_1/\sqrt{M}$ and $w_2 = (\widehat{y}_2 - \sqrt{M}\cdot\bar{x}\cdot w_1)/\sigma = (\widehat{y}_2 - \bar{x}\cdot\widehat{y}_1)/\sigma$. Using backward substitution to solve $R\begin{bmatrix} \alpha \\ \beta \end{bmatrix} = \mathbf{w}$ yields the least squares solution

$$\beta = w_2/\sigma = \frac{(\widehat{y}_2 - \bar{x}\cdot\widehat{y}_1)}{\sigma^2}\text{; and} \tag{6.21}$$

$$\alpha = \frac{(w_1 - \sqrt{M}\cdot\bar{x}\beta)}{\sqrt{M}} = \left(\frac{1}{M} + \frac{\bar{x}^2}{\sigma^2}\right)\cdot\widehat{y}_1 - \frac{\bar{x}\cdot\widehat{y}_2}{\sigma^2}.$$

This agrees with (3.29).

Example 6.11 Apply the method of Example 6.10 to compute a regression for air temperatures and cricket chirp rates, discussed in Example 3.13. We have

$$A = \begin{bmatrix} 1 & 88.6 \\ 1 & 93.3 \\ 1 & 75.2 \\ 1 & 80.6 \\ 1 & 82.6 \end{bmatrix} \text{ and } \mathbf{y} = \begin{bmatrix} 20.0 \\ 19.8 \\ 15.5 \\ 17.1 \\ 17.2 \end{bmatrix}.$$

Thus,

$$\bar{x} = 84.06\,,\ \sigma = 14.09227\,,\ \text{ and } \begin{bmatrix} \widehat{y}_1 \\ \widehat{y}_2 \end{bmatrix} = A^T\mathbf{y} = \begin{bmatrix} 89.60 \\ 7583.92 \end{bmatrix}.$$

The least squares solution, from formula (6.21), is

$$\begin{bmatrix} \alpha \\ \beta \end{bmatrix} = \begin{bmatrix} -4.1515066 \\ 0.2625685 \end{bmatrix}.$$

This agrees with the solution we found in Example 3.13.

Example 6.12 Return to the econometric multiple regression problem from Example 5.6. The goal is to model the effects of personal disposable income (PDI) and time on personal consumption expenditure (PCE) using data for the years 1956 to 1970 (listed as years 1 through 15). We do this by solving the normal equation $A^TA\widehat{\beta} = A^T\mathbf{y}$, where A is the matrix that has all 1s in the first column, the PDI data in the second column, and the year numbers 1–15 in the third column. The vector $\mathbf{y}$ records the PCE data.

We now solve this using QR factorization and equation (6.16). The matrix R is

$$R = \begin{bmatrix} 3.873 & 1559.88 & 30.984 \\ 0 & 291.299 & 16.464 \\ 0 & 0 & 2.988 \end{bmatrix}.$$

Solving $R^T\mathbf{w} = A^T\mathbf{y}$ gives $\mathbf{w} = \begin{bmatrix} 1424.064 \\ 256.7528 \\ 7.9388 \end{bmatrix}$. Solving $R\widehat{\beta} = \mathbf{w}$ yields the effects parameters $\widehat{\beta} = \begin{bmatrix} 53.16 \\ 0.727 \\ 2.736 \end{bmatrix}$, as before.

Equation (6.16) does not require us to write down the matrix Q. If we *do* have Q available, there is another way to solve the normal equation (6.15). Indeed, with $A = QR$, we have two different expressions for the projection of $\mathbf{b}$ into $Col(A)$: $A\widehat{\mathbf{x}}$, where $\widehat{\mathbf{x}}$ is the least squares solution, and $QQ^T\mathbf{b}$. Multiplying both sides of the equation $A\widehat{\mathbf{x}} = QQ^T\mathbf{b}$ by Q^T yields $R\widehat{\mathbf{x}} = Q^T\mathbf{b}$. The matrix R is upper triangular, so we solve this last equation for $\widehat{\mathbf{x}}$ by backward substitution. For the regression problem in Example 6.12, for instance, we find that $Q\mathbf{y}$ is the same as $\mathbf{w} = R\widehat{\beta}$. Thus, we can remove the forward substitution step of solving for $\mathbf{w}$.

Remark When the columns of a matrix A are linearly independent, the matrix A^TA is invertible. In that case, the matrix $A(A^TA)^{-1}A^T$ represents projection into $Col(A)$. With the QR factorization $A = QR$, the matrix QQ^T also represents projection into $Col(A)$. Fortunately, in this case, we have

$$\begin{aligned} A(A^TA)^{-1}A^T &= (QR)(R^TQ^TQR)^{-1}(R^TQ^T) \\ &= (QR)(R^TR)^{-1}(R^TQ^T) \\ &= (QR)(R^{-1}(R^T)^{-1})(R^TQ^T) \\ &= Q(RR^{-1})((R^T)^{-1}R^T)Q^T \\ &= QQ^T. \end{aligned}$$

6.5 Orthogonality and Matrix Norms

Both the operator norm and Frobenius norm of a matrix A are preserved if A is multiplied on either side by an orthogonal matrix. To see this, consider the product of an $M \times M$ orthogonal matrix $Q = \left[\,\mathbf{q}_1 \middle| \cdots \middle| \mathbf{q}_M\,\right]$ and a vector $\mathbf{x} = \left[\,x_1 \cdots x_M\,\right]^T$ in $\mathbb{R}^M$. We have

$$Q\mathbf{x} = x_1\mathbf{q}_1 + x_2\mathbf{q}_2 + \cdots + x_M\mathbf{q}_M.$$

The column vectors of Q form an orthonormal set, so

$$||Q\mathbf{x}||^2 = x_1^2 + x_2^2 + \cdots + x_M^2 = ||\mathbf{x}||^2 . \tag{6.22}$$

For any $M \times N$ matrix A and any nonzero vector $\mathbf{x}$ in $\mathbb{R}^N$, applying equation (6.22) to $A\mathbf{x}$ in place of $\mathbf{x}$ shows that $||QA\mathbf{x}|| = ||A\mathbf{x}||$. Thus, the operator norm of QA satisfies

$$||QA||_{op} = \max_{\mathbf{x}\neq\mathbf{0}} \left\{ \frac{||QA\mathbf{x}||}{||\mathbf{x}||} \right\} = \max_{\mathbf{x}\neq\mathbf{0}} \left\{ \frac{||A\mathbf{x}||}{||\mathbf{x}||} \right\} = ||A||_{op} . \tag{6.23}$$

For the Frobenius norm, we note that the jth column vector of the product QA is equal to $Q\mathbf{a}_j$, where $\mathbf{a}_j$ is the jth column vector of A. Thus,

$$||QA||_F^2 = \sum_{j=1}^{N} ||Q\mathbf{a}_j||^2 = \sum_{j=1}^{N} ||\mathbf{a}_j||^2 = ||A||_F^2 . \tag{6.24}$$

We know that a matrix and its transpose have the same norm and the transpose of an orthogonal matrix is also an orthogonal matrix. Thus, when A is $M \times N$ and Q is an $N \times N$ orthogonal matrix, we have

$$||AQ|| = ||(AQ)^T|| = ||Q^T A^T|| = ||A^T|| = ||A|| ,$$

where $||\cdot||$ is either of the norms we have studied.

6.6 Exercises

1. In each part, use formula (6.5) to compute the volume of the parallelepiped generated by the given vectors.

(a) $\mathbf{a} = \begin{bmatrix} -1 \\ 2 \\ -1 \end{bmatrix}, \mathbf{b} = \begin{bmatrix} 2 \\ -3 \\ 3 \end{bmatrix}, \mathbf{c} = \begin{bmatrix} 4 \\ 1 \\ 2 \end{bmatrix}.$

(b) $\mathbf{a} = \begin{bmatrix} 1 \\ 0 \\ 1 \end{bmatrix}, \mathbf{b} = \begin{bmatrix} 1 \\ -2 \\ 3 \end{bmatrix}, \mathbf{c} = \begin{bmatrix} 2 \\ 1 \\ 0 \end{bmatrix}.$

(c) $\mathbf{a} = \begin{bmatrix} 1 \\ 1 \\ 1 \end{bmatrix}, \mathbf{b} = \begin{bmatrix} 1 \\ 1 \\ 2 \end{bmatrix}, \mathbf{c} = \begin{bmatrix} 1 \\ 2 \\ 3 \end{bmatrix}.$

2. Consider the vectors

$$\begin{bmatrix} 1/2 \\ 1/2 \\ 1/2 \\ 1/2 \end{bmatrix}, \begin{bmatrix} 1/2 \\ 1/2 \\ -1/2 \\ -1/2 \end{bmatrix}, \begin{bmatrix} 1/2 \\ -1/2 \\ -1/2 \\ 1/2 \end{bmatrix}, \text{ and } \begin{bmatrix} 1/2 \\ -1/2 \\ 1/2 \\ -1/2 \end{bmatrix}.$$

(a) Verify that these vectors form an *orthonormal basis* of $\mathbb{R}^4$.

(b) Express the vector $\begin{bmatrix} 4 & -1 & 2 & -3 \end{bmatrix}^T$ as a linear combination of these basis vectors.

(c) Express the generic vector $\begin{bmatrix} b_1 & b_2 & b_3 & b_4 \end{bmatrix}^T$ as a linear combination of these basis vectors.

3. Let $A = \begin{bmatrix} 1 & 1 & 1 \\ 1 & 1 & 2 \\ 1 & 2 & -4 \\ 1 & 4 & 3 \end{bmatrix}$.

(a) Apply the Gram–Schmidt algorithm 6.6 to find an orthonormal set of vectors that generates the column space of A.

(b) Using the answer to part (a), write down the matrix QQ^T for projection into the column space of A.

(c) Using the work from part (a), write down the QR factorization of A. (Note: Q will be a 4×3 matrix while R will be 3×3.)

4. Let $\mathbf{a}_1 = \begin{bmatrix} 1 \\ 0 \\ 1 \end{bmatrix}$ and $\mathbf{a}_2 = \begin{bmatrix} 2 \\ 1 \\ 0 \end{bmatrix}$. The plane generated by $\mathbf{a}_1$ and $\mathbf{a}_2$ is the same as the column space of the matrix $A = \begin{bmatrix} 1 & 2 \\ 0 & 1 \\ 1 & 0 \end{bmatrix}$.

(a) Show that $\begin{bmatrix} 5/6 & 1/3 & 1/6 \\ 1/3 & 1/3 & -1/3 \\ 1/6 & -1/3 & 5/6 \end{bmatrix}$ is the matrix for projection into $Col(A)$.

(b) Compute the projection of the vector $\mathbf{b} = \begin{bmatrix} 1 \\ -2 \\ 3 \end{bmatrix}$ into the plane generated by $\mathbf{a}_1$ and $\mathbf{a}_2$.

(c) Compute the volume of the parallelepiped formed by $\mathbf{a}_1$, $\mathbf{a}_2$, and $\mathbf{b}$.

5. Determine "by hand" the QR factorization of $A = \begin{bmatrix} 1 & 0 \\ 0 & 1 \\ 1 & 0 \end{bmatrix}$.

6. Determine "by hand" the QR factorization of $B = \begin{bmatrix} 1 & 2 \\ 0 & 1 \\ 1 & 0 \end{bmatrix}$.

7. Given: In the QR factorization of $A = \begin{bmatrix} -1 & 2 \\ 2 & -3 \\ -1 & 3 \end{bmatrix}$, we have

$$R = \begin{bmatrix} \sqrt{6} & -11\sqrt{6}/6 \\ 0 & \sqrt{66}/6 \end{bmatrix}.$$

   (a) Solve "by hand" the equation $R^T\mathbf{w} = A^T\mathbf{b}$, where $\mathbf{b} = \begin{bmatrix} 4 \\ 1 \\ 2 \end{bmatrix}$. Use forward substitution.
   (b) Solve "by hand" the equation $R\mathbf{x} = \mathbf{w}$, where $\mathbf{w}$ is the result from part (a). Use backward substitution.

8. Let Q be the matrix in Example 6.4.

   (a) Show that $\mathbf{b} = \begin{bmatrix} 1 \\ 2 \\ 3 \\ 4 \end{bmatrix}$ is in $Col(Q)$ by computing $QQ^T\mathbf{b} = \mathbf{b}$.
   (b) Let $\mathbf{u} = \begin{bmatrix} 3 \\ 1 \\ 2 \\ 1 \end{bmatrix}$. Compute $QQ^T\mathbf{u}$ and $\mathbf{u} - QQ^T\mathbf{u}$. Verify that these are orthogonal to each other.

9. Set $A = \left[\mathbf{a}_1 \middle| \mathbf{a}_2\right] = \begin{bmatrix} -4 & 2 \\ 2 & 1 \\ 4 & -2 \end{bmatrix}$ and $\mathbf{c} = \begin{bmatrix} -3 \\ 4 \\ 1 \end{bmatrix}$.

   (a) Show that $\begin{bmatrix} 1/2 & 0 & -1/2 \\ 0 & 1 & 0 \\ -1/2 & 0 & 1/2 \end{bmatrix}$ is the matrix for projection into $Col(A)$, the plane generated by $\mathbf{a}_1$ and $\mathbf{a}_2$.
   (b) Compute the projection of $\mathbf{c}$ into $Col(A)$.
   (c) Find the volume of the parallelepiped formed by $\mathbf{a}_1$, $\mathbf{a}_2$, and $\mathbf{c}$.

10. For each given inconsistent system $A\mathbf{x} = \mathbf{b}$, do the following:

   (i) Compute $A = QR$, the QR factorization of A.
   (ii) Solve $R^T\mathbf{w} = A^T\mathbf{b}$ for $\mathbf{w}$ using forward substitution.

(iii) Solve $R\widehat{\mathbf{x}} = \mathbf{w}$ using backward substitution, where $\mathbf{w}$ is the answer from part (ii). Note that $\widehat{\mathbf{x}}$ is the least squares solution to the original inconsistent system, as indicated in equation (6.16).

(a) $\begin{bmatrix} -1 & 2 \\ 2 & -3 \\ -1 & 3 \end{bmatrix} \begin{bmatrix} x_1 \\ x_2 \end{bmatrix} = \begin{bmatrix} 4 \\ 1 \\ 2 \end{bmatrix}$;

(b) $\begin{bmatrix} 2 & 4 \\ 1 & 3 \\ 5 & 8 \end{bmatrix} \begin{bmatrix} x_1 \\ x_2 \end{bmatrix} = \begin{bmatrix} 3 \\ -1 \\ 1 \end{bmatrix}$;

(c) $\begin{bmatrix} 1 & 2 \\ -2 & 3 \\ 7 & 4 \end{bmatrix} \begin{bmatrix} x_1 \\ x_2 \end{bmatrix} = \begin{bmatrix} 3 \\ -5 \\ 1 \end{bmatrix}$;

(d) $\begin{bmatrix} 2 & 1 \\ 8 & 3 \\ 3 & 2 \end{bmatrix} \begin{bmatrix} x_1 \\ x_2 \end{bmatrix} = \begin{bmatrix} 4 \\ -2 \\ 1 \end{bmatrix}$.

11. For each given inconsistent system $A\mathbf{x} = \mathbf{b}$, do the following:

(i) Compute $A = QR$, the QR factorization of A.
(ii) Solve $R^T\mathbf{w} = A^T\mathbf{b}$ for $\mathbf{w}$ using forward substitution.
(iii) Solve $R\widehat{\mathbf{x}} = \mathbf{w}$ using backward substitution, where $\mathbf{w}$ is the answer from part (ii). Note that $\widehat{\mathbf{x}}$ is the least squares solution to the original inconsistent system, as indicated in equation (6.16).

(a) $\begin{bmatrix} 1 & 2 & 3 \\ -4 & -6 & -1 \\ -3 & -5 & -4 \\ -3 & -5 & -5 \end{bmatrix} \begin{bmatrix} x_1 \\ x_2 \\ x_3 \end{bmatrix} = \begin{bmatrix} 3 \\ 2 \\ 1 \\ 2 \end{bmatrix}$;

(b) $\begin{bmatrix} 2 & 3 & 4 \\ -2 & 0 & -2 \\ -5 & 1 & -3 \\ 0 & 1 & 1 \end{bmatrix} \begin{bmatrix} x_1 \\ x_2 \\ x_3 \end{bmatrix} = \begin{bmatrix} 3 \\ 2 \\ 1 \\ 2 \end{bmatrix}$;

(c) $\begin{bmatrix} 2 & 4 & 3 \\ -2 & 1 & 0 \\ -5 & 2 & 1 \\ 1 & 1 & 1 \end{bmatrix} \begin{bmatrix} x_1 \\ x_2 \\ x_3 \end{bmatrix} = \begin{bmatrix} 2 \\ -1 \\ 3 \\ 1 \end{bmatrix}$;

(d) $\begin{bmatrix} 4 & 5 & -2 \\ 1 & 0 & 2 \\ 3 & 2 & 2 \\ 5 & 3 & 3 \end{bmatrix} \begin{bmatrix} x_1 \\ x_2 \\ x_3 \end{bmatrix} = \begin{bmatrix} 4 \\ 1 \\ 2 \\ 1 \end{bmatrix}$;

(e) $$\begin{bmatrix} 2 & 1 & 5 \\ 4 & 3 & 8 \\ -1 & 2 & -7 \\ -2 & -7 & 6 \\ 2 & 3 & 1 \end{bmatrix} \begin{bmatrix} x_1 \\ x_2 \\ x_3 \end{bmatrix} = \begin{bmatrix} 1 \\ 3 \\ -1 \\ 2 \\ 2 \end{bmatrix}.$$

12. For each set of paired measurements, do the following:

(i) Determine the equation of the least squares regression line for y_i as a function of x_i. Use formulas (6.20) and (6.21).
(ii) Graph the regression line and the (x_i, y_i) data in the same picture.

(a)

i	1	2	3	4	5	6	7	8
x_i	30	20	25	15	35	40	37	27
y_i	3	1.5	1.5	1	4	6	5.5	3.5

(b)

i	1	2	3	4	5	6	7	8	9	10
x_i	77	95	30	45	85	50	65	60	63	82
y_i	1.5	4.0	0.5	1.4	2.0	0.8	2.5	2.0	1.7	2.8

(c)

i	1	2	3	4	5
x_i	3	4	8	12	17
y_1	4	7	7	11	10

13. The share prices of nine growth stocks were measured in January 1981 (x_i) and January 1982 (y_i), as shown in the table.

i stock	x_i share price \$ Jan. 1981	y_i share price \$ Jan. 1982
1	22.40	28.15
2	45.70	54.75
3	15.65	20.05
4	48.25	56.95
5	32.31	40.06
6	55.14	65.45
7	17.65	21.75
8	25.78	30.95
9	13.17	18.10

(a) Determine the equation of the least squares regression line for the 1982 share price as a function of the 1981 share price. Use formulas (6.20) and (6.21).
(b) Graph the regression line and the share price data in the same picture.

14. *(Proof Problem)* Suppose the columns of a matrix A form an orthogonal set of nonzero vectors. Show that every column of $RREF(A)$ contains the leading 1 of some row. Thus, there will be no free variables when we solve $A\mathbf{x} = \mathbf{b}$.
15. *(Proof Problem)* Prove the following statement. *An $M \times N$ matrix Q, with real number entries, satisfies the condition $Q^T Q = \mathcal{I}_N$ if, and only if, the columns of Q are mutually orthogonal unit vectors in $\mathbb{R}^M$.* (Note: $\mathcal{I}_N$ denotes the $N \times N$ identity matrix.)
16. *(Proof Problem)* Prove that an $M \times M$ orthogonal matrix V has the property that

$$V^T = V^{-1}. \tag{6.25}$$

That is, for an orthogonal matrix, the transpose is equal to the inverse. (See Definition 6.5.)

6.7 Projects

Project 6.1 (Parallelepipeds in $\mathbb{R}^3$)

Given two vectors $\mathbf{a} = \begin{bmatrix} a_1 \; a_2 \; a_3 \end{bmatrix}^T$ and $\mathbf{b} = \begin{bmatrix} b_1 \; b_2 \; b_3 \end{bmatrix}^T$ in $\mathbb{R}^3$, define the *cross product* $\mathbf{a} \times \mathbf{b}$ to be the vector

$$\mathbf{a} \times \mathbf{b} = \begin{bmatrix} (a_2 b_3 - a_3 b_2) \\ (a_3 b_1 - a_1 b_3) \\ (a_1 b_2 - a_2 b_1) \end{bmatrix}. \tag{6.26}$$

See also Example 1.28 and Exercise #27 in Chapter 1.

1. Show that

$$||\mathbf{a} \times \mathbf{b}||^2 = ||\mathbf{a}||^2 \cdot ||\mathbf{b}||^2 - (\mathbf{a}^T \mathbf{b})^2.$$

2. Conclude that the area of the parallelogram generated by $\mathbf{a}$ and $\mathbf{b}$ is equal to $||\mathbf{a} \times \mathbf{b}||$. (See equation (1.34).)
3. Show that $\mathbf{a} \times \mathbf{b}$ is orthogonal to both $\mathbf{a}$ and $\mathbf{b}$.
4. Now let $\mathbf{c} = \begin{bmatrix} c_1 \; c_2 \; c_3 \end{bmatrix}^T$, in $\mathbb{R}^3$. Using the result of the previous step, explain why the *altitude* of the parallelepiped generated by $\mathbf{a}$, $\mathbf{b}$, and $\mathbf{c}$ is equal to the norm of the projection of $\mathbf{c}$ along $\mathbf{a} \times \mathbf{b}$. That is,

$$\text{altitude} = ||\text{proj}_{\mathbf{a} \times \mathbf{b}} \mathbf{c}||.$$

5. Use the previous steps to show that the *volume* of the parallelepiped generated by $\mathbf{a}$, $\mathbf{b}$, and $\mathbf{c}$ is equal to

$$\text{Volume} = \left|(\mathbf{a}\times\mathbf{b})^T\,\mathbf{c}\right| = \left|\mathbf{c}^T\,(\mathbf{a}\times\mathbf{b})\right|. \tag{6.27}$$

(Compare this to formula (6.5).)

6. Using formula (6.27), compute the volume of the parallelepiped generated by the vectors

$$\mathbf{a} = \begin{bmatrix} -1 \\ 2 \\ -1 \end{bmatrix},\ \mathbf{b} = \begin{bmatrix} 2 \\ -3 \\ 3 \end{bmatrix},\ \text{and } \mathbf{c} = \begin{bmatrix} 4 \\ 1 \\ 2 \end{bmatrix}.$$

Project 6.2 (Multiple Regression with QR)
Part I This project revisits the multiple regression example from Project 5.1, where we examined the effect on the quantity of roses sold of the prices of both roses and carnations, a competing product. We will use the QR factorization, as in Example 6.12. A table with the data can be found in the earlier project. The variables are the wholesale price of roses (X_1 \$ per dozen), the wholesale price of carnations (X_2 \$ per dozen), and the quantity of roses sold (Y dozen).

1. Create the matrix A that has all 1s in the first column, the X_1 data in the second column, and the X_2 values in the third column.
2. Form the vector $\mathbf{y}$ containing the Y data. Compute $A^T\mathbf{y}$.
3. Compute the QR factorization $A = QR$.
4. Referring to equation (6.20), do the following:
   (a) Solve $R^T\mathbf{w} = A^T\mathbf{y}$ for $\mathbf{w}$ using forward substitution.
   (b) Solve $\widehat{\beta} = R\mathbf{w}$ to find the least squares solution $\widehat{\boldsymbol{\beta}}$. (Here $\mathbf{w}$ is the output from part (a).)
5. Write down the model $Y \approx \beta_1 + \beta_2 \cdot X_1 + \beta_3 \cdot X_2$.
6. Compute $||A\widehat{\beta} - \mathbf{y}||^2$, the sum of square errors.

Part II. Repeat Part I above for a different example of a multiple regression model. Your example must examine the relationship between a single output variable and at least two input variables. Use the QR factorization and equation (6.20) to find the least squares model that best fits the data.

Project 6.3 (Information Retrieval Using QR)
In this project, we revisit the basic approach to information retrieval introduced in Sections 1.6 and 3.2 and explored in Project 3.2. Given a term–document matrix L and a query vector $\mathbf{q}$, we compute the cosine of the angle between $\mathbf{q}$ and each column vector of L. If a cosine value is above a certain threshold, then the corresponding document is recommended as relevant to the search query. The modification proposed here is to replace the query $\mathbf{q}$ by its projection into $Col(L)$, the column space of L. When $L = QR$ is the QR factorization, the projection into $Col(L)$ is given by the matrix QQ^T. Thus, we substitute the modified query $\widetilde{\mathbf{q}} = QQ^T\mathbf{q}$ in place of $\mathbf{q}$.

We must have $||\widetilde{\mathbf{q}}|| \leq ||\mathbf{q}||$ because $\widetilde{\mathbf{q}}$ is a projection of $\mathbf{q}$. Also, the vector $\mathbf{q} - \widetilde{\mathbf{q}}$ is orthogonal to $Col(L)$ and, hence, belongs to $Null(L^T)$. That means that $L^T\widetilde{\mathbf{q}} = L^T\mathbf{q}$. Now set θ_j to be the angle between $\mathbf{q}$ and the jth column of L and $\widetilde{\theta_j}$ to be the angle between $\widetilde{\mathbf{q}}$ and the jth column of L. The relationship between $\cos(\widetilde{\theta_j})$ and $\cos(\theta_j)$ is like so.

$$\begin{aligned}
\cos(\widetilde{\theta_j}) &= \frac{(L\mathbf{e}_j)^T\widetilde{\mathbf{q}}}{||L\mathbf{e}_j||\,||\widetilde{\mathbf{q}}||} \\
&\geq \frac{(L\mathbf{e}_j)^T\widetilde{\mathbf{q}}}{||L\mathbf{e}_j||\,||\widetilde{\mathbf{q}}||} \cdot \frac{||\widetilde{\mathbf{q}}||}{||\mathbf{q}||} \\
&= \frac{(L\mathbf{e}_j)^T\mathbf{q}}{||L\mathbf{e}_j||\,||\mathbf{q}||} \\
&= \cos(\theta_j)\,.
\end{aligned}$$

It follows that using the modified query $\widetilde{\mathbf{q}}$ can lead to an increase in the number of documents returned in a given search since, for any threshold α, at least as many documents will satisfy $\cos(\widetilde{\theta_j}) > \alpha$ as $\cos(\theta_j) > \alpha$.

Part I. Use the library of documents and dictionary of terms shown in Table 1.3, in Section 1.6.

1. Compute the QR factorization of the term–document matrix. (We only need the Q factor.)
2. Create a query vector $\mathbf{q}$ with terms 3, 8, 12, and 18.
3. Create the modified query $\widetilde{\mathbf{q}} = QQ^T\mathbf{q}$.
4. Compute the *cosine* of the angle between $\widetilde{\mathbf{q}}$ and each column vector of the term–document matrix.
5. Using a threshold of *cosine* > 0.5, determine which documents are relevant to the query.
6. Using a threshold of *cosine* > 0.7, determine which documents are relevant to the query.
7. Create two additional queries of your own and repeat the analysis for each one.
8. If you worked on Project 3.2, you may wish to compare these new results to the earlier ones.

Part II. *(Optional)* If you worked on Project 3.2, Part II, then repeat the analysis of Part I of this project using the library you created in the earlier project. Compare the new results with what you got before.

Chapter 7
Eigenvalues

7.1 Eigenvalues and Eigenvectors

Whenever a system is in motion, some aspects of the motion may be stable or in equilibrium somehow. For example, the earth's axis is (sort of) stable even as the planet spins around it. The motion of a mass–spring system may be in an oscillatory steady state. The birth and death rates of two interacting animal populations may achieve a balance that allows the circle of life to continue. When we roll out a sheet of cookie dough with a rolling pin, the direction of motion is fixed even as the dough stretches out. In this chapter, we investigate the property of stability in linear systems whose dynamics are governed by matrices.

As an example, consider again the population dynamics of the fictional metropolitan area of Shamborough, from Section 3.3. Each resident lives in either the city, the suburbs, or the exurbs. The proportions of the total population living in each part in the kth year are C_k, S_k, and E_k. Our model is that the population proportions for the next year are determined by a transition matrix T, like so.

$$\begin{bmatrix} C_{k+1} \\ S_{k+1} \\ E_{k+1} \end{bmatrix} = T \begin{bmatrix} C_k \\ S_k \\ E_k \end{bmatrix} = \begin{bmatrix} 0.8 & 0.1 & 0 \\ 0.15 & 0.8 & 0.15 \\ 0.05 & 0.1 & 0.85 \end{bmatrix} \begin{bmatrix} C_k \\ S_k \\ E_k \end{bmatrix} . \tag{7.1}$$

We found that

$$T \begin{bmatrix} 3/14 \\ 6/14 \\ 5/14 \end{bmatrix} = \begin{bmatrix} 3/14 \\ 6/14 \\ 5/14 \end{bmatrix} .$$

Supplementary Information The online version contains supplementary material available at https://doi.org/10.1007/978-3-031-39562-8_7.

T. G. Feeman, *Applied Linear Algebra and Matrix Methods*,
Springer Undergraduate Texts in Mathematics and Technology,
https://doi.org/10.1007/978-3-031-39562-8_7

In other words, the relative populations of the three parts of Shamborough are stable when $C_k = 3/14$, $S_k = 6/14$, and $E_k = 5/14$. Moreover, it seemed like we would approach this stable population distribution over time no matter what distribution between city, suburbs, and exurbs we started with.

As a second example, consider the interaction between the populations of a group of owls and of the wood rats upon which the owls prey. Suppose we are able to estimate the sizes of these populations each month. Let x_k denote the number of owls and $y_k \cdot 1000$ denote the number of wood rats in month number k. Perhaps our observations suggest the following linear model:

$$\begin{bmatrix} x_{k+1} \\ y_{k+1} \end{bmatrix} = \begin{bmatrix} 0.5 & 0.4 \\ -0.104 & 1.1 \end{bmatrix} \cdot \begin{bmatrix} x_k \\ y_k \end{bmatrix} . \tag{7.2}$$

Effectively, this model suggests that each owl eats about $104 = (0.104) \cdot 1000$ rats per month. About half of the owls would die off each month with no rats available for food. The rat population would grow by about 10% per month with no owls around.

Let us say that an equilibrium is achieved if there is always the same number of rats available for each owl month after month. That means that the *ratio* of the two populations remains constant. For that to happen, we need $y_{k+1}/x_{k+1} = y_k/x_k$. Equivalently, $x_{k+1}/x_k = y_{k+1}/y_k = \lambda$, for some number λ. Applying the linear model (7.2), this means we are looking for a number λ and values for x_k and y_k such that

$$\begin{bmatrix} 0.5 & 0.4 \\ -0.104 & 1.1 \end{bmatrix} \cdot \begin{bmatrix} x_k \\ y_k \end{bmatrix} = \begin{bmatrix} x_{k+1} \\ y_{k+1} \end{bmatrix} = \lambda \cdot \begin{bmatrix} x_k \\ y_k \end{bmatrix} .$$

In other words, the vector $\begin{bmatrix} x_k \\ y_k \end{bmatrix}$ gets mapped to *a numerical multiple of itself* when we multiply by the coefficient matrix.

We will soon see how to find such an equilibrium state. For now, notice that, in both this problem and the stability problem for Shamborough, we are seeking a vector that remains stable under multiplication by a given matrix.

Definition 7.1 For a given $M \times M$ matrix A, a number λ is called an **eigenvalue** of A if there is a nonzero vector $\mathbf{v}$ such that $A\mathbf{v} = \lambda \cdot \mathbf{v}$. Such a vector $\mathbf{v}$, if it exists, is called an **eigenvector** of A, corresponding to the eigenvalue λ.

Remark The concepts of eigenvalue and eigenvector only apply to square matrices, with the same number of rows as columns. That's because $A\mathbf{v}$ and $\mathbf{v}$ must lie in the same space.

Remark The zero vector $\mathbf{0}$ satisfies $T\mathbf{0} = \lambda\mathbf{0}$ for every value of λ. That's why we do not allow $\mathbf{0}$ to qualify as an eigenvector.

Example 7.2 For the matrix T in (7.1),

$$T\begin{bmatrix}3/14\\6/14\\5/14\end{bmatrix}=\begin{bmatrix}3/14\\6/14\\5/14\end{bmatrix}.$$

Thus, the vector $\mathbf{p}_* = \begin{bmatrix}3/14 & 6/14 & 5/14\end{bmatrix}^T$ is an eigenvector of T corresponding to the eigenvalue $\lambda = 1$. The coordinates of $\mathbf{p}_*$ add up to 1 and represent stable proportions of the different component populations. If the total population of Shamborough is 1 million $= 10^6$ people, then the vector

$$10^6 \cdot \mathbf{p}_* \approx \begin{bmatrix}214,286\\428,571\\357,143\end{bmatrix}$$

gives the total populations of each subregion within Shamborough.

Example 7.3 The matrix $A = \begin{bmatrix}4 & 2\\3 & -1\end{bmatrix}$ has eigenvalues $\lambda_1 = -2$ and $\lambda_2 = 5$, with respective eigenvectors $\mathbf{v}_1 = \begin{bmatrix}1\\-3\end{bmatrix}$ and $\mathbf{v}_2 = \begin{bmatrix}2\\1\end{bmatrix}$. Confirm this by computing

$$A\mathbf{v}_1 = \begin{bmatrix}4 & 2\\3 & -1\end{bmatrix}\begin{bmatrix}1\\-3\end{bmatrix} = \begin{bmatrix}-2\\6\end{bmatrix} = -2\cdot\mathbf{v}_1\,;$$

and

$$A\mathbf{v}_2 = \begin{bmatrix}4 & 2\\3 & -1\end{bmatrix}\begin{bmatrix}2\\1\end{bmatrix} = \begin{bmatrix}10\\5\end{bmatrix} = 5\cdot\mathbf{v}_2\,.$$

Example 7.4 For every natural number N and every vector $\mathbf{v}$ in $\mathbb{R}^N$, we have $\mathcal{I}_N\mathbf{v} = \mathbf{v}$. So $\lambda = 1$ is the only eigenvalue of $\mathcal{I}_N$, and every nonzero vector is an eigenvector.

Example 7.5 A matrix with constant row sums has that sum as an eigenvalue. The "all 1s" vector $\mathbf{1}$ is a corresponding eigenvector.

- The matrix $A = \begin{bmatrix}0&1&0&1\\1&0&1&0\\0&1&0&1\\1&0&1&0\end{bmatrix}$ has constant row sums equal to 2. Compute

$$A\mathbf{1} = \begin{bmatrix}0&1&0&1\\1&0&1&0\\0&1&0&1\\1&0&1&0\end{bmatrix}\begin{bmatrix}1\\1\\1\\1\end{bmatrix} = \begin{bmatrix}2\\2\\2\\2\end{bmatrix} = 2\cdot\mathbf{1}\,.$$

Thus, $\lambda = 2$ is an eigenvalue of A with $\mathbf{1}$ as a corresponding eigenvector.

- The matrix $L = \begin{bmatrix} 2 & -1 & 0 & -1 \\ -1 & 2 & -1 & 0 \\ 0 & -1 & 2 & -1 \\ -1 & 0 & -1 & 2 \end{bmatrix}$ has constant row sums equal to 0. Compute

$$L\mathbf{1} = \begin{bmatrix} 2 & -1 & 0 & -1 \\ -1 & 2 & -1 & 0 \\ 0 & -1 & 2 & -1 \\ -1 & 0 & -1 & 2 \end{bmatrix} \begin{bmatrix} 1 \\ 1 \\ 1 \\ 1 \end{bmatrix} = \begin{bmatrix} 0 \\ 0 \\ 0 \\ 0 \end{bmatrix} = 0 \cdot \mathbf{1} .$$

Thus, $\lambda = 0$ is an eigenvalue of L with $\mathbf{1}$ as a corresponding eigenvector.

Remark The matrix L, in Example 7.5, shows that $\lambda = 0$ can be an eigenvalue. However, the zero vector $\mathbf{0}$ is not allowed to be an eigenvector.

Example 7.6 For a given nonzero vector $\mathbf{a}$ in $\mathbb{R}^M$, the matrix for projection along $\mathbf{a}$ is

$$P_{\mathbf{a}} = (1/||\mathbf{a}||^2) \cdot (\mathbf{a}\mathbf{a}^T) .$$

Since $P_{\mathbf{a}}\mathbf{a} = \mathbf{a}$, it follows that $\lambda = 1$ is an eigenvalue of $P_{\mathbf{a}}$ with $\mathbf{a}$ as an eigenvector. Any nonzero scalar multiple of $\mathbf{a}$ will also be an eigenvector corresponding to $\lambda = 1$. If $M \geq 2$ and $\mathbf{v}$ is any nonzero vector orthogonal to $\mathbf{a}$, then $P_{\mathbf{a}}\mathbf{v} = \mathbf{0}$. Therefore, $\lambda = 0$ is also an eigenvalue of $P_{\mathbf{a}}$.

The matrix for the action of reflection across $\mathbf{a}$ is

$$R_{\mathbf{a}} = 2 \cdot P_{\mathbf{a}} - \mathcal{I}_M .$$

The vector $\mathbf{a}$ is its own reflection: $R_{\mathbf{a}}\mathbf{a} = \mathbf{a}$. Thus, $\lambda = 1$ is an eigenvalue of $R_{\mathbf{a}}$ with eigenvector $\mathbf{a}$. Also, any nonzero vector $\mathbf{v}$ orthogonal to $\mathbf{a}$ satisfies $R_{\mathbf{a}}\mathbf{v} = -\mathbf{v}$. Therefore, $\mathbf{v}$ is an eigenvector of $R_{\mathbf{a}}$ with eigenvalue $\lambda = -1$.

Example 7.7 Suppose Q is an $M \times M$ matrix whose column vectors are an orthonormal set in $\mathbb{R}^M$. According to equation (6.22), $||Q\mathbf{x}|| = ||\mathbf{x}||$ for every vector $\mathbf{x}$ in $\mathbb{R}^M$. It follows that the only possible real number eigenvalues of Q are ± 1. Also, any complex number eigenvalues must have *modulus* equal to 1.

Example 7.8 The matrix $A = \begin{bmatrix} \cos(\theta) & -\sin(\theta) & 0 \\ \sin(\theta) & \cos(\theta) & 0 \\ 0 & 0 & 1 \end{bmatrix}$ corresponds to a rotation of $\mathbb{R}^3$ about the z-axis by the angle θ. Every nonzero multiple of $\begin{bmatrix} 0 & 0 & 1 \end{bmatrix}^T$ is fixed by the rotation and, so, is an eigenvector of A with eigenvalue $\lambda = 1$. Provided that $0 < \theta < \pi$, any vector that does not lie on the z-axis will change its direction, because of the rotation. So no such vector can be an eigenvector. For $\theta = 0$, this matrix is $\mathcal{I}_3$, which we discussed above. For $\theta = \pi$, every vector lying in the xy-plane gets mapped to its own negative. Hence, $\lambda = -1$ is an eigenvalue in this special case.

Example 7.9 The matrix $A_\theta = \begin{bmatrix} \cos(\theta) & -\sin(\theta) \\ \sin(\theta) & \cos(\theta) \end{bmatrix}$ implements rotation of the xy-plane by the angle θ. Unless θ is an integer multiple of π, then $A_\theta\mathbf{v}$ and $\mathbf{v}$ lie on different lines, for every nonzero vector $\mathbf{v}$ in $\mathbb{R}^2$. Thus, the matrix A_θ has no real number eigenvalues.

7.2 Computing Eigenvalues

At this point, we might wonder, *How do we find the eigenvalues and eigenvectors of a matrix?* By Definition 7.1, an eigenvalue is always linked to the presence of an eigenvector. We are looking for *both* a number λ *and* a nonzero vector $\mathbf{v}$ for which $A\mathbf{v} = \lambda\mathbf{v}$. We can rewrite this equation as $(A - \lambda \cdot \mathcal{I})\mathbf{v} = \mathbf{0}$, where $\mathcal{I}$ denotes the identity matrix. We now see that λ is an eigenvalue of A if, and only if, there is a nonzero vector in the nullspace of the matrix $(A - \lambda \cdot \mathcal{I})$. For that to happen, a free variable must emerge when we row reduce the augmented matrix $[(A - \lambda \cdot \mathcal{I}) \mid \mathbf{0}]$. That means that $A - \lambda \cdot \mathcal{I}$ does not row reduce to the identity matrix. In other words, $A - \lambda \cdot \mathcal{I}$ **does not have an inverse**. We record this observation as an alternate definition for the concept of an eigenvalue.

Definition 7.10 For a given $M \times M$ matrix A, a number λ is an **eigenvalue** of A if, and only if, the matrix $(A - \lambda \cdot \mathcal{I}_M)$ does not have an inverse. In this case, every nonzero vector $\mathbf{v}$ such that

$$(A - \lambda \cdot \mathcal{I}_M)\mathbf{v} = \mathbf{0} \tag{7.3}$$

is an **eigenvector** of A, corresponding to the eigenvalue λ.

Remark Even for an $M \times M$ matrix with real number entries, some or all of the eigenvalues may turn out to be complex numbers. In that case, the corresponding eigenvectors will necessarily have at least some complex number coordinates. That is, the eigenvectors will not be in $\mathbb{R}^M$, but, rather, in the space $\mathbb{C}^M$ of complex vectors. Most of the applications we will look at involve real eigenvalues and real eigenvectors. But at this initial stage, we must allow for all possibilities.

Example 7.11 Take $A = \begin{bmatrix} 4 & 2 \\ 3 & -1 \end{bmatrix}$. Then

$$A - \lambda \cdot \mathcal{I} = \begin{bmatrix} 4 - \lambda & 2 \\ 3 & -1 - \lambda \end{bmatrix}.$$

This fails to have an inverse when $(4-\lambda)\cdot(-1-\lambda)-6=0$. That is, $\lambda^2-3\lambda-10=0$, the roots of which are $\lambda=-2$ and $\lambda=5$. Therefore, the eigenvalues of A are $\lambda=-2$ and $\lambda=5$.

Example 7.12 For a 3×3 example, take $A=\begin{bmatrix}6 & -5 & 2\\ 4 & -3 & 2\\ 2 & -2 & 3\end{bmatrix}$. So

$$A-\lambda\cdot\mathcal{I}=\begin{bmatrix}6-\lambda & -5 & 2\\ 4 & -3-\lambda & 2\\ 2 & -2 & 3-\lambda\end{bmatrix}.$$

Carry out these four elementary row operations: (1) switch rows 1 and 3; (2) subtract 2·(row 1) from (row 2); (3) subtract $\frac{(6-\lambda)}{2}$·(Row 1) from (row 3); and (4) subtract the new (row 2) from the new (row 3). This transforms $A-\lambda\cdot\mathcal{I}$ into

$$\begin{bmatrix}2 & -2 & 3-\lambda\\ 0 & 1-\lambda & -4+2\lambda\\ 0 & 0 & -(\lambda^2-5\lambda+6)/2\end{bmatrix}.$$

We see now that this will not row reduce to the identity if either $(1-\lambda)=0$ or $(\lambda^2-5\lambda+6)=0$. Thus, the eigenvalues of A are $\lambda=1$, $\lambda=2$ and $\lambda=3$.

Example 7.13 If A is either upper triangular or lower triangular, then the eigenvalues are exactly the numbers that appear on the diagonal of the matrix. To see this, suppose A is upper triangular and has an entry of λ_* on its diagonal. Then $(A-\lambda_*\cdot\mathcal{I})$ is also upper triangular and has a 0 on its diagonal. Thus, we cannot transform $(A-\lambda_*\cdot\mathcal{I})$ into $\mathcal{I}$ via row reduction. That is, $(A-\lambda_*\cdot\mathcal{I})$ does not have an inverse. Therefore, λ_* is an eigenvalue of A. Conversely, if the number μ does not appear on the diagonal of A, then $(A-\mu\cdot\mathcal{I})$ is an upper triangular matrix with all nonzero entries on its diagonal. This is invertible, so μ is not an eigenvalue of A.

For a lower triangular matrix, the transpose is upper triangular with the same entries on the diagonal. Since a square matrix is invertible only if its transpose is, too, these diagonal entries are the eigenvalues.

Formula 7.14 The eigenvalues of the 2×2 matrix $A=\begin{bmatrix}a & b\\ c & d\end{bmatrix}$ are the roots of the quadratic equation

$$\lambda^2-(a+d)\lambda+(ad-bc)=0\,. \tag{7.4}$$

Proof Take $A=\begin{bmatrix}a & b\\ c & d\end{bmatrix}$. Then $A-\lambda\cdot\mathcal{I}_2=\begin{bmatrix}a-\lambda & b\\ c & d-\lambda\end{bmatrix}$. By Formula 4.23, this *fails* to have an inverse when $(a-\lambda)\cdot(d-\lambda)-bc=0$. This is the same as $\lambda^2-(a+d)\lambda+(ad-bc)=0$. □

Example 7.15 (i) For $A = \begin{bmatrix} 4 & 2 \\ 3 & -1 \end{bmatrix}$, the quadratic $\lambda^2 - 3\lambda - 10 = 0$ has roots $\lambda_1 = -2$ and $\lambda_2 = 5$, as we found in Example 7.11.

(ii) Consider the rotation matrix $A_\theta = \begin{bmatrix} \cos(\theta) & -\sin(\theta) \\ \sin(\theta) & \cos(\theta) \end{bmatrix}$, from Example 7.9. The eigenvalues are the roots of the quadratic equation $\lambda^2 - 2\cos(\theta)\lambda + 1 = 0$. These are $\lambda = \cos(\theta) \pm \sqrt{-\sin^2(\theta)}$, which are not real numbers, unless θ is an integer multiple of π.

7.3 Computing Eigenvectors

Using Definition 7.10, once we have an eigenvalue λ in hand, the corresponding eigenvectors are the nonzero solutions to the homogeneous system $(A - \lambda \cdot \mathcal{I})\mathbf{v} = \mathbf{0}$. This prompts another bit of terminology.

Definition 7.16 Suppose the real or complex number λ is an eigenvalue of the $M \times M$ matrix A. The nullspace

$$Null(A - \lambda \cdot \mathcal{I}_M) = \left\{ \mathbf{v} \text{ in } \mathbb{C}^M \,:\, (A - \lambda \cdot \mathcal{I}_M)\mathbf{v} = \mathbf{0} \right\}$$

is called the **eigenspace** of A corresponding to λ. Thus, the eigenspace corresponding to an eigenvalue λ comprises all the corresponding eigenvectors together with the zero vector $\mathbf{0}$. If λ is a real number and A is a real matrix, then the eigenspace is a subset of $\mathbb{R}^M$.

The dimension of the eigenspace corresponding to the eigenvalue λ is called the **geometric multiplicity** of the eigenvalue. This is equal to the number of free variables that emerge when we row reduce the augmented matrix $\left[A - \lambda \cdot \mathcal{I}_M \middle| \mathbf{0} \right]$.

Remark We will mainly focus on problems involving real number eigenvalues and eigenvectors with real number coordinates. The basic definitions, though, allow for both real and complex number possibilities.

The term *eigenspace* is motivated by the observation that every nonzero linear combination of eigenvectors corresponding to the same eigenvalue is again an eigenvector for that eigenvalue. Indeed, suppose $\mathbf{v}_1$ and $\mathbf{v}_2$ are eigenvectors of A with eigenvalue λ, and suppose c1 and c2 are any constants. Then,

$$\begin{aligned} A(c_1 \cdot \mathbf{v}_1 + c_2 \cdot \mathbf{v}_2) &= c_1 \cdot A\mathbf{v}_1 + c_2 \cdot A\mathbf{v}_2 \\ &= c_1 \cdot (\lambda \mathbf{v}_1) + c_2 \cdot (\lambda \mathbf{v}_2) \\ &= \lambda \cdot (c_1 \cdot \mathbf{v} + c_2 \cdot \mathbf{v}_2) \,. \end{aligned}$$

It follows that the eigenspace of A corresponding to an eigenvalue λ contains all linear combinations of its elements.

Example 7.17 Take $A = \begin{bmatrix} 4 & 2 \\ 3 & -1 \end{bmatrix}$, from Example 7.11. One of the eigenvalues is $\lambda = -2$. To find an eigenvector, look at

$$A - (-2) \cdot \mathcal{I} = \begin{bmatrix} 4-(-2) & 2 \\ 3 & -1-(-2) \end{bmatrix} = \begin{bmatrix} 6 & 2 \\ 3 & 1 \end{bmatrix}.$$

Row reduction gives us

$$\left[\begin{array}{cc|c} 6 & 2 & 0 \\ 3 & 1 & 0 \end{array}\right] \xrightarrow{\text{RREF}} \left[\begin{array}{cc|c} 1 & 1/3 & 0 \\ 0 & 0 & 0 \end{array}\right].$$

Thus, any vector of the form $\mathbf{v} = t \cdot \begin{bmatrix} -1/3 \\ 1 \end{bmatrix}$, for $t \neq 0$, is an eigenvector. The corresponding eigenspace also includes $\mathbf{0}$. This eigenspace defines a line, which is one-dimensional. So this eigenvalue has geometric multiplicity 1.

Example 7.18 Take $A = \begin{bmatrix} 6 & -5 & 2 \\ 4 & -3 & 2 \\ 2 & -2 & 3 \end{bmatrix}$, from Example 7.12. One of the eigenvalues is $\lambda = 2$. To find an eigenvector, we solve the system

$$(A - 2 \cdot I_3)\mathbf{v} = \mathbf{0} \implies \begin{bmatrix} 4 & -5 & 2 \\ 4 & -5 & 2 \\ 2 & -2 & 1 \end{bmatrix} \begin{bmatrix} v_1 \\ v_2 \\ v_3 \end{bmatrix} = \begin{bmatrix} 0 \\ 0 \\ 0 \end{bmatrix}.$$

The augmented matrix is

$$\left[\begin{array}{ccc|c} 4 & -5 & 2 & 0 \\ 4 & -5 & 2 & 0 \\ 2 & -2 & 1 & 0 \end{array}\right] \xrightarrow{\text{RREF}} \left[\begin{array}{ccc|c} 1 & 0 & 1/2 & 0 \\ 0 & 1 & 0 & 0 \\ 0 & 0 & 0 & 0 \end{array}\right].$$

Thus, the third coordinate v_3 is a free variable, $v_1 = -(1/2)v_3$, and $v_2 = 0$. If we take $v_3 = -2$, say, we get the eigenvector $\begin{bmatrix} 1 \\ 0 \\ -2 \end{bmatrix}$. We check our work by verifying that

$$\begin{bmatrix} 4 & -5 & 2 \\ 4 & -5 & 2 \\ 2 & -2 & 1 \end{bmatrix} \begin{bmatrix} 1 \\ 0 \\ -2 \end{bmatrix} = \begin{bmatrix} 0 \\ 0 \\ 0 \end{bmatrix}.$$

The eigenspace is the line generated by this eigenvector.

7.4 Transformation of Eigenvalues

In this section, we explore connections between the eigenvalues of a matrix A and those of certain matrices related to A.

For starters, we know that a matrix is invertible if, and only if, its transpose is invertible. Since $(A - \lambda \cdot \mathcal{I})^T = (A^T - \lambda \cdot \mathcal{I})$, it follows that $(A - \lambda \cdot \mathcal{I})$ is invertible if, and only if, its transpose $(A^T - \lambda \cdot \mathcal{I})$ is invertible. That is, A **and** A^T **have the same eigenvalues**. The corresponding eigenvectors are usually different, however.

Next, suppose A is invertible. Since $A = (A - 0 \cdot \mathcal{I})$, this means that 0 is *not* an eigenvalue of A. Now suppose that $A\mathbf{v} = \lambda\mathbf{v}$ for some nonzero vector $\mathbf{v}$ and some number $\lambda \neq 0$. We compute

$$A^{-1}\mathbf{v} = \frac{1}{\lambda}A^{-1}(\lambda\mathbf{v}) = \frac{1}{\lambda}A^{-1}(A\mathbf{v}) = \frac{1}{\lambda}(A^{-1}A)\mathbf{v} = \frac{1}{\lambda}\mathbf{v}.$$

In other words, $1/\lambda$ is an eigenvalue of A^{-1} and the eigenvector $\mathbf{v}$ is the same. **The eigenvalues of** A^{-1} **are the reciprocals of those of** A. The eigenvectors are the same.

Formula 7.19 Suppose A is an $M \times M$ matrix.

(i) The matrices A and A^T have the same eigenvalues. The corresponding eigenvectors may be different.
(ii) Suppose A is invertible. Then the number λ is an eigenvalue of A if, and only if, the number $1/\lambda$ is an eigenvalue of A^{-1}. Moreover, the corresponding eigenvectors are the same.

Example 7.20 If every row of the $M \times M$ matrix A sums to λ, then λ is an eigenvalue of A, with eigenvector $\mathbf{1}_M$. (See Example 7.5.) Taking the transpose, it follows that, if every *column* of a matrix B sums to λ, then λ is an eigenvalue of B. This applies, for example, to the transition matrix of a Markov process, where each column sums to 1. A corresponding eigenvector depends on the matrix and must be computed separately.

Example 7.21 Let $A = \begin{bmatrix} 4 & 2 \\ 3 & -1 \end{bmatrix}$. In Examples 7.11 and 7.17, we found that A has the eigenvalues $\lambda_1 = -2$ and $\lambda_2 = 5$, with respective eigenvectors $\mathbf{v}_1 = \begin{bmatrix} 1 \\ -3 \end{bmatrix}$ and $\mathbf{v}_2 = \begin{bmatrix} 2 \\ 1 \end{bmatrix}$. The inverse of A is

$$A^{-1} = \frac{1}{10}\begin{bmatrix} 1 & 2 \\ 3 & -4 \end{bmatrix}.$$

We can check that $\mathbf{v}_1$ and $\mathbf{v}_2$ are eigenvectors of A^{-1} as well, with corresponding eigenvalues $-1/2 = 1/\lambda_1$ and $1/5 = 1/\lambda_2$.

Eigenvalues behave nicely when we take powers of a matrix. Suppose $A\mathbf{v} = \lambda\mathbf{v}$, with $\mathbf{v} \neq \mathbf{0}$. Then

$$A^2\mathbf{v} = A(A\mathbf{v}) = A(\lambda\mathbf{v}) = \lambda A\mathbf{v} = \lambda^2\mathbf{v}. \tag{7.5}$$

Thus, λ^2 is an eigenvalue of A^2 with the same eigenvector $\mathbf{v}$. Similarly, for any integer $k \geq 2$, we have

$$A^k\mathbf{v} = A^{k-1}(A\mathbf{v}) = \lambda \cdot A^{k-1}\mathbf{v} = \cdots = \lambda^k\mathbf{v}. \tag{7.6}$$

That is, λ^k is an eigenvalue of A^k with the same vector $\mathbf{v}$ as an eigenvector. We highlight these facts here.

Formula 7.22 Suppose A is a square matrix with eigenvalue λ and corresponding eigenvector $\mathbf{v}$. Then, for every integer $k \geq 2$, λ^k is an eigenvalue of A^k with $\mathbf{v}$ as a corresponding eigenvector.

Example 7.23 Let $A = \begin{bmatrix} 4 & 2 \\ 3 & -1 \end{bmatrix}$, as in Examples 7.11 and 7.17. The eigenvalues of A are $\lambda_1 = -2$ and $\lambda_2 = 5$, with respective eigenvectors $\mathbf{v}_1 = \begin{bmatrix} 1 \\ -3 \end{bmatrix}$ and $\mathbf{v}_2 = \begin{bmatrix} 2 \\ 1 \end{bmatrix}$. Squaring gives us

$$A^2 = \begin{bmatrix} 22 & 6 \\ 9 & 7 \end{bmatrix}.$$

Thus, $\mathbf{v}_1$ and $\mathbf{v}_2$ are eigenvectors of A^2. Checking, we have

$$A^2\mathbf{v}_1 = \begin{bmatrix} 22 & 6 \\ 9 & 7 \end{bmatrix}\begin{bmatrix} 1 \\ -3 \end{bmatrix} = \begin{bmatrix} 4 \\ -12 \end{bmatrix} = 4 \cdot \begin{bmatrix} 1 \\ -3 \end{bmatrix}, \text{ and}$$

$$A^2\mathbf{v}_2 = \begin{bmatrix} 22 & 6 \\ 9 & 7 \end{bmatrix}\begin{bmatrix} 2 \\ 1 \end{bmatrix} = \begin{bmatrix} 50 \\ 25 \end{bmatrix} = 25 \cdot \begin{bmatrix} 2 \\ 1 \end{bmatrix}.$$

The eigenvalues of A^2 are $\mu_1 = 4 = (-2)^2$ and $\mu_2 = 25 = (5)^2$.

Example 7.24 The matrix $P_{\mathbf{a}}$, representing projection along a vector $\mathbf{a}$, is idempotent, meaning that $P_{\mathbf{a}}^2 = P_{\mathbf{a}}$. According to Formula 7.22, every eigenvalue λ of $P_{\mathbf{a}}$ satisfies $\lambda^2 = \lambda$. Thus, the eigenvalues are 0 and 1. This agrees with our finding in Example 7.6.

Similarly, the matrix $R_{\mathbf{a}}$ for reflection across $\mathbf{a}$ satisfies $R_{\mathbf{a}}^2 = \mathcal{I}$. The only eigenvalue of $\mathcal{I}$ is 1. (See Example 7.4.) Thus, the eigenvalues of $R_{\mathbf{a}}$ must satisfy $\lambda^2 = 1$. That is, the eigenvalues of $R_{\mathbf{a}}$ are 1 and -1.

Formula 7.22 extends to any polynomial in A.

Formula 7.25 Let A be an $M \times M$ matrix. For a polynomial $p(x) = c_0 + c_1 x + \cdots + c_n x^n$, we define

$$p(A) = c_0 \cdot \mathcal{I}_M + c_1 \cdot A + \cdots + c_n \cdot A^n . \tag{7.7}$$

If λ is an eigenvalue of A, then $\mu = p(\lambda)$ is an eigenvalue of $p(A)$ with the same corresponding eigenvectors.

Example 7.26 Let $p(x) = x^3 + x^2 - 4x$, so $p(A) = A^3 + A^2 - 4A$. If A is the matrix from Example 7.11, with eigenvalues $\lambda_1 = -2$ and $\lambda_2 = 5$, then $p(A)$ has eigenvalues $\mu_1 = p(-2) = 4$ and $\mu_2 = p(5) = 130$.

Example 7.27 The matrix exponential $\exp(A)$, defined in (3.35), is a limit of polynomials in A. The eigenvalues transform accordingly, with e^{λ} as an eigenvalue of $\exp(A)$ whenever λ is an eigenvalue of A. The eigenvectors are the same.

7.5 Eigenvalue Decomposition

We now look at a matrix factorization based on eigenvalues and eigenvectors.

Suppose the numbers $\lambda_1, \ldots, \lambda_K$ are eigenvalues of a matrix A, with corresponding eigenvectors $\mathbf{v}_1, \ldots, \mathbf{v}_K$. Then we calculate

$$\begin{aligned} A \left[\begin{array}{c|c|c} \mathbf{v}_1 & \cdots & \mathbf{v}_K \end{array} \right] &= \left[\begin{array}{c|c|c} A\mathbf{v}_1 & \cdots & A\mathbf{v}_K \end{array} \right] \\ &= \left[\begin{array}{c|c|c} \lambda_1 \mathbf{v}_1 & \cdots & \lambda_K \mathbf{v}_K \end{array} \right] \\ &= \left[\begin{array}{c|c|c} \mathbf{v}_1 & \cdots & \mathbf{v}_K \end{array} \right] \begin{bmatrix} \lambda_1 & \cdots & 0 \\ \vdots & \ddots & \vdots \\ 0 & \cdots & \lambda_K \end{bmatrix} . \end{aligned} \tag{7.8}$$

In short, the relation $AV = V\Lambda$ holds whenever Λ is a diagonal matrix whose diagonal entries are eigenvalues of A and V is a matrix whose columns are corresponding eigenvectors.

If the matrix V is *invertible*, then the equation $AV = V\Lambda$ implies $A = V\Lambda V^{-1}$. For this to happen, V and Λ must be of the same size as A, say $M \times M$. Also, the geometric multiplicities of all the eigenvalues must add up to M. In that case, the diagonal entries of Λ are the eigenvalues of A, each repeated as many times as its geometric multiplicity. This factorization of A is not always possible.

Formula 7.28 (Eigenvalue Decomposition) An $M \times M$ matrix A is **diagonalizable** when the sum of the geometric multiplicities of its eigenvalues is equal to M. In that case,

(i) let $\Lambda = \begin{bmatrix} \lambda_1 & \cdots & 0 \\ \vdots & \ddots & \vdots \\ 0 & \cdots & \lambda_M \end{bmatrix}$ be an $M \times M$ diagonal matrix whose diagonal entries are the eigenvalues of A, each repeated according to its geometric multiplicity, and

(ii) let $V = \left[\mathbf{v}_1 \middle| \cdots \middle| \mathbf{v}_M \right]$ be a matrix whose columns are linearly independent eigenvectors corresponding to $\lambda_1, \ldots, \lambda_M$, in order.

Then V is invertible and A can be factored as

$$A = V\Lambda V^{-1}. \tag{7.9}$$

This factorization is called the **eigenvalue decomposition** of A.

Example 7.29 Suppose A is an $M \times M$ matrix with M *distinct* eigenvalues. Then any set consisting of one eigenvector for each eigenvalue is linearly independent. Since the number of vectors in a linearly independent set cannot exceed M, it follows that each eigenvalue has geometric multiplicity 1 and that A is diagonalizable.

Example 7.30 The matrix $A = \begin{bmatrix} 4 & 2 \\ 3 & -1 \end{bmatrix}$, from Examples 7.11 and 7.17, has distinct eigenvalues $\lambda_1 = -2$ and $\lambda_2 = 5$. The eigenvalue decomposition is

$$\begin{aligned} A &= \begin{bmatrix} 1 & 2 \\ -3 & 1 \end{bmatrix} \begin{bmatrix} -2 & 0 \\ 0 & 5 \end{bmatrix} \begin{bmatrix} 1 & 2 \\ -3 & 1 \end{bmatrix}^{-1} \\ &= \begin{bmatrix} 1 & 2 \\ -3 & 1 \end{bmatrix} \begin{bmatrix} -2 & 0 \\ 0 & 5 \end{bmatrix} \begin{bmatrix} 1/7 & -2/7 \\ 3/7 & 1/7 \end{bmatrix}. \end{aligned}$$

Example 7.31 The matrix $\begin{bmatrix} 1 & 3 \\ 0 & 1 \end{bmatrix}$ is *not* diagonalizable. The only eigenvalue is $\lambda = 1$, and its geometric multiplicity is 1, not 2.

The eigenvalue decomposition, if it is available, is a big help in computing powers of a matrix. From $A = V\Lambda V^{-1}$, we get

$$A^2 = (V\Lambda V^{-1})(V\Lambda V^{-1}) = V\Lambda(V^{-1}V)\Lambda V^{-1} = V\Lambda^2 V^{-1}.$$

More generally, for any positive integer k, $A^k = V\Lambda^k V^{-1}$. The reason this helps is that

$$\Lambda^k = \begin{bmatrix} \lambda_1 & 0 & \cdots & 0 \\ 0 & \lambda_2 & \cdots & 0 \\ \vdots & \vdots & \ddots & \vdots \\ 0 & 0 & \cdots & \lambda_M \end{bmatrix}^k = \begin{bmatrix} \lambda_1^k & 0 & \cdots & 0 \\ 0 & \lambda_2^k & \cdots & 0 \\ \vdots & \vdots & \ddots & \vdots \\ 0 & 0 & \cdots & \lambda_M^k \end{bmatrix},$$

where each diagonal entry is raised to the kth power. We still have to multiply Λ^k by V and V^{-1} to get A^k, but that requires two matrix multiplications instead of k, a huge savings when k is large.

Example 7.32 For $A = \begin{bmatrix} 4 & 2 \\ 3 & -1 \end{bmatrix}$, the eigenvalue decomposition is

$$A = \begin{bmatrix} 1 & 2 \\ -3 & 1 \end{bmatrix} \begin{bmatrix} -2 & 0 \\ 0 & 5 \end{bmatrix} \begin{bmatrix} 1/7 & -2/7 \\ 3/7 & 1/7 \end{bmatrix}.$$

$$\begin{aligned} \text{Thus, } A^4 &= \begin{bmatrix} 1 & 2 \\ -3 & 1 \end{bmatrix} \begin{bmatrix} (-2)^4 & 0 \\ 0 & (5)^4 \end{bmatrix} \begin{bmatrix} 1/7 & -2/7 \\ 3/7 & 1/7 \end{bmatrix} \\ &= \begin{bmatrix} 1 & 2 \\ -3 & 1 \end{bmatrix} \begin{bmatrix} 16 & 0 \\ 0 & 625 \end{bmatrix} \begin{bmatrix} 1/7 & -2/7 \\ 3/7 & 1/7 \end{bmatrix} \\ &= \begin{bmatrix} 538 & 174 \\ 261 & 103 \end{bmatrix}. \end{aligned}$$

7.6 Population Models

In the introduction to this chapter, we considered the interaction between owls and wood rats. More generally, we may look at any two groups whose members interact in a way that affects the sizes of both groups. A first-order, linear model for the interaction is to assume that the current population of each group has a proportional impact on the size of both groups at the next time check. In symbols, suppose x_k and y_k are the populations of the two groups after k months (or some appropriate time unit), with x_0 and y_0 representing the initial populations at time 0. We assume there are constants, a, b, c, and d, such that, for all $k \geq 0$,

$$\begin{cases} x_{k+1} = a \cdot x_k + b \cdot y_k \\ y_{k+1} = c \cdot x_k + d \cdot y_k \end{cases} ; \text{I.e., } \begin{bmatrix} x_{k+1} \\ y_{k+1} \end{bmatrix} = \begin{bmatrix} a & b \\ c & d \end{bmatrix} \cdot \begin{bmatrix} x_k \\ y_k \end{bmatrix}. \tag{7.10}$$

For short, let $\mathbf{p}_k = \begin{bmatrix} x_k \\ y_k \end{bmatrix}$ and $A = \begin{bmatrix} a & b \\ c & d \end{bmatrix}$. Thus, for all $k \geq 0$, $\mathbf{p}_{k+1} = A\mathbf{p}_k = A^{k+1}\mathbf{p}_0$. Computing A^k for large values of k will enable us to forecast the long-term behavior of the two populations. If the eigenvalue decomposition $A = V \Lambda V^{-1}$ is available, then $A^k = V \Lambda^k V^{-1}$. Hence, $\mathbf{p}_k = V \Lambda^k V^{-1} \mathbf{p}_0$. This formulation is called the *power method* for computing $\mathbf{p}_k$.

Example 7.33 For the interaction between owls and wood rats, suppose there are x_k owls and $y_k \cdot 1000$ wood rats in month number k. Suppose our observations suggest the following:

- Each owl eats about $r \cdot 1000$ rats per month.
- Without any rats to eat, about half of the owls would die off each month.
- Without any owls to eat them, the rat population would grow by about 10% per month.

This gives the following basic linear model:

$$\mathbf{p}_{k+1} = \begin{bmatrix} x_{k+1} \\ y_{k+1} \end{bmatrix} = \begin{bmatrix} 0.5 & 0.4 \\ -r & 1.1 \end{bmatrix} \cdot \begin{bmatrix} x_k \\ y_k \end{bmatrix} = A\mathbf{p}_k \,. \tag{7.11}$$

With $r = 0.104$, as in the introduction to the chapter, the eigenvalues are $\lambda_1 = 1.02$ and $\lambda_2 = 0.58$. The eigenvalue decomposition is

$$A = \begin{bmatrix} 0.5 & 0.4 \\ -0.104 & 1.1 \end{bmatrix} = \begin{bmatrix} 10 & 5 \\ 13 & 1 \end{bmatrix} \begin{bmatrix} 1.02 & 0 \\ 0 & 0.58 \end{bmatrix} \begin{bmatrix} -1/55 & 1/11 \\ 13/55 & -2/11 \end{bmatrix}.$$

Since $(0.58)^k \to 0$ in the limit as $k \to \infty$, we see that

$$\begin{bmatrix} 1.02 & 0 \\ 0 & 0.58 \end{bmatrix}^k \approx \begin{bmatrix} (1.02)^k & 0 \\ 0 & 0 \end{bmatrix},$$

when k is large. The corresponding population vector is

$$\mathbf{p}_k \approx \begin{bmatrix} -2(1.02)^k/11 & 10(1.02)^k/11 \\ -13(1.02)^k/55 & 13(1.02)^k/11 \end{bmatrix} \mathbf{p}_0 \,.$$

Breaking this down into the two component populations, we get

$$\begin{bmatrix} x_k \\ y_k \end{bmatrix} \approx (1.02)^k \cdot \left(\frac{(5 \cdot y_0 - x_0)}{55} \right) \cdot \begin{bmatrix} 10 \\ 13 \end{bmatrix}.$$

In the long run, both populations will thrive with about 13 thousand wood rats for every 10 owls. This population ratio is determined by the eigenvector for the dominant eigenvalue. However, notice that the populations will die off if $5 \cdot y_0 \leq x_0$.

Example 7.34 In the year 1202, the Italian mathematician, Fibonacci, proposed a seemingly simple scenario involving rabbits.

- Rabbits always occurs in pairs of one female and one male;
- After 2 months, a new pair of rabbits matures into a breeding pair;
- A breeding pair of rabbits produces one pair of new rabbits each month;
- Rabbits live—and breed—forever!

Now, suppose we start with one pair of new rabbits in month number 1. In month number 2, we still have one pair of rabbits. In month 3, the original pair of rabbits becomes a breeding pair; so we have 2 pairs of rabbits. In general, in any subsequent month, we have all pairs of rabbits from the previous month. Plus, all pairs from two months earlier are breeding pairs and will give rise to an equal number of new pairs. Thus, the total number of pairs of rabbits in any given month is the sum of the totals from the preceding 2 months. Letting F_n denote the number of pairs of rabbits in month n, we get

$$F_1 = 1\,;\;\; F_2 = 1\,;\;\; \text{and, for } n \geq 3,\;\; F_n = F_{n-1} + F_{n-2}\,. \tag{7.12}$$

The numbers generated by this formula are called **Fibonacci numbers**. The first few are 1, 1, 2, 3, 5, 8, 13, 21, 34, ... In matrix terms,

$$\begin{bmatrix} F_n \\ F_{n-1} \end{bmatrix} = \begin{bmatrix} 1 & 1 \\ 1 & 0 \end{bmatrix} \cdot \begin{bmatrix} F_{n-1} \\ F_{n-2} \end{bmatrix}, \;\text{ for } n \geq 3. \tag{7.13}$$

In particular,

$$\begin{bmatrix} F_n \\ F_{n-1} \end{bmatrix} = \begin{bmatrix} 1 & 1 \\ 1 & 0 \end{bmatrix}^{n-2} \cdot \begin{bmatrix} 1 \\ 1 \end{bmatrix}, \;\text{ for } n \geq 3\,. \tag{7.14}$$

The matrix $\begin{bmatrix} 1 & 1 \\ 1 & 0 \end{bmatrix}$ has the eigenvalue decomposition $V \Lambda V^{-1}$, where

$$\Lambda = \begin{bmatrix} \frac{1+\sqrt{5}}{2} & 0 \\ 0 & \frac{1-\sqrt{5}}{2} \end{bmatrix} \text{ and } V = \begin{bmatrix} \frac{1+\sqrt{5}}{2} & \frac{1-\sqrt{5}}{2} \\ 1 & 1 \end{bmatrix}.$$

This decomposition, along with equation (7.14), leads to *Binet's formula* for the Fibonacci numbers:

$$F_n = \frac{1}{\sqrt{5}} \cdot \left[\left(\frac{1+\sqrt{5}}{2} \right)^n - \left(\frac{1-\sqrt{5}}{2} \right)^n \right] \;\text{ for } n \geq 1\,. \tag{7.15}$$

See Exercise #12.

The Fibonacci numbers arise so often in mathematics and nature that there is a journal, *The Fibonacci Quarterly*, devoted entirely to studying them. This sequence of numbers appears in mathematical writings from India as much as 1400 years before Fibonacci's work, in a study of patterns for Sanskrit poetry. The number $(1+\sqrt{5})/2$, in Binet's formula (7.15), is called the *golden ratio* and has a fascinating history of its own.

7.7 Rotations of $\mathbb{R}^3$

In this section, we use an eigenvalue decomposition to construct rigid rotations of the unit sphere in $\mathbb{R}^3$. This is of interest to computer graphics designers and cartographers, among others.

For a given angle of rotation θ, with $0 \leq \theta < 2\pi$, set

$$R_\theta = \begin{bmatrix} \cos(\theta) & -\sin(\theta) & 0 \\ \sin(\theta) & \cos(\theta) & 0 \\ 0 & 0 & 1 \end{bmatrix}, \tag{7.16}$$

as in Example 7.8. This matrix implements a rotation of $\mathbb{R}^3$ by the angle θ about the z-axis. The column vectors of R_θ form an orthonormal set, so, for every vector $\mathbf{x}$ in $\mathbb{R}^3$, we have $||R_\theta \mathbf{x}|| = ||\mathbf{x}||$, according to equation (6.22). Thus, the unit sphere in $\mathbb{R}^3$ gets mapped onto the unit sphere, with the points $(0,\ 0,\ \pm 1)$ remaining fixed.

To implement a rigid rotation of the unit sphere with an axis of rotation of our own choosing, first select a unit vector $\mathbf{u}$ that lies along the desired axis. Then find unit vectors $\mathbf{v}_1$ and $\mathbf{v}_2$ to complete an orthonormal set in $\mathbb{R}^3$. Set $V = \left[\mathbf{v}_1 \middle| \mathbf{v}_2 \middle| \mathbf{u}\right]$. Notice that we put our new axis in the third column of V. That's because the z-axis is in the third column of our basic rotation matrix R_θ. The matrix $A = V R_\theta V^T$ will now do the trick. Notice that $A^T A = \mathcal{I}$. Therefore, the column vectors of A form an orthonormal set and $||A\mathbf{x}|| = ||\mathbf{x}||$ for every $\mathbf{x}$ in $\mathbb{R}^3$. In other words, multiplication by A maps the unit sphere onto the unit sphere.

Example 7.35 For a counterclockwise rotation by an angle θ about the positive x-axis, we may take

$$V = \begin{bmatrix} 0 & 0 & 1 \\ 1 & 0 & 0 \\ 0 & 1 & 0 \end{bmatrix}.$$

Notice that the third column of V defines our new axis of rotation. We now compute

$$V R_\theta V^T = \begin{bmatrix} 0 & 0 & 1 \\ 1 & 0 & 0 \\ 0 & 1 & 0 \end{bmatrix} \cdot \begin{bmatrix} \cos(\theta) & -\sin(\theta) & 0 \\ \sin(\theta) & \cos(\theta) & 0 \\ 0 & 0 & 1 \end{bmatrix} \cdot \begin{bmatrix} 0 & 1 & 0 \\ 0 & 0 & 1 \\ 1 & 0 & 0 \end{bmatrix}$$

$$= \begin{bmatrix} 1 & 0 & 0 \\ 0 & \cos(\theta) & -\sin(\theta) \\ 0 & \sin(\theta) & \cos(\theta) \end{bmatrix}.$$

Example 7.36 The column vectors of $V = \begin{bmatrix} 1/\sqrt{2} & 1/\sqrt{6} & 1/\sqrt{3} \\ -1/\sqrt{2} & 1/\sqrt{6} & 1/\sqrt{3} \\ 0 & -2/\sqrt{6} & 1/\sqrt{3} \end{bmatrix}$ form an orthonormal set in $\mathbb{R}^3$. To rotate $\mathbb{R}^3$ about the third column vector $\mathbf{u} = \begin{bmatrix} 1/\sqrt{3} & 1/\sqrt{3} & 1/\sqrt{3} \end{bmatrix}^T$ by the angle $\theta = \pi/3$, we use the matrix

$$A = V R_{\pi/3} V^T = \begin{bmatrix} 2/3 & -1/3 & 2/3 \\ 2/3 & 2/3 & -1/3 \\ -1/3 & 2/3 & 2/3 \end{bmatrix}.$$

Instead of having a particular axis of rotation in mind, suppose we have a specific point that we would like to rotate into a preferred position. In cartography, this could mean reconfiguring a map projection formula centered at $\mathbf{e}_1 = \begin{bmatrix} 1 & 0 & 0 \end{bmatrix}^T$, where the equator and prime meridian meet, to have a different center.

Suppose $\mathbf{p} = \begin{bmatrix} x_0 & y_0 & z_0 \end{bmatrix}^T$ is the unit vector in $\mathbb{R}^3$ whose terminal point is the desired center on the map. Our goal is to rotate the sphere so that $\mathbf{p}$ moves to $\mathbf{e}_1$. The angle of rotation is the angle θ between $\mathbf{p}$ and $\mathbf{e}_1$. Thus, $\cos(\theta) = x_0$, and $\sin(\theta) = \sqrt{1 - x_0^2} = \sqrt{y_0^2 + z_0^2}$. Next, compute

$$\mathbf{p} - \text{proj}_{\mathbf{e}_1} \mathbf{p} = \begin{bmatrix} x_0 \\ y_0 \\ z_0 \end{bmatrix} - x_0 \cdot \mathbf{e}_1 = \begin{bmatrix} 0 \\ y_0 \\ z_0 \end{bmatrix}.$$

The unit vector in this direction is $\mathbf{q}_1 = (1/\sqrt{y_0^2 + z_0^2}) \cdot \begin{bmatrix} 0 & y_0 & z_0 \end{bmatrix}^T$. The vectors $\mathbf{e}_1$ and $\mathbf{q}_1$ generate the same plane as $\mathbf{e}_1$ and $\mathbf{p}$. The circle where this plane intersects the sphere is the new equator. To complete a new orthonormal basis, set

$$\mathbf{q}_2 = \begin{bmatrix} 0 \\ -z_0/\sqrt{y_0^2 + z_0^2} \\ y_0/\sqrt{y_0^2 + z_0^2} \end{bmatrix}.$$

(In fact, $\mathbf{q}_2$ is equal to the *cross product* of $\mathbf{e}_1$ and $\mathbf{q}_1$.) We wish to construct a matrix that implements rotation by θ about the axis determined by $\mathbf{q}_2$.

Following our discussion above, let $V = \left[\mathbf{e}_1 \middle| \mathbf{q}_1 \middle| \mathbf{q}_2\right]$. This has our axis of rotation $\mathbf{q}_2$ in the third column. The desired rotation by θ will actually appear to be clockwise from the perspective of $\mathbf{q}_2$, so we use the rotation matrix for $-\theta$ instead of θ. That is,

$$R_{-\theta} = \begin{bmatrix} \cos(\theta) & \sin(\theta) & 0 \\ -\sin(\theta) & \cos(\theta) & 0 \\ 0 & 0 & 1 \end{bmatrix} = \begin{bmatrix} x_0 & \sqrt{y_0^2 + z_0^2} & 0 \\ -\sqrt{y_0^2 + z_0^2} & x_0 & 0 \\ 0 & 0 & 1 \end{bmatrix}.$$

Finally, set $A = V R_{-\theta} V^T$. Thus, $A\mathbf{q}_2 = \mathbf{q}_2$ and $A\mathbf{p} = \mathbf{e}_1$. Indeed,

$$V^T \mathbf{p} = \begin{bmatrix} \mathbf{e}_1^T \mathbf{p} \\ \mathbf{q}_1^T \mathbf{p} \\ \mathbf{q}_2^{Tp} \end{bmatrix} = \begin{bmatrix} x_0 \\ \sqrt{y_0^2 + z_0^2} \\ 0 \end{bmatrix}. \tag{7.17}$$

Next,

$$\begin{aligned} R_{-\theta} V^T \mathbf{p} &= \begin{bmatrix} x_0 & \sqrt{y_0^2 + z_0^2} & 0 \\ -\sqrt{y_0^2 + z_0^2} & x_0 & 0 \\ 0 & 0 & 1 \end{bmatrix} \begin{bmatrix} x_0 \\ \sqrt{y_0^2 + z_0^2} \\ 0 \end{bmatrix} \\ &= \begin{bmatrix} x_0^2 + y_0^2 + z_0^2 \\ 0 \\ 0 \end{bmatrix} = \begin{bmatrix} 1 \\ 0 \\ 0 \end{bmatrix}, \text{ since } ||\mathbf{p}|| = 1. \end{aligned}$$

And, finally,

$$A\mathbf{p} = V R_{-\theta} V^T \mathbf{p} = \left[\mathbf{e}_1 \middle| \mathbf{q}_1 \middle| \mathbf{q}_2\right] \begin{bmatrix} 1 \\ 0 \\ 0 \end{bmatrix} = \mathbf{e}_1 ,$$

as claimed.

Example 7.37 The location of Baghdad, Iraq, on a model globe of radius 1 is not far from the vector $\mathbf{p} = \left[\, 1/\sqrt{3}\ 1/\sqrt{3}\ 1/\sqrt{3} \,\right]^T$. This gives us

$$\mathbf{q}_1 = \begin{bmatrix} 0 \\ 1/\sqrt{2} \\ 1/\sqrt{2} \end{bmatrix} \text{ and } \mathbf{q}_2 = \begin{bmatrix} 0 \\ -1/\sqrt{2} \\ 1/\sqrt{2} \end{bmatrix}.$$

The desired angle of rotation θ satisfies $\cos(\theta) = 1/\sqrt{3}$. That is, $\theta \approx 54.74°$. The matrix that implements this rotation of the sphere is

$$A = \begin{bmatrix} 1 & 0 & 0 \\ 0 & \frac{1}{\sqrt{2}} & -\frac{1}{\sqrt{2}} \\ 0 & \frac{1}{\sqrt{2}} & \frac{1}{\sqrt{2}} \end{bmatrix} \cdot \begin{bmatrix} \frac{1}{\sqrt{3}} & \frac{\sqrt{2}}{\sqrt{3}} & 0 \\ -\frac{\sqrt{2}}{\sqrt{3}} & \frac{1}{\sqrt{3}} & 0 \\ 0 & 0 & 1 \end{bmatrix} \cdot \begin{bmatrix} 1 & 0 & 0 \\ 0 & \frac{1}{\sqrt{2}} & \frac{1}{\sqrt{2}} \\ 0 & -\frac{1}{\sqrt{2}} & \frac{1}{\sqrt{2}} \end{bmatrix}$$

$$= \begin{bmatrix} \frac{1}{\sqrt{3}} & \frac{1}{\sqrt{3}} & \frac{1}{\sqrt{3}} \\ -\frac{1}{\sqrt{3}} & \frac{1+\sqrt{3}}{2\cdot\sqrt{3}} & \frac{1-\sqrt{3}}{2\cdot\sqrt{3}} \\ -\frac{1}{\sqrt{3}} & \frac{1-\sqrt{3}}{2\cdot\sqrt{3}} & \frac{1+\sqrt{3}}{2\cdot\sqrt{3}} \end{bmatrix}.$$

Notice that the column vectors of A form an orthonormal set in $\mathbb{R}^3$.

In Figure 7.1, part (a) is a portrait of a portion of the earth's surface made with an **equirectangular projection**, in which the grid of meridians and parallels on the globe becomes an evenly spaced rectangular grid of vertical and horizontal lines on

Fig. 7.1 (**a**) An equirectangular projection centered where the prime meridian meets the equator; (**b**) An equirectangular projection centered near Baghdad, Iraq

the flat map. In part (b) of the figure, the sphere has been rotated to put Baghdad at the center before applying the equirectangular map formula.

Some important map projections place the north or south pole at the center. To apply these projection formulas with a different center, we could rework the analysis presented above to rotate any unit vector $\mathbf{p}$ to the appropriate vector $\pm\mathbf{e}_3$.

7.8 Existence of Eigenvalues

We have now looked at numerous examples of eigenvalues and eigenvectors for different matrices. In many cases, the eigenvalues and eigenvectors were all real-valued. The rotation matrix in Example 7.9, by contrast, has only complex number eigenvalues. We have not addressed the general question of whether a given square matrix has to have any eigenvalues at all. In fact, this question has an affirmative answer.

Theorem 7.38 *Every $M \times M$ real matrix has an eigenvalue, which may be a complex number. Every real eigenvalue has real-valued eigenvectors.*

A full proof of this fact is beyond our scope here, but can be found in texts, such as [1], that offer a more theoretical treatment of linear algebra. To give the briefest hint of what is involved, consider the problem of using row reduction to find the inverse of a matrix $(A - \lambda \cdot \mathcal{I})$, where λ is an unknown. As in Formula 7.14, generalized to any $M \times M$ matrix, any eigenvalues will have to be roots of a certain polynomial of degree at most M. The Fundamental Theorem of Algebra asserts that such roots exist, at least within the complex number system. In fact, any complex number eigenvalues will occur in so-called conjugate pairs. Hence, every $M \times M$ real matrix where M is an odd number must have a real number eigenvalue.

Theorem 7.38 asserts that every matrix has at least one eigenvalue. Stronger results are known that say more about the whole collection of eigenvalues, but we will not attempt to prove those here, either. In general, finding the eigenvalues of a matrix can be quite difficult, especially if the matrix is large. Developing efficient algorithms to compute eigenvalues is an important and active area of research.

7.9 Exercises

1. Let $A = \begin{bmatrix} 19 & -10 \\ 21 & -10 \end{bmatrix}$.

   (a) Verify that $\begin{bmatrix} 5 \\ 7 \end{bmatrix}$ is an *eigenvector* of A by finding a corresponding *eigenvalue*.
   (b) Verify that $\lambda = 4$ is an *eigenvalue* of A by finding a corresponding *eigenvector* with integer coordinates.
   (c) Write down an invertible matrix V and a diagonal matrix Λ such that $A = V \Lambda V^{-1}$.

2. For each given 2×2 matrix, do the following:

   (i) Determine all eigenvalues of the matrix.
   (ii) For each eigenvalue, find an eigenvector with integer coordinates.

   (a) $\begin{bmatrix} 6 & -4 \\ 3 & -1 \end{bmatrix}$; (b) $\begin{bmatrix} 7 & -6 \\ 12 & -10 \end{bmatrix}$; (c) $\begin{bmatrix} 5 & -6 \\ 3 & -4 \end{bmatrix}$; (d) $\begin{bmatrix} 4 & -2 \\ 1 & 1 \end{bmatrix}$.

3. Let $A = \begin{bmatrix} 1 & 0 & 0 \\ -6 & 8 & 2 \\ 12 & -15 & -3 \end{bmatrix}$.

 (a) Find an *eigenvector* with integer coordinates corresponding to the *eigenvalue* $\lambda = 1$.

 (b) Find *eigenvalues* of A for the *eigenvectors* $\begin{bmatrix} 0 \\ -1 \\ 3 \end{bmatrix}$ and $\begin{bmatrix} 0 \\ -2 \\ 5 \end{bmatrix}$.

 (c) Write down an invertible matrix V and a diagonal matrix Λ such that $A = V \Lambda V^{-1}$.

4. Each of the following 2×2 matrices has distinct eigenvalues $\lambda_1 = 2$ and $\lambda_2 = 1$. For each matrix, do the following:

 (i) For each eigenvalue, find an eigenvector with integer coordinates.
 (ii) Write down an eigenvalue decomposition for the matrix.
 (iii) Compute the matrix power A^5 using the eigenvalue decomposition of A.
 (iv) Write down an eigenvalue decomposition for $(A + \mathcal{I}_2)$.
 (v) Compute the matrix $(A+\mathcal{I}_2)^4$ using the eigenvalue decomposition of $(A+\mathcal{I}_2)$.

 (a) $\begin{bmatrix} 4 & -3 \\ 2 & -1 \end{bmatrix}$; (b) $\begin{bmatrix} 5 & -4 \\ 3 & -2 \end{bmatrix}$; (c) $\begin{bmatrix} 3 & -2 \\ 1 & 0 \end{bmatrix}$; (d) $\begin{bmatrix} 6 & -10 \\ 2 & -3 \end{bmatrix}$.

5. Each of the following 3×3 matrices has distinct eigenvalues $\lambda_1 = 1$, $\lambda_2 = 0$, and $\lambda_3 = -1$. For each matrix, do the following:

 (i) For each eigenvalue, find an eigenvector with integer coordinates.
 (ii) Write down an eigenvalue decomposition for the matrix.
 (iii) Compute the matrix powers A^{99} and A^{100} using the eigenvalue decomposition.

 (a) $\begin{bmatrix} 5 & -5 & -3 \\ 2 & -2 & -1 \\ 4 & -4 & -3 \end{bmatrix}$; (b) $\begin{bmatrix} 1 & 0 & 0 \\ 6 & 5 & 2 \\ 21 & -15 & -6 \end{bmatrix}$; (c) $\begin{bmatrix} 1 & -1 & 1 \\ 2 & -2 & 1 \\ 4 & -4 & 1 \end{bmatrix}$.

6. Each of the following 3×3 matrices has distinct eigenvalues $\lambda_1 = 3$, $\lambda_2 = 2$, and $\lambda_3 = 1$. For each matrix, do the following:

 (i) For each eigenvalue, find an eigenvector with integer coordinates.
 (ii) Write down an eigenvalue decomposition for the matrix.
 (iii) Show that $A^3 - 6 \cdot A^2 + 11 \cdot A - 6 \cdot \mathcal{I}_3 = \mathbf{0}$.

 (a) $\begin{bmatrix} 3 & 5 & -2 \\ 0 & 2 & 0 \\ 0 & 2 & 1 \end{bmatrix}$; (b) $\begin{bmatrix} 1 & 1 & -1 \\ -2 & 4 & -1 \\ -4 & 4 & 1 \end{bmatrix}$; (c) $\begin{bmatrix} 1 & 0 & 0 \\ -6 & 8 & 2 \\ 12 & -15 & -3 \end{bmatrix}$.

7. Each of the following 3×3 matrices has constant row sums λ. Hence, λ is an eigenvalue. Show that the corresponding eigenspace is the line generated by the vector $\mathbf{1}_3 = \begin{bmatrix} 1 & 1 & 1 \end{bmatrix}^T$.

$$\text{(a)} \begin{bmatrix} -2 & 4 & -1 \\ -3 & 5 & -1 \\ -1 & 1 & 1 \end{bmatrix}; \text{ (b)} \begin{bmatrix} 2 & -2 & 1 \\ -1 & 2 & 0 \\ -5 & 7 & -1 \end{bmatrix}; \text{ (c)} \begin{bmatrix} 3 & -2 & 1 \\ 1 & 0 & 1 \\ -1 & 1 & 2 \end{bmatrix}.$$

8. Each of the following 3×3 matrices has constant column sums λ. Hence, λ is an eigenvalue. Find a corresponding eigenvector with integer coordinates.

$$\text{(a)} \begin{bmatrix} 1 & 1 & 1 \\ 1 & 0 & -1 \\ 1 & 2 & 3 \end{bmatrix}; \text{ (b)} \begin{bmatrix} 2 & 0 & -2 \\ 2 & -2 & -1 \\ -2 & 4 & 5 \end{bmatrix}; \text{ (c)} \begin{bmatrix} 5 & -6 & 3 \\ 6 & 7 & 3 \\ -6 & 4 & -1 \end{bmatrix}.$$

9. Let $A = \begin{bmatrix} 6 & -5 & 2 \\ 4 & -3 & 2 \\ 2 & -2 & 3 \end{bmatrix}$.

   (a) Transform $(A - \lambda \cdot \mathcal{I})$ by elementary row operations into the upper triangular matrix

$$U = \begin{bmatrix} 2 & -2 & 3 - \lambda \\ 0 & 1 - \lambda & -4 + 2\lambda \\ 0 & 0 & -(\lambda^2 - 5\lambda + 6)/2 \end{bmatrix}.$$

   (b) Conclude that the eigenvalues of A are 1, 2, and 3. Briefly explain your reasoning.

   (c) Find a corresponding *eigenvector* of A for each eigenvalue indicated in part (b).

   (d) Let $\Lambda = \begin{bmatrix} 3 & 0 & 0 \\ 0 & 2 & 0 \\ 0 & 0 & 1 \end{bmatrix}$. Write down an invertible 3×3 matrix V for which $A = V \Lambda V^{-1}$. Verify by matrix multiplication that your answer is correct.

10. Referring to Example 7.33, suppose that, in month number k, there are x_k owls and $y_k \cdot 1000$ wood rats living in a forest. With $\mathbf{p}_k = \begin{bmatrix} x_k \\ y_k \end{bmatrix}$, assume that

$$\mathbf{p}_{k+1} = \begin{bmatrix} x_{k+1} \\ y_{k+1} \end{bmatrix} = \begin{bmatrix} 0.5 & 0.4 \\ -r & 1.1 \end{bmatrix} \cdot \begin{bmatrix} x_k \\ y_k \end{bmatrix} = A\mathbf{p}_k\,.$$

(a) Take $r = 0.125$. Show that $\lambda = 1$ is an eigenvalue of A. Conclude that, in the long term, the populations will tend toward a stable constant population of about 1250 wood rats per owl.
(b) Take $r = 0.2$. Show that both the owl and wood rat populations will die out in the long term.
(c) Take $r = 0.081$. Show that, in the long term, both populations will increase at a rate of about 4% per month. Determine the long-term ratio of wood rats to owls for this scenario.

11. Suppose x_k is the number of foxes and $y_k \cdot 100$ is the number of rabbits living in a forest in month number k. With $\mathbf{p}_k = \begin{bmatrix} x_k \\ y_k \end{bmatrix}$, assume that

$$\mathbf{p}_{k+1} = \begin{bmatrix} x_{k+1} \\ y_{k+1} \end{bmatrix} = \begin{bmatrix} 0.6 & 0.5 \\ -r & 1.2 \end{bmatrix} \cdot \begin{bmatrix} x_k \\ y_k \end{bmatrix} = A\mathbf{p}_k \,.$$

(a) Set $r = 0.16$. Show that $\lambda = 1$ is an eigenvalue of the matrix A. Conclude that, in the long term, the populations will tend toward a stable constant population of about 80 rabbits per fox.
(b) Set $r = 0.175$. Show that both the fox and rabbit populations will die out in the long term.
(c) Set $r = 0.135$. Show that, in the long term, both populations will increase at a rate of about 5% per month. Determine the long-term ratio of rabbits to foxes in this case.

12. *(Proof Problem)* Refer to Example 7.34 about Fibonacci's rabbits.

(a) Verify that the matrix $\begin{bmatrix} 1 & 1 \\ 1 & 0 \end{bmatrix}$ has the eigenvalue decomposition $V \Lambda V^{-1}$, where

$$\Lambda = \begin{bmatrix} \frac{1+\sqrt{5}}{2} & 0 \\ 0 & \frac{1-\sqrt{5}}{2} \end{bmatrix} \text{ and } V = \begin{bmatrix} \frac{1+\sqrt{5}}{2} & \frac{1-\sqrt{5}}{2} \\ 1 & 1 \end{bmatrix}.$$

(b) Derive *Binet's formula* for the Fibonacci numbers:

$$F_n = \frac{1}{\sqrt{5}} \cdot \left[\left(\frac{1+\sqrt{5}}{2} \right)^n - \left(\frac{1-\sqrt{5}}{2} \right)^n \right] \quad \text{for } n \geq 1 \,. \tag{7.18}$$

(*Hint:* Use part (a) together with equation (7.14).)

13. Multiplication by the matrix $A = \begin{bmatrix} 1 & 0 & 0 \\ 0 & 1/\sqrt{2} & 1/\sqrt{2} \\ 0 & -1/\sqrt{2} & 1/\sqrt{2} \end{bmatrix}$ implements a rotation of the unit sphere in $\mathbb{R}^3$.

    (a) Write down a unit vector along the axis of rotation.
    (b) Determine the angle θ of the rotation.
    (c) Following Example 7.35, find a matrix V such that $A = V R_\theta V^T$, where θ is your answer to part (b), and R_θ is as given in (7.16). Check your work.

14. The matrix $A = \begin{bmatrix} 0 & 1 & 0 \\ 0 & 0 & 1 \\ 1 & 0 & 0 \end{bmatrix}$ can be obtained by first rotating the unit sphere in $\mathbb{R}^3$ by $-\pi/2$ about the positive z-axis and then rotating by $-\pi/2$ about the positive x-axis. Alternatively, A is the matrix that implements a single rotation.

    (a) Show that $\mathbf{u} = \begin{bmatrix} 1/\sqrt{3} & 1/\sqrt{3} & 1/\sqrt{3} \end{bmatrix}^T$ is a unit eigenvector for A with eigenvalue 1.
    (b) Let V be the orthogonal matrix from Example 7.36. Determine the angle θ for which $R_\theta = V^T A V$. This is the angle of the rotation about $\mathbf{u}$.
    (c) The matrices for rotation by $-\pi/2$ about the positive z-axis and for rotation by $-\pi/2$ about the positive x-axis, respectively, are

$$R_z = \begin{bmatrix} 0 & 1 & 0 \\ -1 & 0 & 0 \\ 0 & 0 & 1 \end{bmatrix} \text{ and } R_x = \begin{bmatrix} 1 & 0 & 0 \\ 0 & 0 & 1 \\ 0 & -1 & 0 \end{bmatrix}.$$

    Show that $A = R_x \cdot R_z$. Thus, this combination of two rotations also corresponds to a single rotation.

15. In each part of this exercise, use Example 7.37 and the discussion that precedes it to find a rotation matrix A such that $A\mathbf{p} = \mathbf{e}_1$, for the given unit vector $\mathbf{p}$. (In both cases, these are unit vectors rounded to two decimal places.)

    (a) $\mathbf{p} = \begin{bmatrix} -0.62 & 0.52 & -0.59 \end{bmatrix}^T$. This corresponds approximately to the location of Tokyo, Japan, on a model globe of radius 1.
    (b) $\mathbf{p} = \begin{bmatrix} 0.67 & -0.63 & -0.39 \end{bmatrix}^T$. This corresponds approximately to the location of Rio de Janeiro, Brazil, on a model globe of radius 1.

16. In each part of this exercise, rework the analysis that precedes Example 7.37 to find a rotation matrix A such that $A\mathbf{p} = \mathbf{e}_3$, for the given unit vector $\mathbf{p}$. (In both cases, these are unit vectors rounded to two decimal places.)

    (a) $\mathbf{p} = \begin{bmatrix} -0.62 & 0.52 & -0.59 \end{bmatrix}^T$. This corresponds approximately to the location of Tokyo, Japan, on a model globe of radius 1.

(b) $\mathbf{p} = \begin{bmatrix} 0.67 & -0.63 & -0.39 \end{bmatrix}^T$. This corresponds approximately to the location of Rio de Janeiro, Brazil, on a model globe of radius 1.

17. *(Proof Problem)* Prove the following. If $\Lambda = \begin{bmatrix} \lambda_1 & \cdots & 0 \\ 0 & \ddots & 0 \\ 0 & \cdots & \lambda_M \end{bmatrix}$ is a diagonal matrix and $A = V \Lambda V^{-1}$, for some invertible $M \times M$ matrix V, then the numbers $\lambda_1, \ldots, \lambda_M$ are the eigenvalues of A, and the columns of V are the corresponding eigenvectors.
18. *(Proof Problem)* Suppose A is an $M \times M$ matrix with eigenvalue λ and corresponding eigenvector $\mathbf{v}$. Suppose S is an invertible $M \times M$ matrix, and set $B = SAS^{-1}$. Show that λ is an eigenvalue of B with $S\mathbf{v}$ as a corresponding eigenvector.

Chapter 8
Markov Processes

8.1 Stochastic Matrices

A **Markov process** is a dynamical system where movement within the system consists of transitions between a finite set of states. These transitions are governed by prescribed probabilistic rules and are memoryless in the sense that the transition to a new state depends only on the current state of the system and not on the entire transition history. We consider only finite-state, discrete-time Markov processes.

We first looked at this concept in Section 3.3, with the fictional metropolis of Shamborough. There, we described yearly movements between three states: living in the city, the suburbs, or the exurbs. More generally, suppose there are M different states for the system, numbered 1 through M. Then, for each pair $[i, \ j]$ of integers between 1 and M, let $p_{i,\,j}$ denote the probability that the system will transition to state i at the next time check, given that the system is currently in state j. That is, $p_{i,\,j}$ is the probability of moving *from* state j *to* state i on the next turn. The $M \times M$ matrix T with entries $p_{i,\,j}$ is called the **transition matrix** for the Markov process.

Example 8.1 In a system known as the *Ehrenfest diffusion model*, there are two jars containing a total of N balls. We play a game where, on each turn, we select one ball at random and move it to the other jar. For instance, we could number the balls 1 through N and randomly select a number from this list. Then locate the ball with the selected number and move it. The state of the system is the number of balls in the first jar. Thus, there are $N + 1$ possible states, including the possibility that the first jar is empty. The next state depends only on the current state: If there are k balls in the first jar now, then there will be either $k - 1$ or $k + 1$ balls after one move.

Supplementary Information The online version contains supplementary material available at https://doi.org/10.1007/978-3-031-39562-8_8.

T. G. Feeman, *Applied Linear Algebra and Matrix Methods*,
Springer Undergraduate Texts in Mathematics and Technology,
https://doi.org/10.1007/978-3-031-39562-8_8

The transition matrix for this process has dimension $(N+1) \times (N+1)$. For the case with $N = 4$ balls, the 5×5 transition matrix is

$$T = \begin{array}{c|ccccc} & \multicolumn{5}{l}{\leftarrow \text{ from}} \\ \text{to} \downarrow & 0 & 1 & 2 & 3 & 4 \\ \hline 0 & 0 & 1/4 & 0 & 0 & 0 \\ 1 & 1 & 0 & 1/2 & 0 & 0 \\ 2 & 0 & 3/4 & 0 & 3/4 & 0 \\ 3 & 0 & 0 & 1/2 & 0 & 1 \\ 4 & 0 & 0 & 0 & 1/4 & 0 \end{array}. \tag{8.1}$$

Example 8.2 A mouse is placed in a maze with nine rooms arranged in a 3×3 square grid

1	2	3
4	5	6
7	8	9

. Doors connect the rooms in the same row or column, but not diagonally. The mouse randomly selects its next room from the doors available in its current room. The state of the system is the number of the room the mouse is in. The system is memoryless since the mouse's next room depends only on its current room and not on what happened before. The reader is encouraged to write down the transition matrix for this Markov process.

Example 8.3 The portion of the world wide web that has been indexed contains close to six billion web pages. An automated web crawler could potentially find itself on any one of these pages. If the next page to be visited is selected at random from the links available on the current page, without the use of the `[back]` button, then we have the framework of a Markov process. If page j has m_j available links to other pages, then, for each i, we get

$$p_{i,j} = \begin{cases} 1/m_j, & \text{if page } j \text{ links to page } i; \\ 0, & \text{if page } j \text{ does not link to page } i. \end{cases}$$

The transition matrix is $T = \left[p_{i,j} \right]$. It is effectively impossible to comprehend this transition matrix, though Google attempts it.

In the transition matrix for a Markov process, each entry $p_{i,j}$ represents a probability. So, $0 \leq p_{i,j} \leq 1$, for all i and j. If the system is currently in state j, then, at the next time check, the system must be in *some* state. Therefore, the sum of the entries in column j of the transition matrix is equal to 1, for every j. That is, for every $j = 1, \ldots, M$,

$$p_{1,j} + p_{2,j} + \cdots + p_{M,j} = \sum_{i=1}^{M} p_{i,j} = 1. \tag{8.2}$$

Definition 8.4 (i) A vector whose coordinates are nonnegative numbers that add up to 1 is called a **probability vector**. (ii) A **column stochastic matrix** is a square matrix where each column is a probability vector.

Thus, the transition matrix T associated with a finite-state, discrete-time Markov process is a column stochastic matrix. To explore the long-term behavior of the process, we track the transitions of a large collection of particles as they navigate through the system, randomly selecting one state after another according to the probabilities shown in the transition matrix. The initial distribution of particles among the different states is a probability vector $\mathbf{v}^{(0)}$, where each entry $v_j^{(0)}$ is the proportion of all particles that start in state j of the system. Now suppose that every particle makes a transition at the same time. To find the new distribution of the particles, consider state i. A particle that was in state j moves to state i with probability $p_{i,j}$. Thus, the proportion of *all* particles that move from state j to state i is $p_{i,j}v_j^{(0)}$. Every particle that moves to state i must move there from somewhere. Hence, the proportion of particles in state i after the transition is the sum

$$p_{i,1}v_1^{(0)} + p_{i,2}v_2^{(0)} + \cdots + p_{i,M}v_M^{(0)} = \sum_{j=1}^{M} p_{i,j}v_j^{(0)} .$$

This is exactly the ith coordinate of the vector $T\mathbf{v}^{(0)}$. Denote this by $\mathbf{v}^{(1)}$. That is,

$$\begin{aligned} \mathbf{v}^{(1)} &= T\mathbf{v}^{(0)} \\ &= \begin{bmatrix} \vdots & \vdots & \vdots & \vdots \\ p_{i,1} & p_{i,2} & \cdots & p_{i,M} \\ \vdots & \vdots & \vdots & \vdots \end{bmatrix} \begin{bmatrix} v_1^{(0)} \\ v_2^{(0)} \\ \vdots \\ v_M^{(0)} \end{bmatrix} \\ &= \begin{bmatrix} \vdots \\ p_{i,1}v_1^{(0)} + p_{i,2}v_2^{(0)} + \cdots + p_{i,M}v_M^{(0)} \\ \vdots \end{bmatrix} . \end{aligned}$$

Since T is a column stochastic matrix, the coordinates of $\mathbf{v}^{(1)}$ have the same sum as those of $\mathbf{v}^{(0)}$. In particular, if $\mathbf{v}^{(0)}$ is a probability vector, then so is $\mathbf{v}^{(1)}$. This makes sense since the coordinates of both $\mathbf{v}^{(0)}$ and $\mathbf{v}^{(1)}$ are proportions that show how a total population is distributed. These add up to 1 because every member of the population is accounted for. Symbolically,

$$\sum_{i=1}^{M} v_i^{(1)} = \sum_{i=1}^{M} \left(\sum_{j=1}^{M} p_{i,j}v_j^{(0)} \right) = \sum_{j=1}^{M} \left(\sum_{i=1}^{M} p_{i,j} \right) v_j^{(0)} = \sum_{j=1}^{M} v_j^{(0)} . \tag{8.3}$$

The last step of this calculation uses (8.2).

With the particles now distributed among the different states according to the entries of $\mathbf{v}^{(1)}$, another round of transitions leads to the distribution $\mathbf{v}^{(2)} = T\mathbf{v}^{(1)} =$

$T^2\mathbf{v}^{(0)}$. After any positive number k of transitions, the distribution of particles is given by the vector $\mathbf{v}^{(k)} = T^k\mathbf{v}^{(0)}$. Each of the vectors $\mathbf{v}^{(k)}$ is a probability vector given that $\mathbf{v}^{(0)}$ is, as shown in (8.3). This suggests that a picture of the long-term behavior of a Markov process will emerge as higher powers of the transition matrix are applied to some initial distribution of a population of particles navigating randomly through the system.

8.2 Stationary Distributions

As the particles repeatedly redistribute themselves among the various states of the system, it is natural to wonder if there is a probability vector $\mathbf{v}^*$ for which $T\mathbf{v}^* = \mathbf{v}^*$. Such a distribution is called a *stationary distribution*, not because the particles have stopped moving, but because the overall distribution of the particles remains the same after each transition. At each step, the number of particles leaving a given state will be replaced by the same number moving to that state.

Definition 8.5 A **stationary distribution** for a Markov process with transition matrix T is a probability vector $\mathbf{v}^*$ such that $T\mathbf{v}^* = \mathbf{v}^*$. That is, a stationary distribution is a probability vector that is also an eigenvector of T with eigenvalue 1.

From Example 7.20, every column stochastic matrix does indeed have $\lambda = 1$ as an eigenvalue. The question we face now is whether a corresponding eigenvector can be a probability vector. The existence of such an eigenvector is an important theoretical result.

To work backward, suppose $\mathbf{v}$ is a probability vector that is also an eigenvector of T, with $T\mathbf{v} = \lambda\mathbf{v}$. Since $T\mathbf{v}$ is also a probability vector, according to (8.3), the coordinates of $\lambda\mathbf{v}$ sum to 1. But those same coordinates also sum to λ, since $\mathbf{v}$ is a probability vector. Consequently, $\lambda = 1$. The following Proposition records a slightly more general version of this observation.

Proposition 8.6 *If T is a column stochastic matrix and $\mathbf{v}$ is an eigenvector of T with all nonnegative entries, then the corresponding eigenvalue is* 1.

Unfortunately, this result by itself does not guarantee that a column stochastic matrix T has a probability vector as an eigenvector for the eigenvalue 1. It says only that *if* a probability vector is an eigenvector for T, then 1 is the corresponding eigenvalue. For the result we really want, we appeal to the following powerful theorem, which we only state here. (For a proof, see [11].)

Theorem 8.7 (Perron–Frobenius Theorem) *Suppose A is an $M \times M$ matrix with nonnegative entries. Suppose also that there is some positive integer r such that all entries of the matrix A^r are positive numbers. Then:*

(i) *A has a positive eigenvalue* λ_1 *with a corresponding eigenvector whose entries are all positive;*
(ii) *If* λ *is any other eigenvalue of* A, *then* $|\lambda| < \lambda_1$;
(iii) λ_1 *has multiplicity* 1.

The transition matrix T for a Markov process is column stochastic and has all nonnegative entries. Conclusion (i) in the theorem guarantees that there is a probability vector $\mathbf{v}^*$ that is an eigenvector of T. From Proposition 8.6, we know that 1 is the corresponding eigenvalue. Thus, there really is a stationary distribution for a Markov process. Conclusion (iii) means that all eigenvectors corresponding to the eigenvalue 1 must be constant multiples of the stationary distribution vector. It follows that the stationary distribution is unique. Conclusion (ii) says that all other eigenvalues of T are smaller than 1 in absolute value. That will come in handy when we discuss the *power method.*

Before we jump to conclusions, let's look more closely at the fine print. The statement of the theorem requires that *some power of the matrix should have all positive entries*. For a transition matrix T, an entry of 0 in the $[i,\ j]$ position means that no particle currently in state j will transition to state i in the next step. The matrix T^k represents the transitions over the course of k steps. So an entry of 0 in the $[i,\ j]$ position of T^k means that it is not possible to reach state i from state j in exactly k steps. Thus, the requirement in the statement of the theorem is that there is some number r where it is possible to reach every state from every other state in exactly r steps. If T^r has all positive entries, then so does T^k for all $k \geq r$. A Markov process whose transition matrix has this property is said to be **irreducible**. Thus, the Perron–Frobenius theorem applies to irreducible Markov processes.

Theorem 8.8 *Suppose* T *is the transition matrix of a finite-state, discrete-time Markov process. Suppose also that the Markov process is irreducible. That is, suppose there is some positive integer* r *such that all entries of* T^r *are positive. Then the Markov process has a unique stationary distribution.*

8.3 The Power Method

We now wish to find the stationary distribution of a given irreducible Markov process with transition matrix T. The direct approach is to find a nonzero solution to $(T - \mathcal{I})\mathbf{v} = \mathbf{0}$. This may be difficult in practice, especially if the number of states is large. An alternate approach called the **power method** allows us to approximate the stationary distribution using only matrix-times-vector multiplication. Here is how it works.

Write the eigenvalue decomposition of T as $T = V\Lambda V^{-1}$, where the diagonal matrix $\Lambda = \text{diag}(1,\ \lambda_2,\ \ldots,\ \lambda_M)$ lists the eigenvalue 1 first. Then the first column of V is an eigenvector of T corresponding to the eigenvalue $\lambda_1 = 1$. Denote this column of V by $\mathbf{v}_1$. In other words, $\mathbf{v}_1$ is a constant multiple of the stationary distribution we wish to find.

From the Perron–Frobenius theorem 8.7, we know that 1 is the largest eigenvalue of T in absolute value. Therefore,

$$\lim_{k\to\infty} \Lambda^k = \lim_{k\to\infty} \operatorname{diag}(1,\, \lambda_2^k,\, \ldots,\, \lambda_M^k) = \operatorname{diag}(1,\, 0,\, \ldots,\, 0).$$

Take any probability vector $\mathbf{v}^{(0)}$ as the initial distribution, and let $\mathbf{w} = V^{-1}\mathbf{v}^{(0)}$. Then, for every k,

$$T^k\mathbf{v}^{(0)} = V\Lambda^k V^{-1}\mathbf{v}^{(0)} = V\Lambda^k\mathbf{w} = V\begin{bmatrix} w_1 \\ \lambda_2^k w_2 \\ \vdots \\ \lambda_M^k w_M \end{bmatrix}.$$

It follows that

$$\lim_{k\to\infty} T^k\mathbf{v}^{(0)} = \lim_{k\to\infty} V\begin{bmatrix} w_1 \\ \lambda_2^k w_2 \\ \vdots \\ \lambda_M^k w_M \end{bmatrix} = V\begin{bmatrix} w_1 \\ 0 \\ \vdots \\ 0 \end{bmatrix} = w_1\mathbf{v}_1\,. \tag{8.4}$$

Each vector $T^k\mathbf{v}^{(0)} = V\Lambda^k\mathbf{w}$ is a probability vector because $\mathbf{v}^{(0)}$ is. Thus, $w_1\mathbf{v}_1$ is also a probability vector. Moreover,

$$T\,(w_1\mathbf{v}_1) = T\left(\lim_{k\to\infty} T^k\mathbf{v}^{(0)}\right) = \lim_{k\to\infty} T^{k+1}\mathbf{v}^{(0)} = w_1\mathbf{v}_1\,. \tag{8.5}$$

Thus, the limit $w_1\mathbf{v}_1$ is exactly the stationary distribution we are looking for.

Algorithm 8.9 The power method for approximating the stationary distribution of an irreducible, finite-state, discrete-time Markov process.

Compute the transition matrix T.
Step 0: Create a probability vector $\mathbf{v}^{(0)}$.
Step k, for $k \geq 1$: Compute $\mathbf{v}^{(k)} = T\mathbf{v}^{(k-1)}$.
Stop when $k = K$, for some set number K; or
stop when $||\mathbf{v}^{(k)} - \mathbf{v}^{(k-1)}|| < \varepsilon$, for some error ε. (This does not guarantee that our approximation is within ε of the stationary distribution.)

Remark Do not compute the vector $\mathbf{w}$ or its first coordinate w_1. The end result $w_1\mathbf{v}_1$ emerges of its own accord as we compute $T^k\mathbf{v}^{(0)}$. Also, *do not* compute the full matrix T^k for any power $k > 1$. Instead, compute each successive vector $\mathbf{v}^{(k)}$ as $T\mathbf{v}^{(k-1)}$.

The rate at which $T^k\mathbf{v}^{(0)}$ converges depends on the value of the next largest eigenvalue, λ_2. In practice, we may not know this number. To approximate the stationary distribution as close as we want, we compute the successive vectors $\mathbf{v}^{(k)}$ until the output stabilizes to a desired number of decimal places.

The probability vector $\mathbf{v}^{(0)} = (1/M)\mathbf{1}$, where the distribution is spread uniformly across the different states, is often a convenient choice to start the algorithm, especially if we have no prior belief about the stationary distribution.

Example 8.10 The matrix $T = \begin{bmatrix} 0.8 & 0.3 \\ 0.2 & 0.7 \end{bmatrix}$ is the transition matrix for some two-state Markov process. Take $\mathbf{v}^{(0)} = \begin{bmatrix} 0.5 \\ 0.5 \end{bmatrix}$ as the initial distribution. Using 8-digit arithmetic, we compute $T^{10}\mathbf{v}^{(0)} = \begin{bmatrix} 0.5999023 \\ 0.4000977 \end{bmatrix}$ and $T^{20}\mathbf{v}^{(0)} = \begin{bmatrix} 0.5999999 \\ 0.4000001 \end{bmatrix}$. For all $k \geq 25$, we get $T^k\mathbf{v}^{(0)} = \begin{bmatrix} 0.6000000 \\ 0.4000000 \end{bmatrix}$. In fact, $\begin{bmatrix} 0.6 \\ 0.4 \end{bmatrix}$ is the *exact* stationary distribution, as we can check.

The other eigenvalue of T is $\lambda_2 = 0.5$. Since $(0.5)^k < 5 * 10^{-8}$ for $k \geq 25$, it follows that $T^k\mathbf{v}^{(0)}$ stabilizes to 8 digits for $k \geq 25$.

8.4 Two-State Markov Processes

A Markov process with just two states exhibits all of the essential ideas behind the power method and is relatively simple to solve. Label the states as A and B. Let p denote the probability that a particle currently in state A will transition to state B, and let q denote the probability that a particle now in state B will move to state A. The problem is only interesting when p and q are not equal to 0 or 1. The transition matrix is given by

$$T = \begin{bmatrix} 1-p & q \\ p & 1-q \end{bmatrix}. \tag{8.6}$$

The matrix $T - \lambda \cdot \mathcal{I}_2$ fails to have an inverse when λ satisfies the quadratic equation

$$\begin{aligned} 0 &= \lambda^2 + (p+q-2)\,\lambda + (1-(p+q)) \\ &= (\lambda - 1)(\lambda - (1-(p+q)))\,. \end{aligned}$$

Thus, $\lambda_1 = 1$ and $\lambda_2 = 1-(p+q)$ are the eigenvalues. For $\lambda_1 = 1$, the eigenvectors have the form $t \cdot \begin{bmatrix} q \\ p \end{bmatrix}$. Of these, $\mathbf{v}^* = \frac{1}{p+q} \cdot \begin{bmatrix} q \\ p \end{bmatrix}$ is a probability vector. The vector $\begin{bmatrix} 1 \\ -1 \end{bmatrix}$ is an eigenvector for the eigenvalue λ_2. Therefore, the eigenvalue decomposition of T is

$$V\Lambda V^{-1} = \begin{bmatrix} q/(p+q) & 1 \\ p/(p+q) & -1 \end{bmatrix} \begin{bmatrix} 1 & 0 \\ 0 & 1-(p+q) \end{bmatrix} \begin{bmatrix} q/(p+q) & 1 \\ p/(p+q) & -1 \end{bmatrix}^{-1}.$$

Note that $|1-(p+q)| < 1$, since $0 < p+q < 2$. Hence, $(1-(p+q))^k$ tends to 0 as k tends to ∞, and

$$\begin{aligned} T^k &= V\Lambda^k V^{-1} \\ &= \begin{bmatrix} q/(p+q) & 1 \\ p/(p+q) & -1 \end{bmatrix} \begin{bmatrix} 1 & 0 \\ 0 & (1-(p+q))^k \end{bmatrix} \begin{bmatrix} q/(p+q) & 1 \\ p/(p+q) & -1 \end{bmatrix}^{-1} \\ &\to \begin{bmatrix} q/(p+q) & 1 \\ p/(p+q) & -1 \end{bmatrix} \begin{bmatrix} 1 & 0 \\ 0 & 0 \end{bmatrix} \begin{bmatrix} 1 & 1 \\ p/(p+q) & -q/(p+q) \end{bmatrix} \\ &= \begin{bmatrix} q/(p+q) & q/(p+q) \\ p/(p+q) & p/(p+q) \end{bmatrix}. \end{aligned}$$

This confirms that $\mathbf{v}^* = \frac{1}{p+q} \cdot \begin{bmatrix} q \\ p \end{bmatrix}$ is the stationary distribution.

In Example 8.10, we have $p = 0.2$ and $q = 0.3$. The eigenvalues of T are $\lambda_1 = 1$ and $\lambda_2 = 1-(p+q) = 0.5$. We compute $q/(p+q) = 0.6$ and $p/(p+q) = 0.4$. Thus, the stationary distribution is $\begin{bmatrix} 0.6 \\ 0.4 \end{bmatrix}$, as we found.

8.5 Ranking Web Pages

Surfing the world wide web can be viewed as a Markov process. Each web page is a *state* where someone using the web could be located. The transitions from state to state are determined by the out-links available on each page. To explore the dynamics of this system, we deploy a large team of web surfers who move from web page to web page by randomly selecting an out-link from their current page. At any point in time, we form a probability vector recording the proportion of our team of web surfers visiting each page at that time. To ensure a discrete-time process, we require all surfers to click on new links simultaneously at regular time intervals. Thus, if $\mathbf{v}$ is the probability vector showing the current distribution of our team, and T is the transition matrix, then $T\mathbf{v}$ is the distribution vector after each surfer clicks on the next out-link.

A stationary distribution for this system is a probability vector $\mathbf{v}^*$ for which $T\mathbf{v}^* = \mathbf{v}^*$. In this case, the proportion of the team of random web surfers visiting each page remains unchanged even as each individual surfer clicks and moves to a new page. The significance of the stationary distribution emerges when we think of the world wide web as a large community whose members are the authors of all the many web pages. When one page includes a link to another, it signifies that the host of the first page considers the second page to be a worthwhile or useful page to visit. If many pages include links to one particular page, then that page is considered worthwhile by many members of the community. A random web surfer would visit that particular page more often because many other pages link to it. Conversely, a page with just a few links from other pages might not be visited as often. The situation is not quite that simple, since a few links from popular pages might bring in as many visitors as a larger number of links from less-popular pages. Overall, the value of a page to the community depends on the pages that link to it, the pages that link to those pages, and so on. Thus, the stationary distribution for this Markov process shows the relative value that the community as a whole places on each page. We use that information to create a ranking of all the pages on the web. This is the core premise behind Google's proprietary algorithm known as *PageRank*.

The existence of a stationary distribution for the world wide web is a consequence of the Perron–Frobenius theorem 8.7. For that theorem to hold, we need to know that the Markov process is *irreducible*. In the context of the world wide web, irreducibility means that there is some minimum number of clicks by which we can reach any page no matter where we started. In reality, there may be self-contained sub-webs that are essentially closed off from the rest of the web. This could be as simple as a single page to which no other page has a link. We will never visit that page unless we start there. To work around this obstacle, a search engine might confine its search to a subset of the web that is known to be well-connected and relevant to the query. Another possible work-around is to allow each web surfer to randomly select any page at all for its next move, with some tiny probability. That is equivalent to pretending that every page has a link to every other page, but a random web surfer will nearly always select an actual out-link. For our purposes here, we assume that we have dealt with this issue and that the transition matrix is indeed irreducible. We apply the power method of Algorithm 8.9 to approximate the stationary distribution of the world wide web.

Example 8.11 (A Miniature Model Web) Ranking the billions of pages on the web is a huge undertaking. Consider instead a tiny model to illustrate the process. The following miniature web has only nine pages, numbered #1 through #9. Each page has links to some of the other pages, as shown in the diagram.

Table 8.1 Linkages between pages of a miniature model web

Page # $\xrightarrow{\text{links to}}$ Page #	Page # $\xrightarrow{\text{links to}}$ Page #
#1 $\longrightarrow$#4, 7, 9.	#6 $\longrightarrow$#3, 4, 9.
#2 $\longrightarrow$#3, 5, 6.	#7 $\longrightarrow$#1, 2, 8.
#3 $\longrightarrow$#8.	#8 $\longrightarrow$#2, 7, 9.
#4 $\longrightarrow$#1, 7.	#9 $\longrightarrow$#1, 4, 5.
#5 $\longrightarrow$#1, 3, 6.	

Perhaps we can try to guess what the rankings would be, given this linkage information. Page #3 might be the weakest, since it links to only one other page. But that page, #8, links to three other pages, each of which has further links. For ranking, though, what matters is how readily one can get *to* page #3 from other pages. Think again about the value placed on a page by the community of web page authors. If page #3 were the main authority on the issues relevant to your keyword, you would expect other pages to mention #3 and have links to it. It would be less important for #3 to have links to other pages. The value of a page is a judgment bestowed by others and can't be coerced. I can easily add links from my own site to other sites I value, but I can't force others to link to me. That will happen only if others recognize my site as worth visiting. Thus, the links into a given web site from other sites matter more than the links out from the site.

Table 8.1 leads to the following transition matrix:

$$T = \begin{bmatrix} 0 & 0 & 0 & 1/2 & 1/3 & 0 & 1/3 & 0 & 1/3 \\ 0 & 0 & 0 & 0 & 0 & 0 & 1/3 & 1/3 & 0 \\ 0 & 1/3 & 0 & 0 & 1/3 & 1/3 & 0 & 0 & 0 \\ 1/3 & 0 & 0 & 0 & 0 & 1/3 & 0 & 0 & 1/3 \\ 0 & 1/3 & 0 & 0 & 0 & 0 & 0 & 0 & 1/3 \\ 0 & 1/3 & 0 & 0 & 1/3 & 0 & 0 & 0 & 0 \\ 1/3 & 0 & 0 & 1/2 & 0 & 0 & 0 & 1/3 & 0 \\ 0 & 0 & 1 & 0 & 0 & 0 & 1/3 & 0 & 0 \\ 1/3 & 0 & 0 & 0 & 0 & 1/3 & 0 & 1/3 & 0 \end{bmatrix} \tag{8.7}$$

This Markov process is irreducible, since T^4 has all positive entries. To start, set $\mathbf{v}^{(0)} = (1/9)\mathbf{1}$. This distributes the team of web surfers equally among all nine pages. When every surfer selects an out-link at random and clicks, the new distribution of surfers is

$$\mathbf{v}^{(1)} = T\mathbf{v}^{(0)} = \frac{1}{54} \cdot \begin{bmatrix} 9 & 4 & 6 & 6 & 4 & 4 & 7 & 8 & 6 \end{bmatrix}^T . \tag{8.8}$$

Now, more surfers visit site #1 than any other site. After another click, the new configuration is

$$\mathbf{v}^{(2)} = T\mathbf{v}^{(1)} = \frac{1}{162} \cdot \begin{bmatrix} 26 & 15 & 12 & 19 & 10 & 8 & 26 & 25 & 21 \end{bmatrix}^T . \tag{8.9}$$

Sites #1 and #7 are now tied for the most visitors, with site #8 close behind. Site #3, which we thought might be the weakest site, is lagging, but sites #5 and #6 have even fewer visitors. After each web surfer has clicked on 30 randomly selected out-links, the preferences are clear: $\mathbf{v}^{(30)}$ is

$$\frac{1}{9} \cdot \begin{bmatrix} 1.584 & 0.864 & 0.672 & 1.056 & 0.648 & 0.504 & 1.44 & 1.152 & 1.08 \end{bmatrix}^T .$$

In fact, these values are within 10^{-10} of the stationary distribution values, as we can check by comparing $\mathbf{v}^{(30)}$ to $\mathbf{v}^{(29)}$. This application of the power method shows that, in the long run, page #1 gets the most visits of any page, with 50 percent more visitors than page #4 and more than three times as many as page #6. Based on this analysis, the ranking of the pages in this miniature web is

$$\#1, \#7, \#8, \#9, \#4, \#2, \#3, \#5, \#6\,.$$

Google estimates that it gets a good approximation of the stationary distribution of the full world wide web by computing $T^{50}\mathbf{v}^{(0)}$. This computation requires several days of computer time and has been called the world's largest linear algebra problem.

Remark The process just outlined provides a full ranking of every page on the web. When we do a search, we only want to see a ranking of pages that are relevant to certain keywords and phrases. So, it might make sense to find the relevant pages first and then rank them only against each other. In practice, that requires applying the power method to a new transition matrix for each search, thus causing search engine users to wait longer to get their results. Google's approach is to rank everything ahead of time. Then the pages relevant to a given search are listed in the order of their overall ranking. This significantly reduces the wait time for users and is part of the reason Google was able to beat out their early competitors in the last years of the twentieth century.

Remark Google's premise that pages can be accurately ranked based on linkages between pages seems objective. Yet, this approach has been demonstrated to reinforce, and even promote, both implicit and explicit biases against different groups of people and sets of ideas. The system is also vulnerable to what has become known as *Google bombing*, in which malicious, or at least mischievous, users can deploy excessive hyperlinking to effectively take control of certain search terms and move specific pages to the top of search results. This is possible in part because all links to a given page are considered as positive recommendations of equal weight. Google's own commercial interests create another potential source of manipulation. To generate revenue, the company relies on users to click on advertisements. The advertisers, for their part, wish to have their ads appear alongside specific search terms. This exposes the ranking system to distortions that can further promote favoritism and bias. The book [18] provides a thorough analysis of the biases and shortcomings inherent in many search engines and other social media applications.

8.6 The Monte Carlo Method

The transition matrix for the web is constantly changing as new pages and new links are added and old ones are deleted. Therefore, any version of the transition matrix is temporary. Linkage information must be updated regularly in order to keep the rankings as accurate as possible. The *Monte Carlo method* offers a way to approximate the stationary distribution by *simulating* the behavior of a random web surfer as it navigates from page to page. A record is kept of how many times the surfer visits each page. Page ranks are then assigned based on the visit counts. One advantage of this approach is that the transition matrix does not have to be computed. The web surfer simply selects a link at random from whatever is available. We trust it to select links with the correct probability. This procedure requires us to track an immense number of page visits. We need powerful computers to do that. The use of the Monte Carlo method to solve problems related to Markov processes has blossomed in these first decades of the twenty-first century and has been dubbed the *Markov Chain Monte Carlo (MCMC) revolution*.

Example 8.12 To illustrate the principles involved, return to the nine-page miniature web with the links listed above in Example 8.11. Playing the part of a web surfer, we select a page at random to start our journey. Then, with uniform probability, we randomly select one of the out-links that appear on our current page. Click on the selected link to travel to a new page. Repeat this procedure many times. We keep a count of how many visits we make to each of the nine pages. Table 8.2 shows the results for two separate experiments, each with 101 page visits. The visit counts are tallied separately and also combined together. Once the experiment is complete and the results are recorded, the pages are ranked from most visits to fewest. For the results in Table 8.2, the rankings for each tally separately and for the combined tallies are shown in Table 8.3.

Table 8.2 Number of visits by a random web surfer to each page of a miniature web

Page	First run visits	Second run visits	Combined visits
1	20	20	40
2	12	9	21
3	5	8	13
4	15	12	27
5	7	7	14
6	7	4	11
7	17	12	29
8	9	13	22
9	9	16	25

In this small sample, pages #3, #5, and #6 received significantly fewer visits than the other pages. This tentatively suggests that these pages are less useful overall. Page #1 received noticeably more visits than any other page, earning its ranking as the strongest page. Behind page #1, pages #7, #4, and #9 have similar numbers of visits. Next come pages #8 and #2. A few hundred more page visits should help to properly distinguish the pages and get reliable rankings.

When we follow a web surfer through many page visits, we are actually sampling from the power method algorithm. If we visit enough pages, we get a reasonable sample of the distribution $T^k \mathbf{v}^{(0)}$. That's why the Monte Carlo method works. The underlying assumption is that there is a true, correct ranking of the pages, meaning

Table 8.3 Page rankings based on the data from Table 8.2

Rankings from first run	**1, 7, 4, 2, 9, 8, 5, 6, 3.**
Rankings from second run	**1, 9, 8, 7, 4, 2, 3, 5, 6.**
Rankings from combined runs	**1, 7, 4, 9, 8, 2, 5, 3, 6.**

that there is a unique stationary distribution for the Markov process whose dynamics we are simulating. Again, this will be the case provided the web we are navigating is irreducible.

Algorithm 8.13 Markov Chain Monte Carlo

- Create the transition matrix T of the model web;
- Choose the total number of page visits to be made;
- Create a vector to count the number of visits to each page;
- Randomly select a page at which to start;
- Visit a new page by sampling from the out-links available on the current page;
- Update the visit count vector by adding $+1$ to the visit count for the new page;
- Stop when the total number of page visits has been reached;
- Rank the pages in descending order of visit count.

Example with *R* 8.14 With a small example like the nine-page web, it makes a fun game to run a Monte Carlo simulation by hand, using dice to randomly select from the available out-links. That is how the data in Tables 8.2 and 8.3 were generated. We really need a computer, though, to work with a larger web or to run many more trials even on a small web. Here is an *R* script to implement the Markov Chain Monte Carlo algorithm for ranking web pages. In this version, only the cumulative page visit counts are stored. The full record of individual page visits is lost.

```
Nsteps=50000
visit.count=double(ncol(T))
visit.new=sample(1:ncol(T),1)
for (i in 2:(Nsteps+1)){
visit.new=sample(1:ncol(T),1,prob=T[,visit.new])
visit.count[visit.new]=visit.count[visit.new]+1}
visit.count
order(visit.count,decreasing=TRUE)
```

We have now looked at two approaches to finding the stationary distribution for the world wide web. The power method is an *analytical* approach for which we must know the linkages between the different pages and the corresponding transition matrix. The Monte Carlo method is a statistical approach based on *simulation*. It relies on repetition and randomness to *approximate*, rather than compute, a solution.

Remark The idea that we can approximate a correct ranking of web pages by what amounts to playing a game, or gambling for lack of a better word, is the reason for the name *Monte Carlo*. For instance, to determine if a mysterious pair of dice is fair or loaded, you can roll them many times and keep track of the results of the rolls. If each of the possible values 2 through 12 shows up about the number of times you would expect for a fair pair of dice, then the dice are probably fair. If you get too many 2s, say, then the dice are probably loaded. A simulation of web surfing uses the same approach to help reveal which web pages are the "most loaded".

8.7 Random Walks on Graphs

A random web surfer moves from page to page by randomly selecting out-links. This same behavior makes sense on an undirected graph of the type studied in Section 3.4. Indeed, an undirected graph is analogous to a web where out-linking is always mutual. To make a random walk through a graph, imagine we start at the vertex ν_j. From there, we can walk to any neighboring vertex along the corresponding edge. The number of neighbors is equal to the degree of the vertex, $\deg(\nu_j)$. Assuming that we select from among the neighbors with uniform probability, there is probability $1/\deg(\nu_j)$ of choosing each neighbor. Recall that the adjacency matrix for the graph has entries $a_{i,j}$ equal to 1 when ν_i and ν_j are neighbors and equal to 0 if not. It follows that, if we are now at vertex ν_j, then we will travel to vertex ν_i with probability $a_{i,j}/\deg(\nu_j)$. This produces the transition matrix for a Markov process where each vertex in the graph is a state and movement between states has the form of a random walk along the edges of the graph.

Definition 8.15 Let $\mathcal{G}$ be a graph with N vertices, labeled ν_j, for $j = 1, \ldots, N$. Let $A = \left[a_{i,j}\right]$ denote the adjacency matrix for $\mathcal{G}$. Let Δ denote the $N \times N$ degree matrix with diagonal entries $\deg(\nu_1), \deg(\nu_2), \ldots, \deg(\nu_N)$, and all other entries 0. A **random walk** along $\mathcal{G}$ is a Markov process with transition matrix $T = A\Delta^{-1}$, i.e., $T[i, j] = a_{i,j}/\deg(\nu_j)$, for $i, j = 1, \ldots, N$.

Example 8.16 The graph from Example 3.6 has $N = 5$ vertices and adjacency matrix A. The degree matrix is $\Delta = \text{diag}(3, 2, 2, 1, 2)$. The transition matrix for a random walk on this graph is

$$
\begin{aligned}
T &= A\Delta^{-1} \\
&= \begin{bmatrix} 0&1&1&0&1 \\ 1&0&0&0&1 \\ 1&0&0&1&0 \\ 0&0&1&0&0 \\ 1&1&0&0&0 \end{bmatrix} \begin{bmatrix} 1/3&&&& \\ &1/2&&& \\ &&1/2&& \\ &&&1& \\ &&&&1/2 \end{bmatrix} \\
&= \begin{bmatrix} 0&1/2&1/2&0&1/2 \\ 1/3&0&0&0&1/2 \\ 1/3&0&0&1&0 \\ 0&0&1/2&0&0 \\ 1/3&1/2&0&0&0 \end{bmatrix}.
\end{aligned}
$$

Notice that T is indeed a column stochastic matrix.

The stationary distribution for T is the probability vector

$$\mathbf{v} = \begin{bmatrix} 0.3 & 0.2 & 0.2 & 0.1 & 0.2 \end{bmatrix}^T .$$

For instance, a random walker on this graph will spend about 3/10 of their time at vertex ν_1 and only 1/10 of their time at ν_4.

For any graph, the vector whose coordinates are the degrees of the different vertices is an eigenvector with eigenvalue 1 for the transition matrix T corresponding to a random walk on the graph. Indeed, for each choice of i between 1 and N, we have

$$\sum_{j=1}^{N} T[i, j] \cdot \deg(\nu_j) = \sum_{j=1}^{N} \frac{a_{i,j}}{\deg(\nu_j)} \cdot \deg(\nu_j) = \sum_{j=1}^{N} a_{i,j} = \deg(\nu_i) .$$

To make this into a probability vector, we divide each coordinate by the sum of all the degrees, which is twice the number of edges in the graph. This gives us the stationary distribution of the random walk *provided the Markov process is irreducible.*

Proposition 8.17 *Let $\mathcal{G}$ be a connected graph whose vertices, $\nu_1, \ldots, \nu_N$, have degrees $deg(\nu_1), \ldots, deg(\nu_N)$, respectively. Let T be the transition matrix for a random walk on $\mathcal{G}$ as defined in Definition 8.15. Suppose T is irreducible. Then the stationary distribution for T is given by the probability vector*

$$\mathbf{v} = \frac{1}{deg(\nu_1) + \cdots + deg(\nu_N)} \begin{bmatrix} deg(\nu_1) & \cdots & deg(\nu_N) \end{bmatrix}^T .$$

Example 8.18 A random walk on a graph can fail to be irreducible. For example, suppose the graph is not connected, but, instead, consists of two or more separate smaller graphs with no edges between them. In that case, we could consider each subgraph as a separate graph in its own right. Another case that is not irreducible is the "Markov mouse" of Example 8.2. If we think of each room in the maze as a vertex, then the corresponding graph is *bipartite*. That means we can partition the vertices into two sets such that all of the edges connect from a vertex in one set to a vertex in the other. In this example, the odd numbered vertices form one set and the even numbered vertices make up the other set. The resulting random walk is not irreducible because we can't walk from an odd-numbered vertex to an even-numbered vertex in an even number of steps. Similarly, we can't walk between two odd-numbered vertices, or between two even-numbered vertices, in an odd number of steps. So there is no power of the transition matrix that has all positive entries.

If the transition matrix for a random walk is irreducible, we can choose to rank the vertices in the graph using the stationary distribution. That amounts to ranking the vertices in order of their degrees. In that case, it may seem pointless to apply the power method or try a Monte Carlo approach. However, for a large graph, representing an extensive social network, say, we might not be able to identify all of the vertices and the connections between them beforehand. By simulating a random walk with many steps, the Monte Carlo method offers a way to approximate the degree of every vertex and, thereby, enhance our understanding of a complex graph.

8.8 Exercises

1. Returning to the Shamborough metropolitan area of Section 3.3, the transition matrix describing the movement between city, suburbs, and exurbs is

$$T = \begin{bmatrix} 0.8 & 0.1 & 0 \\ 0.15 & 0.8 & 0.15 \\ 0.05 & 0.1 & 0.85 \end{bmatrix}.$$

   (a) Verify that $\mathbf{v} = \begin{bmatrix} 3/14 & 6/14 & 5/14 \end{bmatrix}^T$ satisfies $T\mathbf{v} = \mathbf{v}$. (Thus, $\mathbf{v}$ gives the stationary distribution for the Shamborough population.)
   (b) The other eigenvalues of T are 0.8 and 0.65. Given that $(0.8)^{35} \approx 0.00041$, if we use T^{35} to estimate the stationary distribution by means of the power method, how many decimal places of accuracy should we expect in our result? Explain.

2. Suppose three similar products A, B, and C, compete for market share. Consumers who have purchased one product this month will buy either the same product or one of the competing products next month, according to the following probabilities:

$$T = \left[\begin{array}{c|ccc} & \multicolumn{3}{c}{\leftarrow \text{this month}} \\ \text{next month} \downarrow & A & B & C \\ \hline A & 0.6 & 0.3 & 0.4 \\ B & 0.2 & 0.4 & 0.4 \\ C & 0.2 & 0.3 & 0.2 \end{array}\right].$$

   (a) Suppose this month's purchases are the same for each product. What will be the approximate market share for each product after 3 more months?
   (b) Verify that $\mathbf{v} = \begin{bmatrix} 6/13 & 4/13 & 3/13 \end{bmatrix}^T$ satisfies $T\mathbf{v} = \mathbf{v}$. Thus, $\mathbf{v}$ gives the stationary distribution for the market share of these three products.
   (c) The next largest eigenvalue of T is 0.273. Find the smallest value of k for which T^k can be used to estimate the stationary distribution to within three decimal places. Explain.

3. Consider the same three products from the previous exercise. Now suppose product B introduces a loyalty program to encourage repeat purchases. This results in the following new transition matrix for consumers of these products:

$$T = \left[\begin{array}{c|ccc} & \multicolumn{3}{c}{\leftarrow \text{this month}} \\ \text{next month} \downarrow & A & B & C \\ \hline A & 0.6 & 0.2 & 0.4 \\ B & 0.2 & 0.6 & 0.4 \\ C & 0.2 & 0.2 & 0.2 \end{array}\right].$$

   (a) Determine the long-term market share for each product given these new purchasing patterns.
   (b) Do these values for the market share of each product make sense to you? Explain.

4. For each 2×2 column stochastic matrix below, answer the following.

   (i) Compute the stationary distribution.
   (ii) Suppose state A represents *living in the city*, while state B represents *living in the suburbs* of a metropolitan area. In the long run, what proportion of the total population will live in the city?
   (iii) If 60% of the population live in the city in year 0, then how many years will it take for the city population to be within 0.05 of the proportion given by the stationary distribution?

$$\text{(a)} \begin{bmatrix} 0.7 & 0.2 \\ 0.3 & 0.8 \end{bmatrix}; \text{ (b)} \begin{bmatrix} 0.7 & 0.3 \\ 0.3 & 0.7 \end{bmatrix}; \text{ (c)} \begin{bmatrix} 0.7 & 0.4 \\ 0.3 & 0.6 \end{bmatrix}.$$

5. Show that the Markov mouse maze, described in Example 8.2, is not irreducible. The even/odd parity of the room number changes after any odd number of moves, but stays the same after any even number of moves. Therefore, show that $T^k_{[even,even]} = 0$ for all odd k and $T^k_{[odd,even]} = 0$ for all even k. Now explain why there is no value of k for which all entries of T^k are nonzero. Next, show that there are distinct probability vectors $\mathbf{v}_1$ and $\mathbf{v}_2$ such that $T\mathbf{v}_1 = \mathbf{v}_2$ and $T\mathbf{v}_2 = \mathbf{v}_1$. Thus, the power method does not lead to a stationary distribution.
6. Modify the Markov mouse maze, described in Example 8.2, like so. We now allow the mouse to either move to an adjacent room or stay in its current room, with uniform probability. For example, a mouse in room #2 may find itself next in any of the rooms #1, #2, #3, or #5, each with probability $1/4$.

   (a) Write down the new transition matrix T. (That is, enter T into a computer.)
   (b) Show that T^4 has all positive entries. Thus, the new transition matrix is irreducible.
   (c) Find the stationary distribution.

(d) What percent of the time will a mouse be in the middle room #5? In a corner room?

7. Show that the Ehrenfest diffusion model, described in Example 8.1, is not irreducible. The even/odd parity of the number of balls in the first jar changes after any odd number of moves, but stays the same after any even number of moves. Therefore, show that $T^k_{[even,even]} = 0$ for all odd k and $T^k_{[odd,even]} = 0$ for all even k. Now explain why there is no value of k for which all entries of T^k are nonzero. Next, show that there are distinct probability vectors $\mathbf{v}_1$ and $\mathbf{v}_2$ such that $T\mathbf{v}_1 = \mathbf{v}_2$ and $T\mathbf{v}_2 = \mathbf{v}_1$. Thus, the power method does not lead to a stationary distribution.
8. Revisit the "barbell" graph from Exercise #12 in Chapter 3. This graph consists of two triangles joined by a line segment. The adjacency matrix is shown in the earlier exercise.

   (a) Write down the transition matrix for a random walk on this graph, as described in Definition 8.15.
   (b) Compute the stationary distribution for this Markov process. Does the result make sense to you?

9. Revisit the "pentagonal wheel" graph from Exercise 13 in Chapter 3. This is a five-sided wheel with a central hub and spokes. The adjacency matrix is shown in the earlier exercise.

   (a) Write down the transition matrix for a random walk on this graph, as described in Definition 8.15.
   (b) Compute the stationary distribution for this Markov process. Does the result make sense to you?

10. The graph in Project 3.4 has nine vertices and adjacency matrix as shown in the earlier project.

    (a) Write down the transition matrix for the Markov process associated with a random walk on this graph.
    (b) Compute the stationary distribution for this Markov process.

8.9 Project

Project 8.1 (Ranking Web Pages)
Part I. Here we rank the pages of the miniature nine-page web from Example 8.11. The linkages are shown in Table 8.1. The transition matrix is given in equation (8.7).

1. First, we apply the **power method** for approximating the stationary distribution of this model web web. Use a computer to implement Algorithm 8.9. You will need the following steps:

- Type in the transition matrix T.
- Create a probability vector $\mathbf{v}^{(0)}$ as the initial distribution.
- Select the number of clicks, say N, that each web surfer will make.
- For each k from 1 to N, compute $\mathbf{v}^{(k)} = T\mathbf{v}^{(k-1)}$. This is a recursive algorithm where the output from one step becomes the input for the next step. **Do not** compute the full matrix powers T^k.
- Rank the web pages in descending order of the entries of the final distribution vector $\mathbf{v}^{(N)}$.
- Assess the accuracy of your estimate of the stationary distribution by comparing $\mathbf{v}^{(N)}$ to $\mathbf{v}^{(N-1)}$.

2. Now, we apply the **Monte Carlo method** for ranking the pages in this model web. Using the same linkage matrix as above, apply the procedure outlined in Algorithm 8.13. The *R* script provided in Example 8.14 may be helpful. After all page visits have been made, rank the pages in descending order of the visit counts.

Part II. Create your own model web with *at least 20 pages*. It must be possible to reach every page from every other page in some finite number of steps. Then **repeat Part I above** for this new web to estimate the stationary distribution and the ranking of the pages using both the power method and the Markov Chain Monte Carlo approach. *Note:* You will need to type in the transition matrix of your new web.

Project 8.2 (Ranking Sports Teams with Random Walks)
Part I. Here we rank the teams in the Big East men's basketball conference from the 2019/2020 season. In this matrix, the $[i, j]$ entry represents the fraction of the jth team's losses that were to the ith team. For example, two of team #5's ten losses were to team #2; so the $[2, 5]$ entry is $2/10 = 1/5$. Notice that each column has a column sum of 1. This is a *column stochastic matrix.*

$$T = \begin{bmatrix} 0 & 1/5 & 1/15 & 1/13 & 1/10 & 1/5 & 2/13 & 0 & 1/5 & 1/5 \\ 1/8 & 0 & 2/15 & 1/13 & 1/5 & 1/5 & 1/13 & 2/5 & 1/5 & 1/5 \\ 1/8 & 0 & 0 & 1/13 & 1/10 & 0 & 0 & 0 & 0 & 0 \\ 1/8 & 1/5 & 1/15 & 0 & 0 & 0 & 2/13 & 0 & 0 & 0 \\ 1/8 & 0 & 1/15 & 2/13 & 0 & 0 & 1/13 & 0 & 1/5 & 1/5 \\ 1/8 & 1/5 & 2/15 & 2/13 & 1/5 & 0 & 1/13 & 1/5 & 1/5 & 1/10 \\ 0 & 1/5 & 2/15 & 0 & 1/10 & 1/5 & 0 & 0 & 0 & 1/10 \\ 1/4 & 0 & 2/15 & 2/13 & 1/5 & 1/5 & 2/13 & 0 & 1/5 & 0 \\ 1/8 & 1/5 & 2/15 & 2/13 & 1/10 & 0 & 2/13 & 1/5 & 0 & 1/5 \\ 0 & 0 & 2/15 & 2/13 & 0 & 1/5 & 2/13 & 1/5 & 0 & 0 \end{bmatrix}.$$

(The teams are listed alphabetically: (1)Butler; (2) Creighton; (3) DePaul; (4) Georgetown; (5) Marquette; (6) Providence; (7) Saint John's; (8) Seton Hall; (9) Villanova; (10) Xavier.) In this context, when we "visit" a team, we are in effect voting for them as best team. Then we look to see who beat them and randomly select one of those teams to vote for instead. Then we switch our vote again to a team that beat that team, and so on, for as many steps as we want. The more often we find ourselves voting for a particular team, the better that team is, at least in this scheme.

1. First, we apply the **power method** for ranking the teams. Use a computer to complete the following steps:

    (i) Begin with an initial ratings vector $\mathbf{v}^{(0)}$. For instance, if all teams are ranked equally to start, we can take

    $$\mathbf{v}^{(0)} = (0.1) \cdot \begin{bmatrix} 1 & 1 & 1 & 1 & 1 & 1 & 1 & 1 & 1 & 1 \end{bmatrix}^T .$$

    (ii) Compute the next ratings vector as $\mathbf{v}^{(1)} = T\mathbf{v}^{(0)}$.
    (iii) Select a number of steps to make, say N steps. (Remember: each step amounts to switching votes for best team.) For each number k from 1 to N, compute the ratings vector $\mathbf{v}^{(k)} = T\mathbf{v}^{(k-1)}$. Notice that this is a recursive algorithm where the output from one step becomes the input for the next step. We do not compute the full matrix powers T^k.
    (iv) Compute the final ratings vector. Then rank the teams, with the highest-rated team ranked as #1 and so on in descending order of the ratings.

2. Now, we apply the **Monte Carlo method** for ranking the teams. Using the same transition matrix as above, apply the procedure outlined in Algorithm 8.13. The *R* script provided in Example 8.14 may be helpful. After all votes have been counted, rank the teams in descending order of the vote counts. The procedure involves the following steps:

    (i) Randomly select a team at which to begin.
    (ii) Vote for a new team by randomly selecting one of the teams that beat the current team according to the distribution of the current team's losses.
    (iii) Update the vote count for the new team by adding $+1$ to the current vote count for that team.
    (iv) Repeat this voting procedure some specified number of times. Keep a count of how many votes each of the teams has received.
    (v) Determine the team rankings by listing them in descending order of the number of votes each team received.

Part II. Describe a different ranking scenario you would like to explore. What data would you need to gather? What would the entries in the transition matrix represent? What would it mean to be ranked #1 in your scenario?

Chapter 9
Symmetric Matrices

9.1 The Spectral Theorem

The adjacency matrix of an undirected graph is symmetric. In the normal equation (5.4), the matrix $A^T A$ is symmetric. So, what can we say about the eigenvalues and eigenvectors of a symmetric matrix? The answer, it turns out, is, "A lot!"

Suppose A is a symmetric matrix in $\mathbb{M}_{M\times M}(\mathbb{R})$. So, $A^T = A$. Suppose λ and μ are eigenvalues of A with corresponding nonzero eigenvectors $\mathbf{v}$ and $\mathbf{w}$. Now compute

$$\lambda(\mathbf{v}^T\mathbf{w}) = (A\mathbf{v})^T\mathbf{w} = \mathbf{v}^T A^T\mathbf{w} = \mathbf{v}^T(A\mathbf{w}) = \mu(\mathbf{v}^T\mathbf{w})\,. \tag{9.1}$$

Thus, $(\lambda - \mu)(\mathbf{v}^T\mathbf{w}) = 0$. If $\lambda \neq \mu$, it follows that $\mathbf{v}^T\mathbf{w} = 0$. In other words, **eigenvectors that correspond to different eigenvalues are orthogonal to each other.** This basic fact is the key to showing that all eigenvalues of a symmetric matrix are real numbers and that every symmetric matrix has a full eigenvalue decomposition. The full result is stated here. Part *(i)* of this theorem is a restatement of (9.1). The proofs of parts *(ii)* and *(iii)* may be found in texts, such as [1], that offer a more theoretical treatment of linear algebra.

Theorem 9.1 (Spectral Theorem for Real Symmetric Matrices) *Let A be a symmetric $M \times M$ matrix with real number entries. Then A has the following properties:*

(i) Eigenvectors of A corresponding to different eigenvalues are orthogonal to each other.

Supplementary Information The online version contains supplementary material available at https://doi.org/10.1007/978-3-031-39562-8_9.

T. G. Feeman, *Applied Linear Algebra and Matrix Methods*,
Springer Undergraduate Texts in Mathematics and Technology,
https://doi.org/10.1007/978-3-031-39562-8_9

(ii) *The eigenvalues of A are real numbers with eigenvectors in $\mathbb{R}^M$.*

(iii) *A has an eigenvalue decomposition of the form $A = V\Lambda V^T$, where Λ is a diagonal matrix whose diagonal entries are the eigenvalues of A, repeated according to multiplicity, and V is an orthogonal matrix whose columns are unit eigenvectors of A corresponding to the entries of Λ. In particular, $V^T = V^{-1}$.*

The eigenvalue decomposition $A = V\Lambda V^T$ also has an outer product expansion. Writing everything out, we have

$$\Lambda = \begin{bmatrix} \lambda_1 & 0 & \cdots & 0 \\ 0 & \lambda_2 & 0 & \cdots \\ \vdots & \vdots & \ddots & \vdots \\ 0 & 0 & \cdots & \lambda_M \end{bmatrix} \text{ and } V = \left[\begin{array}{c|c|c|c} \mathbf{v}_1 & \mathbf{v}_2 & \cdots & \mathbf{v}_M \end{array}\right],$$

where $A\mathbf{v}_i = \lambda_i\mathbf{v}_i$, for $1 \le i \le M$. Thus,

$$\begin{aligned} V\Lambda V^T &= \left[\begin{array}{c|c|c|c} \mathbf{v}_1 & \mathbf{v}_2 & \cdots & \mathbf{v}_M \end{array}\right] \begin{bmatrix} \lambda_1 & 0 & \cdots & 0 \\ 0 & \lambda_2 & 0 & \cdots \\ \vdots & \vdots & \ddots & \vdots \\ 0 & 0 & \cdots & \lambda_M \end{bmatrix} \begin{bmatrix} \mathbf{v}_1^T \\ \hline \mathbf{v}_2^T \\ \hline \vdots \\ \hline \mathbf{v}_M^T \end{bmatrix} \\ &= \left[\begin{array}{c|c|c|c} \lambda_1\mathbf{v}_1 & \lambda_2\mathbf{v}_2 & \cdots & \lambda_M\mathbf{v}_M \end{array}\right] \begin{bmatrix} \mathbf{v}_1^T \\ \hline \mathbf{v}_2^T \\ \hline \vdots \\ \hline \mathbf{v}_M^T \end{bmatrix} \\ &= \lambda_1\mathbf{v}_1\mathbf{v}_1^T + \lambda_2\mathbf{v}_2\mathbf{v}_2^T + \cdots + \lambda_M\mathbf{v}_M\mathbf{v}_M^T. \end{aligned} \tag{9.2}$$

Geometrically, the columns of V are mutually orthogonal unit vectors in $\mathbb{R}^M$ and can be used as the coordinate axes in a new coordinate framework for $\mathbb{R}^M$. For any vector $\mathbf{x}$ in $\mathbb{R}^M$, the companion vector $\mathbf{y} = V^T\mathbf{x}$ gives the coordinates of $\mathbf{x}$ relative to this new coordinate framework. We have $||\mathbf{y}|| = ||\mathbf{x}||$, from equation (6.22). To find the angle between $\mathbf{x}$ and $A\mathbf{x}$, we need to compute the inner product $\mathbf{x}^T A\mathbf{x}$. This becomes

$$\begin{aligned} \mathbf{x}^T A\mathbf{x} &= \mathbf{x}^T V\Lambda V^T\mathbf{x} \\ &= (V^T\mathbf{x})^T\Lambda(V^T\mathbf{x}) \\ &= \mathbf{y}^T\Lambda\mathbf{y}. \end{aligned} \tag{9.3}$$

According to the results of section 6.5, we also have

$$\begin{aligned} ||\mathbf{y}|| &= ||V^T\mathbf{x}|| = ||\mathbf{x}|| \text{ and} \\ ||A\mathbf{x}|| &= ||V\Lambda V^T\mathbf{x}|| = ||\Lambda V^T\mathbf{x}|| = ||\Lambda\mathbf{y}||. \end{aligned}$$

Thus, the angle between $\mathbf{x}$ and $A\mathbf{x}$ is the same as that between $\mathbf{y}$ and $\Lambda\mathbf{y}$.

9.2 Norm of a Symmetric Matrix

In Definition 5.16, we defined two norms for any matrix—the Frobenius norm and the operator norm. Equations (6.23) and (6.24) show that, if V is an orthogonal matrix, then $||VA|| = ||AV|| = ||A||$ for both norms. For a symmetric matrix A with eigenvalue decomposition $A = V\Lambda V^T$, it follows that A and Λ have the same norm. This is great, because it is straightforward to compute the norm of a diagonal matrix. If $\Lambda = \text{diag}(\lambda_1 \ldots \lambda_M)$, then

$$||\Lambda||_F = \left(\lambda_1^2 + \cdots + \lambda_M^2\right)^{1/2} \text{ and } ||\Lambda||_{op} = \max\{|\lambda_j| : 1 \le j \le M\}. \tag{9.4}$$

That is, the Frobenius norm of a symmetric matrix is the square root of the sum of the squares of its eigenvalues. The operator norm is the largest absolute value of an eigenvalue.

9.3 Positive Semidefinite Matrices

Every real symmetric $M \times M$ matrix A has real number eigenvalues and a full eigenvalue decomposition

$$A = V\Lambda V^T = \sum_{i=1}^{M} \lambda_i \mathbf{v}_i \mathbf{v}_i^T,$$

as in (9.2). For each $\mathbf{x}$ in $\mathbb{R}^M$, set $\mathbf{y} = V^T\mathbf{x}$. The computation in (9.3) implies that

$$\mathbf{x}^T A\mathbf{x} = \mathbf{y}^T \Lambda \mathbf{y} = \lambda_1 y_1^2 + \lambda_2 y_2^2 + \cdots + \lambda_M y_M^2. \tag{9.5}$$

As $||\mathbf{y}|| = ||\mathbf{x}||$, it follows that

$$\min\{\lambda_i\}||\mathbf{x}||^2 \le \mathbf{x}^T A\mathbf{x} \le \max\{\lambda_i\}||\mathbf{x}||^2. \tag{9.6}$$

In particular, suppose the eigenvalues of A are all *nonnegative*. Then $0 \le \mathbf{x}^T A\mathbf{x}$, for all $\mathbf{x}$. Conversely, suppose $0 \le \mathbf{x}^T A\mathbf{x}$ for all $\mathbf{x}$. Then, for each eigenvector $\mathbf{v}_i$, we get $0 \le \mathbf{v}_i^T A\mathbf{v}_i = \lambda_i$. So, the eigenvalues of A are all nonnegative. We have proven the following.

Proposition 9.2 *For a real symmetric $M \times M$ matrix A, the following statements are equivalent:*

(i) All eigenvalues of A are nonnegative.
(ii) For all $\mathbf{x}$ in $\mathbb{R}^M$, $0 \le \mathbf{x}^T A\mathbf{x}$.

Definition 9.3 A real symmetric matrix A that enjoys the properties in Proposition 9.2 is called **positive semidefinite**.

In the eigenvalue decomposition $A = V\Lambda V^T$ of a positive semidefinite matrix A, it is customary to list the nonnegative eigenvalues of A in decreasing order, $\lambda_1 \geq \lambda_2 \geq \dots \lambda_M \geq 0$, on the diagonal of Λ. Each eigenvalue is repeated according to its multiplicity. The columns of V must be arranged to match the ordering of the eigenvalues.

When A is positive semidefinite, the angle between $\mathbf{x}$ and $A\mathbf{x}$ has a nonnegative *cosine*, since $\mathbf{x}^T A\mathbf{x} \geq 0$. Therefore, the angle itself is between 0 and $\pi/2$. In other words, $\mathbf{x}$ and $A\mathbf{x}$ can't point in "opposite" directions. This is another way of seeing that A has nonnegative eigenvalues.

Example 9.4 For any matrix A, of any size array, the matrix $A^T A$ is symmetric. Moreover, $A^T A$ is *positive semidefinite*. Indeed, for all $\mathbf{x}$,

$$\mathbf{x}^T(A^T A\mathbf{x}) = (\mathbf{x}^T A^T)(A\mathbf{x}) = (A\mathbf{x})^T(A\mathbf{x}) = ||A\mathbf{x}||^2 \geq 0\,.$$

Equivalently, suppose $(A^T A)\mathbf{v} = \lambda\mathbf{v}$ for some nonzero vector $\mathbf{v}$. Then

$$0 \leq ||A\mathbf{v}||^2 = \mathbf{v}^T(A^T A\mathbf{v}) = \lambda\, ||\mathbf{v}||^2\,,$$

which implies that $0 \leq \lambda$. So, the eigenvalues of $A^T A$ are nonnegative.

Example 9.5 Let $A = \begin{bmatrix} 1 & 1 & 0 \\ 1 & 1 & 0 \\ 1 & 0 & 1 \\ 1 & 0 & 1 \end{bmatrix}$. Then $A^T A = \begin{bmatrix} 4 & 2 & 2 \\ 2 & 2 & 0 \\ 2 & 0 & 2 \end{bmatrix}$, which has eigenvalues of 6, 2, and 0. We store these, in decreasing order, in the diagonal matrix Λ. We compute corresponding unit eigenvectors and store them as the columns of the orthogonal matrix V. Thus, the eigenvalue decomposition is $A^T A = V\Lambda V^T$, where

$$\Lambda = \begin{bmatrix} 6 & 0 & 0 \\ 0 & 2 & 0 \\ 0 & 0 & 0 \end{bmatrix} \text{ and } V = \begin{bmatrix} 2/\sqrt{6} & 0 & -1/\sqrt{3} \\ 1/\sqrt{6} & -1/\sqrt{2} & 1/\sqrt{3} \\ 1/\sqrt{6} & 1/\sqrt{2} & 1/\sqrt{3} \end{bmatrix}.$$

9.3.1 Matrix Square Roots

In the eigenvalue decomposition $A = V\Lambda V^T$ of a positive semidefinite matrix A, let $\Lambda^{1/2}$ denote the diagonal matrix whose diagonal entries are the nonnegative square roots of the eigenvalues of A. We then observe that

$$(V\Lambda^{1/2}V^T)(V\Lambda^{1/2}V^T) = V\Lambda V^T = A\,. \tag{9.7}$$

Moreover, the matrix $V\Lambda^{1/2}V^T$ is symmetric and positive semidefinite in its own right. We conclude that *every positive semidefinite matrix has a positive semidefinite square root.*

Example 9.6 As seen in Example 9.5, the matrix $B = \begin{bmatrix} 4 & 2 & 2 \\ 2 & 2 & 0 \\ 2 & 0 & 2 \end{bmatrix}$ is positive semidefinite, with eigenvalues 6, 2, and 0. Thus, B has a square root given by $B^{1/2} = V\Lambda^{1/2}V^T$, where $\Lambda^{1/2} = \begin{bmatrix} \sqrt{6} & 0 & 0 \\ 0 & \sqrt{2} & 0 \\ 0 & 0 & 0 \end{bmatrix}$ and V is the same as in the earlier example. That is,

$$B^{1/2} = V\Lambda^{1/2}V^T = \begin{bmatrix} \frac{2\sqrt{6}}{3} & \frac{\sqrt{6}}{3} & \frac{\sqrt{6}}{3} \\ \frac{\sqrt{6}}{3} & \frac{\sqrt{6}+3\sqrt{2}}{6} & \frac{\sqrt{6}-3\sqrt{2}}{6} \\ \frac{\sqrt{6}}{3} & \frac{\sqrt{6}-3\sqrt{2}}{6} & \frac{\sqrt{6}+3\sqrt{2}}{6} \end{bmatrix}.$$

The reader is invited to check that $B^{1/2}B^{1/2} = B$.

9.4 Clusters in a Graph

A cluster within a graph is a subset of vertices that is well-connected internally, but has relatively few connections to vertices outside the subset. For instance, a person's set of friends might be made up of smaller sets, such as friends from work, from school, from the cycling club, and so on. Some of these people may happen to know each other, such as a friend from work who is also a cyclist. But there are more connections within each subset than between subsets. The challenge for the data scientist is to find these clusters when only the aggregate network of connections is known.

So, suppose we have a graph with N vertices and adjacency matrix $A = [a_{i,j}]$. To identify a cluster in the graph, we need to classify different vertices, and pairs of vertices, as *in* or *out*. One way to quantify this effort is to assign a score or *weight* to each vertex in the graph. Let w_i denote the weight for vertex v_i. We now assign a *cost* to each edge in the graph, where the cost is based on the difference between the weights of the endpoints. To make sure the cost of an edge is the same no matter which direction we go along it, we assign the cost value $(w_i - w_j)^2$ to the edge between vertices v_i and v_j, if that edge exists. Since v_i and v_j are connected by an edge if, and only if, $a_{i,j} = 1$, the cost value of each edge is the same as $a_{i,j}(w_i - w_j)^2$. This quantity is equal to 0 when v_i and v_j are not joined by an edge. Adding up the cost values $a_{i,j}(w_i - w_j)^2$ over all i and j results in counting each edge twice. To fix this, for each i, we need only consider values $j > i$. Therefore, the *total cost* of all edges for the given choice of weights is

$$\text{Cost} = \sum_{i=1}^{N} \sum_{j:j>i} a_{i,j} \left(w_i - w_j\right)^2 . \tag{9.8}$$

Our goal is to find a set of weights that minimizes the sum in (9.8). The clusters then consist of vertices whose weights are close together. When the cost is minimized, there will be relatively few "expensive" edges connecting vertices whose weights are far apart.

A simple way to have the total cost equal to 0 is to make all the weights the same. That amounts to selecting the entire graph as one cluster. Also, a fixed constant can be added to every weight in a given set of weights without changing the cost of any edge. So, there is never just one set of weights for the same cost. Finally, the cost can be reduced artificially by multiplying every weight by some small number, say $1/10$. Then the cost is multiplied by $1/100$. To avoid these problems, add the two requirements that $\sum_1^N w_i^2 = 1$ and $\sum_1^N w_i = 0$. With these technical fixes in place, we now state our problem.

Problem 9.7 For a graph with N vertices and adjacency matrix $A = [a_{i,j}]$, find the weights vector $\mathbf{w} = \begin{bmatrix} w_1 & \cdots & w_N \end{bmatrix}^T$ that satisfies

$$\mathbf{1}^T \mathbf{w} = \sum_{i=1}^{N} w_i = 0 \text{ and } ||\mathbf{w}||^2 = \sum_{i=1}^{N} w_i^2 = 1 ,$$

and minimizes the quantity

$$\text{Cost} = \sum_{i=1}^{N} \sum_{j:j>i} a_{i,j} \left(w_i - w_j\right)^2 .$$

This problem may seem daunting, especially if it means we must try every possible unit vector $\mathbf{w}$ as a choice of weights. Fortunately, there is a different way to express the total cost in (9.8). Look at just one specific vertex, say v_i, and add up the quantities $a_{i,j}(w_i - w_j)$ for all j. We get

$$\begin{aligned} \sum_{j=1}^{N} a_{i,j}(w_i - w_j) &= w_i \cdot \sum_{j=1}^{N} a_{i,j} - \sum_{j=1}^{N} a_{i,j} w_j \\ &= \deg(v_i) \cdot w_i - \sum_{j=1}^{N} a_{i,j} w_j \end{aligned} \tag{9.9}$$

$$= \begin{bmatrix} -a_{i,1} & \cdots & \deg(\nu_i) & \cdots & -a_{i,N} \end{bmatrix} \begin{bmatrix} w_1 \\ \vdots \\ w_i \\ \vdots \\ w_N \end{bmatrix}.$$

Let Δ denote the $N \times N$ diagonal matrix whose diagonal entries are the *degrees* of the respective vertices, with all other entries equal to 0. That is,

$$\Delta = \text{diag}(\deg(\nu_1), \deg(\nu_2), \ldots, \deg(\nu_N)). \tag{9.10}$$

The row vector in the last expression in (9.9) is the ith row of the matrix $(\Delta - A)$. Thus, the sum in (9.9) is exactly the ith entry of the vector $(\Delta - A)\mathbf{w}$. This gives us

$$\mathbf{w}^T(\Delta - A)\mathbf{w} = \begin{bmatrix} w_1 & \cdots & w_N \end{bmatrix} \begin{bmatrix} \sum_{j=1}^N a_{1,j}(w_1 - w_j) \\ \vdots \\ \sum_{j=1}^N a_{N,j}(w_N - w_j) \end{bmatrix}$$

$$= \sum_{i=1}^N w_i \cdot \left(\sum_{j=1}^N a_{i,j}(w_i - w_j) \right)$$

combine the $[i, j]$ and $[j, i]$ terms:

$$= \sum_{i=1}^N \sum_{j:j>i} a_{i,j} \cdot \left(w_i \cdot \left(w_i - w_j\right) + w_j \cdot \left(w_j - w_i\right)\right)$$

$$= \sum_{i=1}^N \sum_{j:j>i} a_{i,j} \cdot \left(w_i - w_j\right)^2. \tag{9.11}$$

This is the total cost from (9.8)! The matrix $(\Delta - A)$ has earned its own name.

Definition 9.8 The **Laplacian**, or **graph Laplacian**, of a graph is the matrix $L = \Delta - A$, where A is the adjacency matrix of the graph and Δ is the diagonal matrix whose diagonal entries are the *degrees* of the respective vertices.

Formulas (9.8) and (9.11) together give us $\text{Cost} = \mathbf{w}^T L \mathbf{w}$. To find the optimal weights vector $\mathbf{w}$, notice that the Laplacian matrix L is *symmetric*. Moreover, equation (9.11) shows that $\mathbf{w}^T L \mathbf{w}$ is a sum of squares and, hence, nonnegative, for every $\mathbf{w}$. That means L is a *positive semidefinite* matrix. Therefore, L has an eigenvalue decomposition $L = V \Lambda V^T$, where $V^T = V^{-1}$ and the diagonal entries of Λ are the nonnegative eigenvalues of L in decreasing order: $\lambda_1 \geq \cdots \geq \lambda_{N-1} \geq \lambda_N \geq 0$. In fact, $\lambda_N = 0$, and the last column of V is $(1/\sqrt{N}) \cdot \mathbf{1}$. This is so because the entries in every row of L add up to 0. The *smallest positive eigenvalue* of L is λ_{N-1}, and the next to last column of V is a corresponding unit eigenvector, $\mathbf{v}_{N-1}$.

Now take $\mathbf{w}$ to be any unit vector whose coordinates add up to 0, as required by the statement of Problem 9.7. That is $\mathbf{1}^T \mathbf{w} = \sum w_i = 0$ and $||\mathbf{w}|| = 1$. It follows that the vector $\mathbf{y} = V^T \mathbf{w}$ is also a unit vector, and $y_N = 0$. The computations in (9.5) and (9.6) imply that

$$\begin{aligned} \mathbf{w}^T L \mathbf{w} &= \mathbf{y}^T \Lambda \mathbf{y} \\ &= \lambda_1 y_1^2 + \cdots + \lambda_{N-1} y_{N-1}^2 \\ &\geq \lambda_{N-1} \left(y_1^2 + \cdots + y_{N-1}^2 \right) \\ &= \lambda_{N-1} \, . \end{aligned} \tag{9.12}$$

The minimum is achieved when $\mathbf{y} = \mathbf{e}_{N-1}$, which, in turn, yields

$$\mathbf{w} = V \mathbf{e}_{N-1} = \mathbf{v}_{N-1} \, ,$$

the next to last column vector of V. This solves Problem 9.7.

Proposition 9.9 *Let $\mathcal{G}$ be a graph with N vertices, adjacency matrix $A = [a_{i,j}]$, and graph Laplacian matrix L. The unit vector $\mathbf{w}$ that satisfies $\mathbf{1}^T \mathbf{w} = 0$ and minimizes the quantity*

$$\mathbf{w}^T L \mathbf{w} = \sum_{i=1}^{N} \sum_{j:j>i} a_{i,j} \left(w_i - w_j \right)^2$$

is the unit eigenvector for the smallest positive eigenvalue of L. This eigenvector is called the ***Fiedler vector****.*

Remark The *Fiedler vector* is so named in honor of the Czech mathematician, Miroslav Fiedler (1926–2015), who first recognized its importance, in [8].

Once we have found the optimal weights vector $\mathbf{w}$, we have to interpret the result to identify good clusters of vertices. The main principle is that two vertices in the cluster should have weights that are relatively close. We could form one cluster of all the vertices with positive weights and another cluster of the vertices with negative weights. Or perhaps we could restrict the first cluster to those vertices with the largest positive weights. Some other suitable scheme might emerge from the solution. The specific application we have in mind can guide us.

Example 9.10 Consider the graph with nine vertices and adjacency matrix

$$A = \begin{bmatrix} 0\,0\,0\,1\,0\,0\,1\,0\,0 \\ 0\,0\,1\,0\,0\,1\,0\,0\,0 \\ 0\,1\,0\,0\,1\,1\,0\,1\,0 \\ 1\,0\,0\,0\,0\,0\,1\,0\,0 \\ 0\,0\,1\,0\,0\,0\,0\,0\,1 \\ 0\,1\,1\,0\,0\,0\,0\,0\,0 \\ 1\,0\,0\,1\,0\,0\,0\,1\,0 \\ 0\,0\,1\,0\,0\,0\,1\,0\,0 \\ 0\,0\,0\,0\,1\,0\,0\,0\,0 \end{bmatrix}.$$

The smallest positive eigenvalue of the Laplacian is $\lambda_8 \approx 0.204$, with Fiedler vector

$$\mathbf{v}_8 = \left[0.46\ -0.23\ -0.18\ 0.46\ -0.34\ -0.23\ 0.37\ 0.10\ -0.42\right]^T.$$

This suggests ranking the vertices as $(v_1,\ v_4),\ v_7,\ v_8,\ v_3,\ (v_2,\ v_6),\ v_5,\ v_9$. To form clusters, we might group v_7 with v_1 and v_4, include v_3 with the $(v_2,\ v_6)$ group, and put v_5 and v_9 together.

Example 9.11 In Example 3.10, we discussed the idea of forming a graph whose vertices are the terms in a dictionary used for evaluating search queries. Two vertices (terms) are joined by an edge when the *cosine* of the angle between their respective document vectors exceeds some threshold value. We can identify clusters of closely related terms using the Fiedler vector of the resulting adjacency matrix. For the terms and documents shown in Table 1.3, and using a threshold of 0.5 for the *cosine* value, we get a term graph with 18 vertices and 93 edges. Table 9.1 shows the Fiedler vector for this graph. The terms *analysis*, *Riemann*, and *sequence* form a cluster. The

Table 9.1 The Fiedler vector for the term graph associated with the term–document array in Table 1.3

algebra 0.110	analysis −0.519	applied 0.123	derivative −0.067	eigenvalue 0.125
factorization 0.125	group 0.125	inner product 0.108	linear 0.110	matrix 0.110
nullspace 0.125	orthogonal 0.110	permutation 0.110	Riemann −0.519	sequence −0.519
symmetric 0.106	transpose 0.125	vector 0.110		

term *derivative* serves as the only connection between those terms and the remaining terms. There is a cluster of six terms with component near 0.125 and another cluster of eight terms at 0.106 to 0.110. This is not surprising, perhaps, given that two of the eight books used for this study are in advanced calculus, while the rest are from the somewhat related fields of linear algebra and abstract algebra.

9.5 Clustering a Graph with k-Means

We can use the method of k-means, introduced in Section 3.6, to identify clusters in a graph. Comparing the rows of the adjacency matrix for two different vertices, we see that the vertices are close when they have many common neighbors. Thus, if we form clusters based on the adjacency matrix, we will group together vertices that tend to have more common neighbors. This is similar to how we clustered terms together in the term–document configuration in Example 3.17. It may be difficult to determine an appropriate value of k, representing the number of separate clusters we want to identify. It is also hard to picture what we are doing because each row of the adjacency matrix has as many entries as there are vertices.

We can improve our results by using our knowledge of the graph Laplacian. The simplest approach is to identify each vertex with the corresponding numerical coordinate of the Fiedler vector.

Example 9.12 For the graph with nine vertices in Example 9.10, running the k-means algorithm, with $k = 2$, on the set of numbers that are the coordinates of the Fiedler vector, $\mathbf{v}_8$, separates vertices $\{\nu_1, \nu_4, \nu_7, \nu_8\}$, whose corresponding coordinates are positive, from $\{\nu_2, \nu_3, \nu_5, \nu_6, \nu_9\}$, which have negative coordinates. With $k = 3$, one output from R had the clusters $\{\nu_1, \nu_4, \nu_7\}$, with larger positive coordinates, $\{\nu_8\}$, with coordinate nearest to 0, and $\{\nu_2, \nu_3, \nu_5, \nu_6, \nu_9\}$, which have negative coordinates. Running the algorithm again yielded clusters $\{\nu_1, \nu_4, \nu_7\}$, $\{\nu_2, \nu_3, \nu_6, \nu_8\}$, and $\{\nu_5, \nu_9\}$.

The Fiedler vector reveals top-level clusters in a graph. Bringing the eigenvector for the second-smallest positive eigenvalue of the Laplacian into the picture can add detail by helping to locate clusters within clusters. We now associate the ith vertex, ν_i, with the pair of numbers $\mathbf{v}_{N-1}[i]$ and $\mathbf{v}_{N-2}[i]$, similar to what we did with the SAT scores example earlier. A scatter plot of the points $(\mathbf{v}_{N-1}[i], \mathbf{v}_{N-2}[i])$ might suggest a value for k.

Example 9.13 For the graph in Example 9.10, the Fiedler vector is $\mathbf{v}_8$, as shown there, and the next eigenvector is

$$\mathbf{v}_7 = \left[-0.12\ 0.46\ 0.21\ -0.12\ -0.29\ 0.46\ -0.06\ 0.11\ -0.64\right]^T.$$

In this case, one pass using $k = 4$ gave ν_8 as its own cluster. This is the only vertex where $\mathbf{v}_8[i]$ and $\mathbf{v}_7[i]$ are both positive. The cluster $\{\nu_5, \nu_9\}$ includes vertices where $\mathbf{v}_8[i]$ and $\mathbf{v}_7[i]$ are both negative. Vertices $\{\nu_1, \nu_4, \nu_7\}$ have the largest positive values of $\mathbf{v}_8[i]$, while $\{\nu_2, \nu_3, \nu_6\}$ have the largest positive values of $\mathbf{v}_7[i]$. Figure 9.1 shows these clusters along with the cluster centers.

With $k = 5$, $\{\nu_1, \nu_4\}$, $\{\nu_7\}$, $\{\nu_8\}$, $\{\nu_2, \nu_3, \nu_6\}$, and $\{\nu_5, \nu_9\}$ were the suggested clusters. In Example 9.10, we speculated this clustering just using the Fiedler vector, but now the k-means algorithm finds it for us analytically.

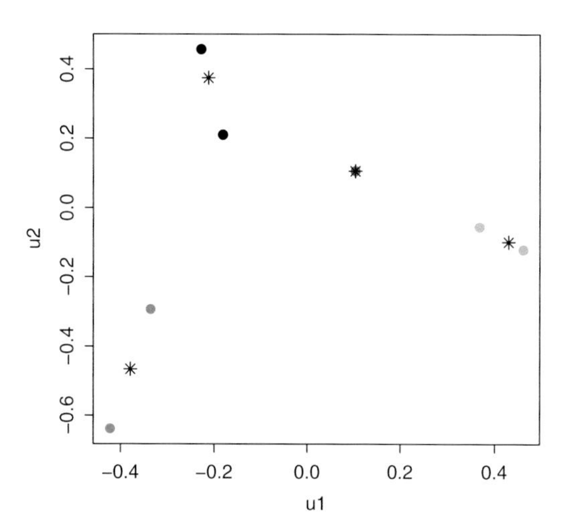

Fig. 9.1 We use two eigenvectors and k-means with $k = 4$ to cluster the vertices in a graph with nine vertices. Each vertex is associated with the point $(\mathbf{v}_8[i], \mathbf{v}_7[i])$. Note that the same point may correspond to more than one vertex

The results of the k-means algorithm may vary from one run to the next, but using the eigenvectors for the smallest few positive eigenvalues of the Laplacian can help give focus to the application of k-means to finding clusters in a graph.

9.6 Drawing a Graph

If we try to draw a graph by randomly placing some vertices and then connecting them according to the entries in the adjacency matrix, the resulting picture may not be very satisfying or explanatory. For instance, it may not be clear where the triangles are in the drawing. Spielman, in [19], suggests using the Fiedler vector to design an attractive presentation for the graph, like so. Put the vertices in order according to the values of the coordinates of $\mathbf{v}_{N-1}$, from high to low, say. Now arrange the vertices in this order around a ring or a horseshoe and connect them accordingly. In this way, most of the edges of the graph will connect vertices that are close to each other in the drawing.

Using a computer, we can make a picture of the adjacency matrix itself as a color grid, where an entry of 0 corresponds to a white square in the grid and an entry of 1 yields a black square. When the vertices are arranged randomly, the resulting grid has no discernible structure apart from symmetry across the diagonal. If we list the vertices in the order of the coordinates of the Fiedler vector and rearrange the rows and columns of the adjacency matrix accordingly, we get a color grid that shows the clusters in the graph near the diagonal of the grid. Figure 9.2 illustrates this for the graph with nine vertices in Example 9.10.

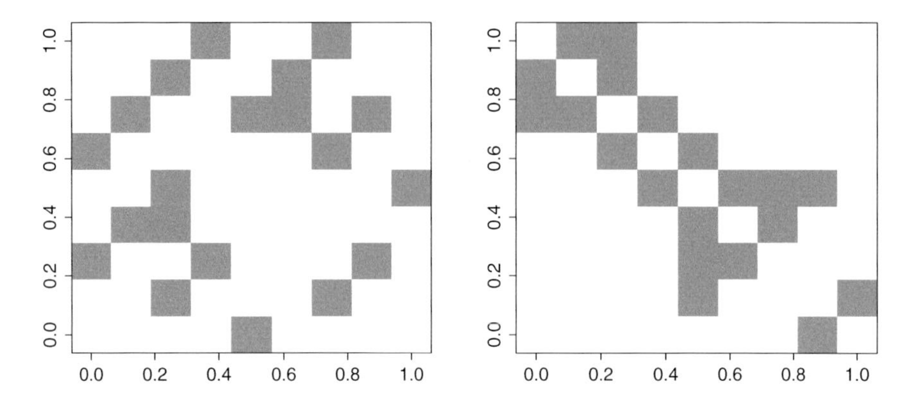

Fig. 9.2 Two renditions of a graph with nine vertices: On the right, the vertices are arranged in the order of the coordinates of the Fiedler vector

9.7 Exercises

1. For each given 2×2 symmetric matrix, do the following:
   (i) Find the eigenvalues of A.
   (ii) Find a unit eigenvector for each eigenvalue.
   (iii) Write down an eigenvalue decomposition $A = V \Lambda V^T$, as described in Theorem 9.1.
   (iv) Express the eigenvalue decomposition as an outer product expansion, as in formula (9.2).

$$\text{(a)} \begin{bmatrix} 3 & 1 \\ 1 & 3 \end{bmatrix}; \text{ (b)} \begin{bmatrix} 6 & -2 \\ -2 & 3 \end{bmatrix}; \text{ (c)} \begin{bmatrix} -1 & 1 \\ 1 & 1 \end{bmatrix}; \text{ (d)} \begin{bmatrix} 1 & 2 \\ 2 & 4 \end{bmatrix}.$$

2. Consider the symmetric matrix $A = \begin{bmatrix} a & b \\ b & a \end{bmatrix}$, where $b \neq 0$.
   (a) Show that $\lambda_1 = a + b$ and $\lambda_2 = a - b$ are the eigenvalues of A by finding corresponding eigenvectors.
   (b) Find a matrix V such that $V^T = V^{-1}$ and

$$A = V \begin{bmatrix} a+b & 0 \\ 0 & a-b \end{bmatrix} V^T.$$

   (c) Apply the previous parts of this exercise to write down the eigenvalue decompositions of the following symmetric matrices:

$$\text{(i)} \begin{bmatrix} 5 & 2 \\ 2 & 5 \end{bmatrix}; \text{ (ii)} \begin{bmatrix} 6 & -2 \\ -2 & 6 \end{bmatrix}; \text{ (iii)} \begin{bmatrix} -1 & 1 \\ 1 & -1 \end{bmatrix}.$$

3. Consider the symmetric matrix $A = \begin{bmatrix} a & b \\ b & -a \end{bmatrix}$, where $b \neq 0$.

   (a) Show that $\begin{bmatrix} b \\ -a + \sqrt{a^2 + b^2} \end{bmatrix}$ and $\begin{bmatrix} -b \\ a + \sqrt{a^2 + b^2} \end{bmatrix}$ are eigenvectors of A by determining the corresponding eigenvalues.

   (b) Use the result of part (a) to find eigenvector/eigenvalue pairs for the following symmetric matrices:

$$\text{(i)} \begin{bmatrix} 1 & 2 \\ 2 & -1 \end{bmatrix}; \ \text{(ii)} \begin{bmatrix} 3 & -4 \\ -4 & -3 \end{bmatrix}; \ \text{(iii)} \begin{bmatrix} -5 & 12 \\ 12 & 5 \end{bmatrix}.$$

4. Let $A = \begin{bmatrix} 1 & 2 \\ 0 & -5 \\ 3 & 4 \end{bmatrix}$.

   (a) Compute the positive semidefinite matrix $A^T A$.

   (b) Compute $\mathbf{x}^T (A^T A\mathbf{x})$ for the generic vector $\mathbf{x} = \begin{bmatrix} x_1 \\ x_2 \end{bmatrix}$.

   (c) Conclude that $10x_1^2 + 28x_1x_2 + 45x_2^2 \geq 0$ for all x_1 and x_2.

5. Let $B = \begin{bmatrix} 3 & 0 \\ -1 & 4 \\ 6 & 5 \end{bmatrix}$.

   (a) Compute the positive semidefinite matrix $B^T B$.

   (b) Compute $\mathbf{x}^T (B^T B\mathbf{x})$ for the generic vector $\mathbf{x} = \begin{bmatrix} x_1 \\ x_2 \end{bmatrix}$.

   (c) Conclude that $46x_1^2 + 52x_1x_2 + 41x_2^2 \geq 0$ for all x_1 and x_2.

6. Find the positive semidefinite square root of each of the following positive semidefinite matrices:

   (a) $A = \begin{bmatrix} 8 & 6 \\ 6 & 17 \end{bmatrix}$.

   (b) $B = \begin{bmatrix} 17 & 4 \\ 4 & 23 \end{bmatrix}$.

   (c) $C = \begin{bmatrix} 10 & 1 & 5 \\ 1 & 10 & 5 \\ 5 & 5 & 6 \end{bmatrix}$.

7. Revisit the graph from Exercise 14 in Chapter 3. This graph consists of a square with one diagonal, connecting vertices v_2 and v_4. The adjacency matrix is $A = \begin{bmatrix} 0&1&0&1 \\ 1&0&1&1 \\ 0&1&0&1 \\ 1&1&1&0 \end{bmatrix}$.

   (a) Verify that the *graph Laplacian* is $L = \begin{bmatrix} 2&-1&0&-1 \\ -1&3&-1&-1 \\ 0&-1&2&-1 \\ -1&-1&-1&3 \end{bmatrix}$.

   (b) Show that $\mathbf{x}^T L\mathbf{x} \geq 0$ for every vector $\mathbf{x}$ in $\mathbb{R}^4$. Conclude that L is *positive semidefinite*.

   (c) Verify, by matrix multiplication, that $L = V \Lambda V^T$, where

$$\Lambda = \begin{bmatrix} 4&0&0&0 \\ 0&4&0&0 \\ 0&0&2&0 \\ 0&0&0&0 \end{bmatrix} \text{ and } V = (1/2) \cdot \begin{bmatrix} 1&0&-\sqrt{2}&1 \\ -1&-\sqrt{2}&0&1 \\ 1&0&\sqrt{2}&1 \\ -1&\sqrt{2}&0&1 \end{bmatrix}.$$

   (d) Identify the *Fiedler vector* for this graph. What clusters of vertices does this vector suggest? Does this result make sense to you?

8. Revisit the "barbell" graph from Exercise 12 in Chapter 3. This graph consists of two triangles joined by a line segment. The adjacency matrix is

$$A = \begin{bmatrix} 0&1&1&0&0&0 \\ 1&0&1&0&0&0 \\ 1&1&0&1&0&0 \\ 0&0&1&0&1&1 \\ 0&0&0&1&0&1 \\ 0&0&0&1&1&0 \end{bmatrix}.$$

   (a) Write down the *graph Laplacian* L for this graph.

   (b) Verify that $\mathbf{x}^T L\mathbf{x} \geq 0$ for every vector $\mathbf{x}$ in $\mathbb{R}^6$.

   (c) Given: $\lambda_5 = (5 - \sqrt{17})/2 \approx 0.43844$ is the smallest positive eigenvalue of L. Verify that the Fiedler vector is

$$\begin{bmatrix} -1 & -1 & -\alpha & \alpha & 1 & 1 \end{bmatrix}^T,$$

   where $\alpha = (\sqrt{17} - 3)/2 \approx 0.56155$.

(d) Interpret the Fiedler vector from part (c) to identify any clusters in this graph. Does this result make sense to you?

9. *(Proof Problem)* Prove that the square of any symmetric matrix is positive semidefinite.

9.8 Projects

Project 9.1 (Finding Clusters in a Graph)

Part I. Here we look at clusters for a graph with twelve (12) vertices and adjacency matrix given by

$$A = \begin{bmatrix} 0&1&0&0&1&1&0&0&1&0&0&0 \\ 1&0&0&0&0&1&1&0&1&0&0&0 \\ 0&0&0&1&0&0&0&1&0&1&0&0 \\ 0&0&1&0&0&0&0&1&0&1&1&1 \\ 1&0&0&0&0&0&0&0&0&0&1&0 \\ 1&1&0&0&0&0&0&0&1&0&0&0 \\ 0&1&0&0&0&0&0&0&1&0&0&0 \\ 0&0&1&1&0&0&0&0&0&1&0&0 \\ 1&1&0&0&0&1&1&0&0&0&0&0 \\ 0&0&1&1&0&0&0&1&0&0&1&1 \\ 0&0&0&1&1&0&0&0&0&1&0&1 \\ 0&0&0&1&0&0&0&0&0&1&1&0 \end{bmatrix}.$$

1. (i) **Create an image** of the adjacency matrix A as given above. The image will be a 12-by-12 grid where each square is either black (if $a_{i,j} = 1$) or white (if $a_{i,j} = 0$). For example, in *R*, an image of a square matrix named `mat` can be formed using the following code:

```
image(t(mat[rev(1:ncol(mat)),1:ncol(mat)]),
col=gray(1-(0.5)*(0:256)/256))
```

   (ii) Write down the Laplacian matrix for this graph.
   (iii) Compute the eigenvalues of the Laplacian for this graph.
   (iv) Identify the smallest positive eigenvalue of the Laplacian; write down the corresponding eigenvector (i.e., the Fiedler vector).
   (v) Re-order the vertices in either ascending or descending order of the coordinates of the eigenvector you just found. Write down this new order.
   (vi) Compute the modified adjacency matrix according to the new ordering of the vertices. Write down the new matrix. For example, in *R*, you might use the code `mat.new=mat[order(vec),order(vec)]`, where `vec` is the eigenvector.

(vii) Create an image of the new adjacency matrix. (As before, the image will be a grid of black and white squares.)

(viii) By hand, draw two sketches of the graph. For the first sketch, arrange 12 dots, labeled with letters ν_1 through ν_{12}, clockwise around a circle and connect them according to the original adjacency matrix A above. For the second sketch, label the dots in the new ordering of the vertices determined by the Fiedler vector; connect the dots accordingly. Comment on the differences you see in these sketches.

Part II. Create your own graph with at least 15 vertices and at least 40 edges. Then repeat the steps of Part I above for this graph. *Note:* You do not have to write down the adjacency matrix or the Laplacian by hand, but you will need to type the adjacency matrix into the computer.

Project 9.2 (Term Clusters in a Term–Document System)
Verify the claims made in Example 9.11. Explore how the adjacency matrix and Fiedler vector change if we raise the *cosine* threshold to 0.65 or to 0.75. Try this out on a term–document system of your own devising.

Chapter 10
Singular Value Decomposition

10.1 Singular Value Decomposition

The eigenvalues and eigenvectors of a square matrix reveal structure and patterns hiding within the numbers. The eigenvalue decomposition of a symmetric or positive semidefinite matrix exposes an orthogonal framework that completely explains the action of the matrix. In this chapter, we extend our eigenvalue analysis to produce a powerful factorization of any non-square matrix.

Begin with an $M \times N$ real matrix A. The $N \times N$ matrix $A^T A$ is positive semidefinite and, so, has an eigenvalue decomposition $A^T A = V \Lambda V^T$. In this decomposition, the diagonal entries of Λ are the nonnegative eigenvalues of $A^T A$, which we now list in decreasing order. So,

$$\Lambda = \begin{bmatrix} \lambda_1 & 0 \cdots & 0 \\ 0 & \lambda_2 \cdots & 0 \\ \vdots & \vdots \ddots & \vdots \\ 0 & 0 \cdots & \lambda_N \end{bmatrix}, \text{ where } \lambda_1 \geq \lambda_2 \geq \ldots \geq \lambda_N \geq 0 .$$

The column vectors $\{\mathbf{v}_j\}$ of the $N \times N$ matrix V are chosen to be mutually orthogonal corresponding unit eigenvectors of $A^T A$. Therefore, $V^T = V^{-1}$. The **rank** of $A^T A$ (and also of A) is the largest number $r \leq N$ for which $\lambda_r > 0$. The nullspace of A has dimension $N - r$.

For each eigenvalue λ_j of $A^T A$, with unit eigenvector $\mathbf{v}_j$, we have

$$||A\mathbf{v}_j|| = \sqrt{(A\mathbf{v}_j)^T A\mathbf{v}_j} = \sqrt{(A^T A\mathbf{v}_j)^T \mathbf{v}_j} = \sqrt{(\lambda_j \mathbf{v}_j)^T \mathbf{v}_j} = \sqrt{\lambda_j} .$$

Supplementary Information The online version contains supplementary material available at https://doi.org/10.1007/978-3-031-39562-8_10.

T. G. Feeman, *Applied Linear Algebra and Matrix Methods*,
Springer Undergraduate Texts in Mathematics and Technology,
https://doi.org/10.1007/978-3-031-39562-8_10

Therefore, for each $j = 1, \ldots, r$, there is a unit vector $\mathbf{u}_j$ such that $A\mathbf{v}_j = \sqrt{\lambda_j}\,\mathbf{u}_j$. Moreover, the vectors $\mathbf{u}_j$ and $\mathbf{u}_k$ are orthogonal if $j \neq k$. This holds even if $\lambda_j = \lambda_k$. Indeed,

$$\begin{aligned}(\mathbf{u}_j)^T\mathbf{u}_k &= \frac{1}{\sqrt{\lambda_j \cdot \lambda_k}}(A\mathbf{v}_j)^T(A\mathbf{v}_k) = \frac{1}{\sqrt{\lambda_j \cdot \lambda_k}}(A^TA\mathbf{v}_j)^T\mathbf{v}_k \\ &= \frac{1}{\sqrt{\lambda_j \cdot \lambda_k}}(\lambda_j\mathbf{v}_j)^T\mathbf{v}_k = \sqrt{\frac{\lambda_j}{\lambda_k}}\,(\mathbf{v}_j^T\mathbf{v}_k) = 0\,,\end{aligned}$$

since the column vectors of V are mutually orthogonal. In this way, we construct an $M \times r$ matrix $\widetilde{U}$ whose columns, $\mathbf{u}_1, \ldots, \mathbf{u}_r$, form an orthonormal set in $\mathbb{R}^M$ and satisfy $A\mathbf{v}_j = \sqrt{\lambda_j}\,\mathbf{u}_j$. This construction also shows that $r \leq M$. If $r < M$, we compute an additional set of orthogonal unit vectors, $\mathbf{u}_{r+1}, \ldots, \mathbf{u}_M$, so that the $M \times M$ matrix $U = \left[\,\widetilde{U}\middle|\mathbf{u}_{r+1}\middle|\cdots\middle|\mathbf{u}_M\,\right]$ is invertible with $U^{-1} = U^T$.

Next, set $\sigma_j = \sqrt{\lambda_j}$, for each eigenvalue λ_j of A^TA. Note that $\sigma_1 \geq \sigma_2 \geq \cdots \geq \sigma_N$ and that $\sigma_j > 0$ for $j \leq r$. For the final ingredient, define the $M \times N$ matrix Σ with diagonal entries $\Sigma_{j,j} = \sigma_j$, for $j = 1, \ldots, \min\{M, N\}$ and all other entries equal to 0. Putting this all together, we have $AV = U\Sigma$, or, equivalently, $A = U\Sigma V^T$.

Definition 10.1 Let A be an $M \times N$ real matrix.

(i) The **singular values** of A are the square roots of the eigenvalues of the positive semidefinite matrix A^TA.
(ii) The **singular value decomposition (SVD)** of A is the factorization

$$A = U\Sigma V^T\,, \tag{10.1}$$

where U is an $M \times M$ orthogonal matrix, V is an $N \times N$ orthogonal matrix, and Σ is an $M \times N$ diagonal matrix whose diagonal entries are the singular values of A, in decreasing order.

The singular value decomposition of A also has the **outer product expansion**

$$A = \sigma_1\mathbf{u}_1\mathbf{v}_1^T + \sigma_2\mathbf{u}_2\mathbf{v}_2^T + \cdots + \sigma_N\mathbf{u}_N\mathbf{v}_N^T\,. \tag{10.2}$$

Each outer product $\mathbf{u}_j\mathbf{v}_j^T$ is an $M \times N$ matrix. As above, r is the largest number for which $\sigma_r > 0$. Since $\sigma_{r+1} = \ldots = \sigma_N = 0$, those terms do not contribute anything to the sum. Each row of $\mathbf{u}_j\mathbf{v}_j^T$ is a scalar multiple of $\mathbf{v}_j^T$, and each column is a scalar multiple of $\mathbf{u}_j$, as we saw in Example 2.29.

Remark When we delete the 0 terms from the outer product expansion in (10.2), we get the **thin SVD** of A, defined by

$$A = \sigma_1\mathbf{u}_1\mathbf{v}_1^T + \sigma_2\mathbf{u}_2\mathbf{v}_2^T + \cdots + \sigma_r\mathbf{u}_r\mathbf{v}_r^T = \widetilde{U}\Sigma_r V_r^T\,, \text{ where} \tag{10.3}$$

$$\widetilde{U} = \left[\mathbf{u}_1 \middle| \cdots \middle| \mathbf{u}_r\right], \ \Sigma_r = \begin{bmatrix} \sigma_1 & & \\ & \ddots & \\ & & \sigma_r \end{bmatrix}, \text{ and } V_r = \left[\mathbf{v}_1 \middle| \cdots \middle| \mathbf{v}_r\right].$$

Example 10.2 Take $A = \begin{bmatrix} 1 & 1 \\ 1 & 2 \\ 1 & 3 \\ 1 & 4 \end{bmatrix}$. Then $A^T A = \begin{bmatrix} 4 & 10 \\ 10 & 30 \end{bmatrix} = V \Lambda V^T$, where

$$V \approx \begin{bmatrix} 0.3220 & -0.9467 \\ 0.9467 & 0.3220 \end{bmatrix} \text{ and } \Lambda \approx \begin{bmatrix} 33.4012 & 0 \\ 0 & 0.5988 \end{bmatrix}.$$

The singular values of A are $\sigma_1 = \sqrt{\lambda_1} \approx 5.7794$ and $\sigma_2 = \sqrt{\lambda_2} \approx 0.7738$. We compute

$$\mathbf{u}_1 = (1/\sigma_1) A \begin{bmatrix} 0.3220 \\ 0.9467 \end{bmatrix} \approx \begin{bmatrix} 0.2195 \\ 0.3833 \\ 0.5472 \\ 0.7110 \end{bmatrix}, \text{ and}$$

$$\mathbf{u}_2 = (1/\sigma_2) A \begin{bmatrix} -0.9467 \\ 0.3220 \end{bmatrix} \approx \begin{bmatrix} -0.8074 \\ -0.3912 \\ 0.0249 \\ 0.4410 \end{bmatrix}.$$

The thin SVD of A is given by

$$A \approx \begin{bmatrix} 0.2195 & -0.8074 \\ 0.3833 & -0.3912 \\ 0.5472 & 0.0249 \\ 0.7110 & 0.4410 \end{bmatrix} \begin{bmatrix} 5.7794 & 0 \\ 0 & 0.7738 \end{bmatrix} \begin{bmatrix} 0.3220 & 0.9467 \\ -0.9467 & 0.3220 \end{bmatrix}.$$

For the full SVD, we need additional unit vectors $\mathbf{u}_3$ and $\mathbf{u}_4$ orthogonal to $\mathbf{u}_1$ and $\mathbf{u}_2$ and to each other, in $\mathbb{R}^4$. For example, set $\mathbf{u}_3 \approx \begin{bmatrix} 0.5472 \\ -0.7120 \\ -0.2176 \\ 0.3824 \end{bmatrix}$ and $\mathbf{u}_4 \approx \begin{bmatrix} 0.0236 \\ -0.4393 \\ 0.8079 \\ -0.3921 \end{bmatrix}$. In this case, $\Sigma = \begin{bmatrix} 5.7794 & 0 \\ 0 & 0.7738 \\ 0 & 0 \\ 0 & 0 \end{bmatrix}$.

Example 10.3 Set $B = \begin{bmatrix} 1 & 1 & -1 \\ 1 & 2 & 0 \\ 1 & 3 & 1 \\ 1 & 4 & 2 \end{bmatrix}$. (This is the matrix in Example 10.2 with an extra column that is a combination of the first two.) We have $B^T B = \begin{bmatrix} 4 & 10 & 2 \\ 10 & 30 & 10 \\ 2 & 10 & 6 \end{bmatrix} = V \Lambda V^T$, where

$$\Lambda \approx \begin{bmatrix} 36.7332 & 0 & 0 \\ 0 & 3.2668 & 0 \\ 0 & 0 & 0 \end{bmatrix} \text{ and } V \approx \begin{bmatrix} 0.2949 & 0.4963 & 0.8165 \\ 0.9028 & 0.1351 & -0.4082 \\ 0.3130 & -0.8576 & 0.4082 \end{bmatrix}.$$

The singular values of B are $\sigma_1 \approx 6.0608$, $\sigma_2 \approx 1.8074$, and $\sigma_3 = 0$. We compute

$$\mathbf{u}_1 = (1/\sigma_1) B \mathbf{v}_1 \approx \frac{1}{6.0608} \begin{bmatrix} 0.8848 \\ 2.1006 \\ 3.3163 \\ 4.5321 \end{bmatrix} \approx \begin{bmatrix} 0.1460 \\ 0.3466 \\ 0.5472 \\ 0.7478 \end{bmatrix} \text{ and}$$

$$\mathbf{u}_2 = (1/\sigma_2) B \mathbf{v}_2 \approx \frac{1}{1.8074} \begin{bmatrix} 1.4890 \\ 0.7666 \\ 0.0441 \\ -0.6783 \end{bmatrix} \approx \begin{bmatrix} 0.8238 \\ 0.4241 \\ 0.0244 \\ -0.3753 \end{bmatrix}.$$

The thin SVD of B is

$$B = \left[\mathbf{u}_1 \middle| \mathbf{u}_2 \right] \cdot \begin{bmatrix} 6.0608 & 0 \\ 0 & 1.8074 \end{bmatrix} \cdot \begin{bmatrix} \mathbf{v}_1^T \\ \hline \mathbf{v}_2^T \end{bmatrix}.$$

For the full SVD, $B = U \Sigma V^T$, we need unit vectors $\mathbf{u}_3$ and $\mathbf{u}_4$ orthogonal to $\mathbf{u}_1$ and $\mathbf{u}_2$ in $\mathbb{R}^4$. Also, $\Sigma = \begin{bmatrix} 6.068 & 0 & 0 \\ 0 & 1.8074 & 0 \\ 0 & 0 & 0 \\ 0 & 0 & 0 \end{bmatrix}$.

The singular value decomposition has a geometric interpretation that comes from thinking of the $M \times N$ matrix A as a linear transformation that maps $\mathbb{R}^N$ into $\mathbb{R}^M$. In this view, A maps the N-dimensional unit sphere of $\mathbb{R}^N$, which is the set $\{\mathbf{x} \in \mathbb{R}^N : ||\mathbf{x}|| = 1\}$, onto an r-dimensional ellipsoid sitting inside $\mathbb{R}^M$, where r is the number of nonzero singular values of A. The principal axes of this ellipsoid lie in the directions of the first r column vectors of the matrix U and have lengths $2 \cdot \sigma_j$, for $j = 1, \ldots, r$. According to the formula $A\mathbf{v}_j = \sigma_j \mathbf{u}_j$, the first r columns of V are the unit vectors that A maps to these principal axes. Thus, using the columns of V as an orthogonal coordinate framework in $\mathbb{R}^N$, and the columns of U as the framework in $\mathbb{R}^M$, the action of A is described by the diagonal matrix Σ. One consequence of this is that the largest singular value, σ_1, is equal to the operator norm of A as a linear transformation. That is, σ_1 is the maximum factor by which the mapping A can stretch the length of any nonzero vector. In symbols,

$$\sigma_1 = ||A||_{op} = \max \{||A\mathbf{x}||/||\mathbf{x}|| \,:\, \mathbf{x} \neq \mathbf{0}\} \,. \tag{10.4}$$

We know that the Frobenius norm of a matrix is unchanged when we multiply the matrix on either side by an orthogonal matrix. Hence,

$$\begin{aligned} ||A||_F &= ||\Sigma||_F \\ &= \sqrt{(\sigma_1)^2 + (\sigma_2)^2 + \cdots + (\sigma_N)^2} \\ &= \sqrt{\lambda_1 + \lambda_2 + \cdots + \lambda_N} \,. \end{aligned} \tag{10.5}$$

In the outer product expansion of A, each of the outer products $\mathbf{u}_j \mathbf{v}_j^T$ has Frobenius norm, and operator norm, equal to $||\mathbf{u}_j|| \cdot ||\mathbf{v}_j|| = 1$.

The singular value decomposition $A = U\Sigma V^T$ also exhibits orthonormal generating sets for all of the four fundamental subspaces associated with A and A^T. As the thin SVD (10.3) shows, the vectors $\mathbf{u}_1, \ldots, \mathbf{u}_r$ generate $Col(A)$, the column space of A. Then the remaining vectors $\mathbf{u}_{r+1}, \ldots, \mathbf{u}_M$ generate $Null(A^T)$, the orthogonal complement to $Col(A)$ in $\mathbb{R}^M$. Similarly, the vectors $\mathbf{v}_{r+1}, \ldots, \mathbf{v}_N$ form an orthonormal basis for $Null(A)$, while $\mathbf{v}_1, \ldots, \mathbf{v}_r$ generate $Col(A^T) = Null(A)^{\perp}$ in $\mathbb{R}^N$.

Example 10.4 For the matrix A in Example 10.2, $Col(A)$ is generated by $\mathbf{u}_1$ and $\mathbf{u}_2$, while $\mathbf{u}_3$ and $\mathbf{u}_4$ are a basis of $Null(A^T)$. The vectors $\mathbf{v}_1$ and $\mathbf{v}_2$ generate $Col(A^T)$, and $Null(A)$ contains only $\mathbf{0}$, the zero vector. In Example 10.3, $Col(B)$ is generated by the vectors $\mathbf{u}_1$ and $\mathbf{u}_2$ of that example. The vectors $\mathbf{v}_1$ and $\mathbf{v}_2$ generate $Col(B^T)$, and $Null(B)$ consists of all numerical multiples of $\mathbf{v}_3$. For a basis of $Null(B^T)$, we have to compute suitable vectors $\mathbf{u}_3$ and $\mathbf{u}_4$ for the full SVD of B.

10.2 Reduced Rank Approximation

In the singular value decomposition $A = U\Sigma V^T$, the column vectors of U form an orthonormal set. Therefore, the outer product expansion of A, shown in (10.2), expresses A as a sum of mutually orthogonal slices. The importance of each slice to the total is determined by the magnitude of the singular value for that slice relative to the other singular values. As equations (10.4) and (10.5) suggest, we can obtain a useful approximation of A by selecting just the biggest slices from the outer product expansion. The smallest slices may be omitted without much loss of information.

For instance, an ellipse with one axis of length 100 and the other axis of length 1 is not much different than a line segment. Truncating the singular value decomposition like this yields a **reduced rank approximation** of the original matrix.

Specifically, the rank of A is equal to the number of nonzero singular values. Now select just the k largest positive singular values and approximate A by

$$A \approx A_k = \sigma_1 \mathbf{u}_1 \mathbf{v}_1^T + \sigma_2 \mathbf{u}_2 \mathbf{v}_2^T + \cdots + \sigma_k \mathbf{u}_k \mathbf{v}_k^T = \sum_{j=1}^{k} \sigma_j \mathbf{u}_j \mathbf{v}_j^T \, . \tag{10.6}$$

The $M \times N$ matrix A_k has rank equal to k and is called the **rank k truncated SVD** of A. To measure how well the truncated SVD approximates the original matrix A, compute the norm $||A - A_k||$. For the Frobenius norm,

$$||A - A_k||_F^2 = ||\sigma_{k+1} \mathbf{u}_{k+1} \mathbf{v}_{k+1}^T + \cdots + \sigma_N \mathbf{u}_N \mathbf{v}_N^T||_F^2 = \sigma_{k+1}^2 + \cdots + \sigma_N^2 \, . \tag{10.7}$$

In the operator norm,

$$||A - A_k||_{op} = ||\sigma_{k+1} \mathbf{u}_{k+1} \mathbf{v}_{k+1}^T + \cdots + \sigma_N \mathbf{u}_N \mathbf{v}_N^T||_{op} = \sigma_{k+1} \, . \tag{10.8}$$

For both norms, the truncated SVD A_k gives a smaller error compared to A than any other matrix of rank k. This is because the difference $(A - A_k)$ has no components in any of the directions corresponding to the k largest singular values of A, while the difference between A and a general matrix of rank k will have such components.

10.3 Image Compression

It has been estimated that as much as 90% of the data currently on the Internet has been created in just the past few years. Every day, millions of images are posted on different web sites, and tens of thousands of hours of video are streamed and uploaded. It is only through *data compression* that the constant transmission of so much information is possible. Researchers have developed a wide range of data compression applications designed to preserve a faithful approximation to the original while substantially reducing the amount of information that must be stored or transmitted. For example, `mpeg` audio files and `jpeg` images are produced using data compression algorithms that have been standardized for worldwide use. Here, we look at a basic compression technique that uses the singular value decomposition and reduced rank approximation.

Every digital image can be represented as a matrix, as described in Section 3.1. The color in each pixel of the image is assigned a numerical value in the corresponding entry of the matrix. Only grayscale images are considered here, though the same principles apply to RGB color images. As a first example, consider the image from Figure 3.1 and its associated 10×10 matrix A, given in (3.1). Using a computer, we learn that A has rank 7, with positive singular values:

$$\sigma_1 = 4.03 > 2.92 > 1.48 > 1.12 > 1.06 > 0.33 > 0.22 = \sigma_7 .$$

(The singular value 0 has multiplicity 3.) In the singular value decomposition $A = \sum_{j=1}^{7} \sigma_j \mathbf{u}_j \mathbf{v}_j^T$, think of each summand as representing a "layer" of the image. The first three layers, with the largest singular values, capture many fundamental features of the image. The first five layers together portray nearly all the essential information. Figure 10.1 illustrates this.

The image on the left in Figure 10.2 shows an idealized CAT scan slice of a brain, depicted in a 256×256 grid of pixels. This image, called a *phantom*, is defined mathematically, not built from actual X-ray data. The phantom shown here has rank 120 when interpreted as a matrix of numerical grayscale values. The rank 30 truncated SVD of the matrix reveals most of the important features of the phantom,

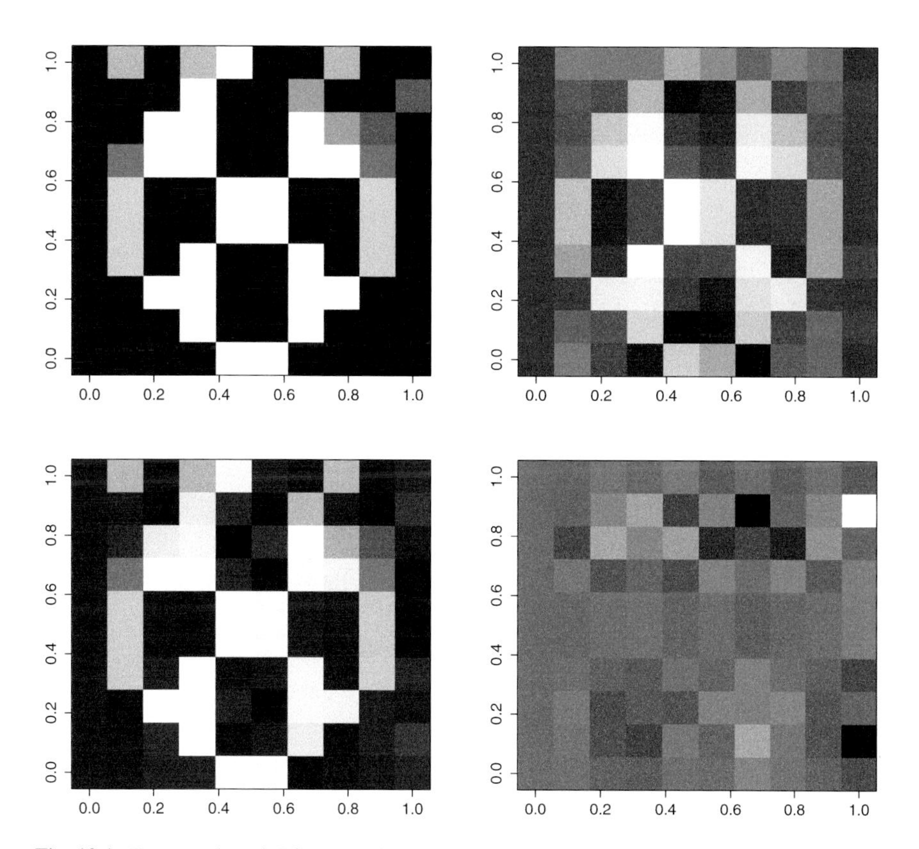

Fig. 10.1 *Top row:* A rank 7 image and its best rank 3 approximation. *Bottom row:* The best rank 5 approximation portrays all fundamental features of the image; the layers corresponding to σ_6 and σ_7 add very little content and could represent noise or distortion

including the “tumor” in the lower right near the skull. This approximation is shown on the right in the figure. To learn more about how the singular value decomposition is used in medical imaging algorithms, see [7].

Example with *R* 10.5 The `imager` package in *R* has a small library of stock digital photographic images that we can use to explore more. Once the package has been installed and loaded, the command `library(imager)` accesses the images. To convert the image `boats` into matrix form and create a grayscale plot, use

```
B=as.matrix(grayscale(boats))
image(B[1:nrow(B),rev(1:ncol(B))],col=gray
     ((0:256)/256))
```

Use `BB=svd(grayscale(boats))` to compute the singular value decomposition. This image matrix has rank 256. The best rank 75 approximation uses less than one-third of the data in the image. Compute and plot this approximation using

```
B75=(BB$u[,1:75]%*%diag(BB$d[1:75])%*%t(BB$v[,1:75]))
image(B75[1:nrow(B75),rev(1:ncol(B75))],col=gray
    ((0:256)/256))
```

The result is a good reproduction of the original, as shown in Figure 10.3.

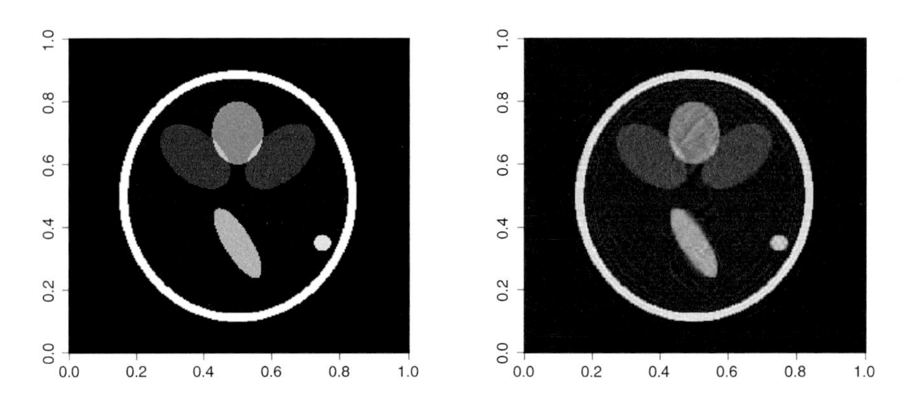

Fig. 10.2 *Left:* An idealized CAT scan image of a brain, with rank 120. *Right:* The rank 30 truncated SVD

10.4 Latent Semantic Indexing

Latent semantic indexing, or LSI, is a method for improving term–document searches and retrieval by using a reduced rank approximation of the original term–document matrix. The description here follows the presentation in [2]. Recall the setup from Section 3.2. For a given dictionary of M terms and a library of N

Fig. 10.3 *Left:* A stock image of sailboats, with rank 256. *Right:* The rank 75 truncated SVD

documents, we form the $M \times N$ matrix L whose $[i, j]$ entry reflects the occurrence of term i in document j. (We have used 0s and 1s, but it is also possible to use weights that indicate the frequency or importance of each occurrence.) To evaluate a search query vector $\mathbf{q}$, we compute the *cosine* of the angle θ_j between $\mathbf{q}$ and each column vector of L. Thus,

$$\cos(\theta_j) = \frac{(L\mathbf{e}_j)^T \mathbf{q}}{||L\mathbf{e}_j|| \cdot ||\mathbf{q}||},$$

where $\mathbf{e}_j$ is the jth column vector of the identity matrix $\mathcal{I}_N$. If $\cos(\theta_j)$ meets a pre-selected threshold, then document j is recommended. The choice of threshold affects the balance between *precision*, which is the proportion of recommended documents that are a good match for the query, and the *yield*, which measures the proportion of recommended documents overall.

Each column vector of the term–document matrix L represents one document in the library. In the singular value decomposition $L = U\Sigma V^T$, the columns of U effectively represent "super–documents" obtained by combining the actual documents in a way that captures the majority of the variation in term usage. Comparing a search query to these super–documents can improve the results by identifying additional relevant documents that may have missed the cutoff before. To make this work, we select a rank k so that the rank k approximation to L captures most of the overall variation between documents. For instance, if the ratio $(\sigma_1^2 + \cdots + \sigma_k^2)/(\sigma_1^2 + \cdots + \sigma_r^2)$ is about 0.9, then about 90% of the variation will be present in the approximation. So, let $L_k = U_k \Sigma_k V_k^T$ denote the rank k truncation of the singular value decomposition of L. Using L_k in place of L to evaluate a search query vector $\mathbf{q}$, we compute the *cosines* of the angles $\widehat{\theta}_j$ between $\mathbf{q}$ and the columns of L_k. Thus,

$$\cos(\widehat{\theta}_j) = \frac{(L_k\mathbf{e}_j)^T \mathbf{q}}{||L_k\mathbf{e}_j|| \cdot ||\mathbf{q}||} = \frac{\mathbf{e}_j^T (V_k \Sigma_k)(U_k^T \mathbf{q})}{||U_k \Sigma_k V_k^T \mathbf{e}_j|| \cdot ||\mathbf{q}||}. \tag{10.9}$$

Now, $||U_k \Sigma_k V_k^T \mathbf{e}_j|| = ||\Sigma_k V_k^T \mathbf{e}_j||$, because the columns of U_k are an orthonormal set. So, let $\mathbf{s}_j = \Sigma_k V_k^T \mathbf{e}_j$ for each j. Notice that these do not depend on the query, so we can compute them once and use them for all queries. Next, set $\widehat{\mathbf{q}} = U_k^T \mathbf{q}$. Then $U_k \widehat{\mathbf{q}} = U_k U_k^T \mathbf{q}$ is the *projection* of $\mathbf{q}$ into the column space of U_k. Thus, $||\widehat{\mathbf{q}}|| \leq ||\mathbf{q}||$. Plugging this back in to (10.9), we have

$$\cos(\widehat{\theta}_j) \leq \frac{\mathbf{s}_j^T \widehat{\mathbf{q}}}{||\mathbf{s}_j|| \cdot ||\widehat{\mathbf{q}}||} = \cos(\theta_j^*), \tag{10.10}$$

where θ_j^* now denotes the angle between the vector $\mathbf{s}_j$ and the modified query vector $\widehat{\mathbf{q}}$. Since $\cos(\widehat{\theta}_j) \leq \cos(\theta_j^*)$, the threshold $\cos(\theta_j^*) \geq \alpha$ will generally return more documents than our previous analysis. By comparing a query to the first k super-documents, the hope is that the search results will be stronger than before.

Example 10.6 Return to the term–document matrix L in Table 1.3. The singular values are

$$\sigma_1 = 7.175 > 3.128 > 2.637 > 1.614 > 1.36 > 0.988 > 0.948 > 0.672 = \sigma_8 .$$

The first three values capture more than 90% of the total variation of 75. Thus, we will use the rank 3 truncated SVD

$$L_3 = U_3 \begin{bmatrix} 7.175 & 0 & 0 \\ 0 & 3.128 & 0 \\ 0 & 0 & 2.637 \end{bmatrix} V_3^T .$$

The vectors $\mathbf{s}_j$ are the columns of the matrix $\Sigma_3 V_3^T$, equal to

$$\begin{bmatrix} -2.18 & -2.91 & -1.26 & -2.76 & -2.32 & -3.39 & -0.39 & -3.47 \\ -1.12 & 0.47 & -2.08 & -0.20 & 0.14 & 0.43 & -1.71 & 0.90 \\ 1.06 & -1.20 & -0.66 & 1.04 & 1.30 & -0.93 & -0.55 & -0.15 \end{bmatrix}.$$

Suppose now that the query vector $\mathbf{q}$ corresponds to terms 3, 8, 12, and 18 (*applied*, *inner product*, *orthogonal*, and *vector*). The modified query vector is $\widehat{\mathbf{q}} = U_3^T \mathbf{q} = \left[-0.924\ 0.462\ -0.559\right]^T$. Normalizing both the columns of $\Sigma_3 V_3^T$ and the modified query $\widehat{\mathbf{q}}$, we compute the list of *cosine* values $\left[\cos(\theta_j^*)\right]$ as

$$\left[0.288\ 0.957\ 0.192\ 0.539\ 0.474\ 0.925\ -0.058\ 0.880\right]^T .$$

Documents 2, 4, 6, and 8 now meet the 0.5 threshold. Our earlier approach only returned documents 6 and 8. Increasing the threshold to 0.8 now returns documents 2, 6, and 8. In our first analysis, no documents reached the 0.8 threshold.

We can also use the truncated SVD of the term–document matrix to analyze term-to-term comparisons. This concept was introduced in Example 3.2. If L is the term–document matrix at hand, then row i shows the occurrences of term i in the various documents. To compare term i and term j, we compute the *cosine* of the angle $\omega_{i,j}$ between rows i and j of L. That is,

$$\cos(\omega_{i,j}) = \frac{(\mathbf{e}_i^T L)(L^T \mathbf{e}_j)}{||L^T \mathbf{e}_i|| \cdot ||L^T \mathbf{e}_j||} .$$

Now we replace L with its rank k best approximation $L_k = U_k \Sigma_k V_k^T$. Let $\widehat{\omega}_{i,j}$ denote the angle between rows i and j of L_k. We compute

$$\begin{aligned}
\cos(\widehat{\omega}_{i,j}) &= \frac{(\mathbf{e}_i^T L_k)(L_k^T \mathbf{e}_j)}{||L_k^T \mathbf{e}_i|| \cdot ||L_k^T \mathbf{e}_j||} \\
&= \frac{(\mathbf{e}_i^T U_k \Sigma_k V_k^T)(V_k \Sigma_k U_k^T \mathbf{e}_j)}{||V_k \Sigma_k U_k^T \mathbf{e}_i|| \cdot ||V_k \Sigma_k U_k^T \mathbf{e}_j||} \\
&= \frac{(\mathbf{e}_i^T U_k \Sigma_k)(\Sigma_k U_k^T \mathbf{e}_j)}{||\Sigma_k U_k^T \mathbf{e}_i|| \cdot ||\Sigma_k U_k^T \mathbf{e}_j||} .
\end{aligned} \tag{10.11}$$

We have used the facts that $V_k^T V_k = \mathcal{I}$ and that $||V_k \mathbf{x}|| = ||\mathbf{x}||$ for every $\mathbf{x}$. If we now set $\mathbf{w}_i = \Sigma_k U_k^T \mathbf{e}_i$, then it follows that

$$\cos(\widehat{\omega}_{i,j}) = \frac{\mathbf{w}_i^T \mathbf{w}_j}{||\mathbf{w}_i|| \cdot ||\mathbf{w}_j||} .$$

That is, $\widehat{\omega}_{i,j}$ is the angle between $\mathbf{w}_i$ and $\mathbf{w}_j$. In matrix form, let W be the matrix obtained from $\Sigma_k U_k^T$ by normalizing the columns to be unit vectors. Then the $[i,\ j]$ entry of $W^T W$ is equal to $\cos(\widehat{\omega}_{i,j})$. In general, $\cos(\widehat{\omega}_{i,j}) \geq \cos(\omega_{i,j})$, so the same threshold will return more significant comparisons between terms.

Example 10.7 Return to the term–document matrix L from Table 1.3. There are 153 pairs of terms to be compared. In Example 3.2, we found that, for 93 of these pairs, the corresponding angle had a *cosine* value of at least 0.5. Using the rank 3 approximation L_3, we now find that 108 pairs of terms compare favorably using the 0.5 threshold. For example, the term *factorization* (#6) now compares favorably to all but terms 2, 14, and 15, where before it compared well with eight other terms. Raising the threshold to 0.65, we have 90 pairs of terms that pass, compared to only 73 pairs that made this cutoff before. In practice, a search query can be augmented with additional terms that compare well to those in the query in order to enhance the results of the search.

10.5 Principal Component Analysis

The outcome of an event or experiment may depend on contributions from a variety of factors. However, it may not be clear how to assess the importance of each individual factor to the overall result. Consider, for example, the athletics competition known as the *heptathlon*, which combines seven events: 100 meter hurdles; high jump; shot put; 200 meter run; long jump; javelin throw; and 800 meter run. In each event, an athlete's performance is converted to a score according to a formula that rewards 1000 points for a standard performance. The seven scores are combined to give the overall score on which rankings are based. The scoring formulas were developed by Karl Ülbrich, an Austrian mathematician. The heptathlon has been contested by women athletes at the international level since about 1980.

Twenty-six women completed all events of the heptathlon at the 2015 IAAF World Championships, held in Beijing. The medalists were Jessica Ennis–Hill (GBR), Brianne Theisen–Eaton (CAN), and Laura Ikauniece–Admidina (LAT). Here, we analyze the performances of the top 10 finishers, whose scores are recorded in the following 10×7 matrix H. The row sums of H give each athlete's official overall score.

$H =$

	hurdles	high jmp	shot	200 m	long jmp	jav.	800 m
1st	1138	1054	776	1037	985	716	963
2nd	1127	978	774	986	1023	724	942
3rd	1093	941	708	984	949	931	910
4th	1043	1054	833	850	912	928	871
5th	1059	978	733	966	1043	692	970
6th	1000	1054	803	848	940	847	897
7th	1114	1016	783	924	1010	647	865
8th	1153	978	738	1002	893	669	911
9th	1021	978	802	862	899	844	916
10th	1026	1016	781	884	902	791	908

To quantify the impact of each event on an athlete's overall result, we use the concepts of *variance* and *covariance*. This is similar to correlation, but without correcting for differences in the range of units. Specifically, suppose we have M observations, say $\{x_i\}$ and $\{y_i\}$, respectively, for each of the random variables X and Y. Let $\bar{x}$ and $\bar{y}$, respectively, denote the average values of these sets of observations. The **covariance** of X and Y is defined by

$$Cov(X, Y) = \frac{(x_1 - \bar{x}) \cdot (y_1 - \bar{y}) + \cdots + (x_M - \bar{x}) \cdot (y_M - \bar{y})}{M - 1}.$$

We divide by $M - 1$ for technical reasons. (See [16], for example, for more details.) A large positive covariance occurs when X and Y tend to be both above or both below their respective means. A negative value of $Cov(X, Y)$ suggests that when one value is above its mean, the other tends to be below. In the special case where $Y = X$, the resulting sum of squares is called the **variance** of X, denoted $Var(X)$.

Notice that the covariance $Cov(X, Y)$ is the inner product of the mean-adjusted vectors $[x_i - \bar{x}]$ and $[y_i - \bar{y}]$, divided by $(M - 1)$. For the matrix

$$A = \begin{bmatrix} x_1 - \bar{x} & y_1 - \bar{y} \\ \vdots & \vdots \\ x_M - \bar{x} & y_M - \bar{y} \end{bmatrix},$$

we have

$$\frac{1}{M-1} \cdot A^T A = \begin{bmatrix} Var(X) & Cov(X, Y) \\ Cov(X, Y) & Var(Y) \end{bmatrix}.$$

This positive semidefinite matrix is called the **variance/covariance matrix** for the given random variables. The same concept extends automatically to the case where there are more than two random variables to be compared.

Returning to the analysis of the heptathlon, we first center the scores matrix around the mean score for each event. So, we obtain H_0 from the scores matrix H by subtracting the average value of the entries in each column from that column. Next, with $M = 10$ athletes, we compute the variance/covariance matrix $(1/9) \cdot H_0^T H_0$ as

$$\left[\begin{array}{ccccccc} \text{hurdles} & \text{high jmp} & \text{shot} & 200\,m & \text{long jmp} & \text{jav.} & 800\,m \\ \hline 2974.0 & -559.0 & -973.3 & 3329.3 & 1036.3 & -3377.0 & 470.6 \\ & 1606.2 & 1125.6 & -1033.6 & -221.1 & 171.9 & -347.2 \\ & & 1396.1 & -1861.9 & -592.3 & 983.0 & -585.7 \\ & & & 4855.1 & 1773.1 & -3815.2 & 1511.9 \\ & & & & 3132.0 & -2926.7 & 902.1 \\ & & & & & 10931.7 & -1201.4 \\ & & & & & & 1212.0 \end{array}\right].$$

The variance in javelin performances (the [6, 6] entry) is noticeably larger than for the other events. The [2, 6] entry is small in absolute value, suggesting that there is no consistent relation between an athlete's performances in the high jump and the javelin. Two events that involve running, the 100 meter hurdles and the 200 meter run, have a high positive covariance, while the 100 meter hurdles and the shot put have a strong negative covariance—it is hard to have the skill and strength sets needed to excel at both events.

The eigenvalues of the variance/covariance matrix are

$$\lambda_1 = 16231.6 > 4742.3 > 2306.8 > 1576.6 > 951.0 > 293.1 > 5.8 = \lambda_7 .$$

This shows that the first eigenvalue accounts for about 62% of the overall variation in the scores and the first two account for about 80%. In each case, we look at how the scores of the different athletes are spread out along the corresponding eigenvector. The first eigenvector is

$$\mathbf{v}_1 = \left[0.343\ -0.072\ -0.150\ 0.444\ 0.273\ -0.751\ 0.139\right]^T .$$

This vector is called the **first principal component direction**. The coordinates of $\mathbf{v}_1$ with the largest absolute values indicate those events where a marginal change in performance can have the largest overall effect. The component for the javelin (event #6) is the largest one, indicating that the javelin may be the single most important event in distinguishing between the athletes. This also suggests that an athlete can improve her overall standing more by making a modest improvement in the javelin throw than in any other event. A fast 200 meter run (#4) is also key to a strong performance. Looking at both the absolute value and the $+/-$ sign of each coordinates suggests that improvements in the hurdles, 200 meters, or long jump may come at the expense of an athlete's performance in the shot put or javelin and vice versa.

The next eigenvector, called the **second principal component direction**, is

$$\mathbf{v}_2 = \left[0.272\ -0.346\ -0.359\ 0.554\ -0.007\ 0.586\ 0.166\right]^T .$$

This vector indicates that the high jump and shot put offer the next level of events where marginal improvements can pay off. The third largest eigenvalue brings the total variance up to 89%, with the long jump providing the dominant coordinate of the corresponding eigenvector.

We use the principal component directions to compute the **principal component scores**. This is the same as using the singular value decomposition to compute a reduced rank approximation to a matrix. In this example, we obtain the best rank 2 approximation $H_0 \approx \sqrt{\lambda_1}\mathbf{u}_1\mathbf{v}_1^T + \sqrt{\lambda_2}\mathbf{u}_2\mathbf{v}_2^T$. That is,

$$H_0 \approx \begin{bmatrix} 49.8 & -18.0 & -28.0 & 69.7 & 33.7 & -78.0 & 21.7 \\ 42.3 & -15.5 & -24.0 & 59.4 & 28.5 & -65.5 & 18.4 \\ 19.3 & -51.8 & -48.2 & 58.3 & -21.5 & 153.1 & 17.0 \\ -69.7 & 18.8 & 33.9 & -93.2 & -52.2 & 135.5 & -29.1 \\ 36.9 & -6.0 & -14.6 & 46.5 & 30.7 & -88.0 & 14.6 \\ -61.1 & 30.1 & 41.1 & -91.2 & -35.3 & 63.3 & -28.2 \\ 14.4 & 22.7 & 15.2 & 0.6 & 31.3 & -136.7 & 0.9 \\ 48.1 & -14.5 & -24.7 & 65.3 & 34.8 & -87.2 & 20.4 \\ -45.4 & 14.7 & 24.1 & -62.4 & -32.1 & 78.1 & -19.4 \\ -34.4 & 19.5 & 25.2 & -53.1 & -17.9 & 25.3 & -16.3 \end{bmatrix} .$$

The row sums of this matrix give the adjusted "scores" of the athletes in the first two principal component directions:

$$\begin{bmatrix} 50.8 \; 43.6 \; 126.2 \; -55.9 \; 20.1 \; -81.3 \; -51.7 \; 42.2 \; -42.3 \; -51.8 \end{bmatrix}.$$

Ranking the athletes by these scores gives the alternate results

$$\begin{bmatrix} 3 \; 1 \; 2 \; 8 \; 5 \; 9 \; 7 \; 10 \; 4 \; 6 \end{bmatrix}$$

for the ten contestants. This obviously differs from the official results. For instance, the athlete who officially finished fourth is now ranked eighth. Looking at her performances, we see that she had the best shot put and second best javelin throw. The javelin features as a principal component of both the first and second eigenvectors, while the shot put is significant in the second eigenvector. However, her weakest event, the 200 meter run, rivals the javelin in importance in the rank 2 approximation. This athlete can use this analysis to see how a small improvement in her 200 meter run can affect her future performances.

We can use the same principal components, $\mathbf{v}_1$ and $\mathbf{v}_2$, to rank the athletes from a different heptathlon competition. For example, let J equal the matrix of scores for the top ten finishers in the heptathlon at the 2017 IAAF World Championships. Then compute $J_2 = (J\mathbf{v}_1)\mathbf{v}_1^T + (J\mathbf{v}_2)\mathbf{v}_2^T$. This gives a rank 2 approximation of J along the two principal component directions of H. Ranking the athletes according to the row sums of J_2 yields

$$\begin{bmatrix} 5 \; 7 \; 3 \; 8 \; 6 \; 2 \; 9 \; 4 \; 1 \; 10 \end{bmatrix}.$$

Again, this differs significantly from the official results and would come as quite a shock to the winner of that competition. The first two principal components used here account for only 80% of the variability in the scores. At the elite level of athletics, the difference between first and second place is often under 2%. Thus, this reduced-rank analysis is not the whole story. But it can be useful for athletes in identifying areas for improvement and planning their training programs.

10.6 Least Squares with SVD

The *least squares problem* for an overdetermined system $A\mathbf{x} = \mathbf{b}$ asks for a vector $\mathbf{x}$ that minimizes the norm of the error $||A\mathbf{x} - \mathbf{b}||$. Such a vector $\mathbf{x}$ must satisfy the normal equation $A^T A\mathbf{x} = A^T\mathbf{b}$. Now, suppose $\mathbf{x}$ is a least squares solution and let $\mathbf{x}_0$ be the projection of $\mathbf{x}$ into the nullspace of A. Then the difference, $(\mathbf{x} - \mathbf{x}_0)$, is orthogonal to $Null(A)$ and also satisfies the normal equation. Indeed,

$$A^T A(\mathbf{x} - \mathbf{x}_0) = A^T A\mathbf{x} - \mathbf{0} = A^T\mathbf{b}.$$

At the same time, $||\mathbf{x} - \mathbf{x}_0|| \leq ||\mathbf{x}||$, because $\mathbf{x}$ is the hypotenuse of a right triangle that has $(\mathbf{x} - \mathbf{x}_0)$ as an adjacent side. Thus, every solution to the normal equation leads to a solution that is orthogonal to $Null(A)$ with a norm that is the same or smaller. However, there cannot be two distinct least squares solutions both orthogonal to $Null(A)$ since their difference would be simultaneously orthogonal to $Null(A)$ and in $Null(A)$. (See exercise #12.) In other words, there is a *unique solution* to the normal equation that is also orthogonal to the nullspace of A. This particular solution, denoted $\mathbf{x}^+$, has the *smallest possible norm* for solutions to the normal equation and is called the **Moore–Penrose solution** to the normal equation.

Definition 10.8 For a given matrix A and vector $\mathbf{b}$, the **Moore–Penrose solution** to the normal equation $A^T A\mathbf{x} = A^T\mathbf{b}$ is the uniquely determined vector $\mathbf{x}^+$ that satisfies the normal equation and is orthogonal to $Null(A)$. The norm of $\mathbf{x}^+$ is the smallest among all solutions to the normal equation. That is, $\mathbf{x}^+$ is the unique vector such that

(i) $A^T A\mathbf{x}^+ = A^T\mathbf{b}$ and
(ii) $\mathbf{x}^+ \in Null(A)^\perp$.

As a consequence of (i) and (ii), $||\mathbf{x}^+|| = \min\{||\mathbf{x}|| \,:\, A^T A\mathbf{x} = A^T\mathbf{b}\}$.

Remark When $A^T A$ is invertible, then $Null(A)$ is just the zero vector. In this case, the normal equation (5.4) yields the unique least squares solution $\mathbf{x}^+ = \left(A^T A\right)^{-1} A^T\mathbf{b}$, which coincides with the Moore–Penrose solution.

Example 10.9 Let $A = \begin{bmatrix} 1 & 1 & 0 \\ 1 & 1 & 0 \\ 1 & 0 & 1 \\ 1 & 0 & 1 \end{bmatrix}$ and $\mathbf{b} = \begin{bmatrix} 1 & 3 & 8 & 2 \end{bmatrix}^T$. In this case, $A^T A$ is not invertible, and $Null(A)$ is generated by the vector $\begin{bmatrix} -1 & 1 & 1 \end{bmatrix}^T$. Every solution to the normal equation has the form $\begin{bmatrix} 5-t & -3+t & t \end{bmatrix}^T$, for some real number t. The norm of these solutions is minimized when $t = 8/3$. Thus, the Moore–Penrose solution is

$$\mathbf{x}^+ = \begin{bmatrix} 7/3 & -1/3 & 8/3 \end{bmatrix}^T.$$

This vector is indeed orthogonal to $\begin{bmatrix} -1 & 1 & 1 \end{bmatrix}^T$ and, hence, to $Null(A)$, as required.

To find the Moore–Penrose solution $\mathbf{x}^+$ in Example 10.9, we first found the general solution to the normal equation $A^T A\mathbf{x} = A^T\mathbf{b}$ and then figured out which solution had the minimum possible norm. The singular value decomposition offers an alternative. First, rewrite the equation $A\mathbf{x} = \mathbf{b}$ as $U\Sigma V^T\mathbf{x} = \mathbf{b}$. If Σ is invertible, then its inverse has along its diagonal the reciprocals of the diagonal entries of Σ. In general, Σ is not a square matrix, but $M \times N$. Also, Σ may have some 0s on its diagonal. So, let Σ^+ be the $N \times M$ matrix whose diagonal entries are the reciprocals of the nonzero diagonal entries of Σ or 0 where Σ has a 0. Specifically,

$$\Sigma = \begin{bmatrix} \sigma_1 & & & \\ & \ddots & & \\ & & \sigma_r & \\ \hline & & & 0 \\ & & & 0 \end{bmatrix} \Rightarrow \Sigma^+ = \begin{bmatrix} 1/\sigma_1 & & & \\ & \ddots & & \\ & & 1/\sigma_r & \\ \hline & & & 0\ 0 \end{bmatrix}. \tag{10.12}$$

Each of the products $\Sigma^+\Sigma$ and $\Sigma\Sigma^+$ has the $r \times r$ identity matrix in its upper left corner, where r is the number of nonzero singular values. For this reason, Σ^+ is called the **pseudoinverse** of Σ.

Now, consider the vector $\mathbf{x} = V\Sigma^+U^T\mathbf{b}$. We compute

$$\begin{aligned} A^TA\mathbf{x} &= (V\Lambda V^T)V\Sigma^+U^T\mathbf{b} \text{ (since } A^TA = V\Lambda V^T) \\ &= V\Sigma^TU^T\mathbf{b} \text{ (since } V^TV = \mathcal{I} \text{ and } \Lambda\Sigma^+ = \Sigma^T) \\ &= A^T\mathbf{b} \text{ (since } A^T = V\Sigma^TU^T). \end{aligned}$$

We have found a solution to the normal equation! Even more, the vector $\mathbf{x} = V\Sigma^+U^T\mathbf{b}$ is, by its definition, a linear combination of those columns of V that correspond to the *nonzero* singular values of A. These columns of V are orthogonal to $Null(A)$. Therefore, the solution we have found is also orthogonal to $Null(A)$. That's the Moore–Penrose solution $\mathbf{x}^+$. We sum up this discussion.

Definition 10.10 Let A be an $M \times N$ matrix with singular value decomposition $A = U\Sigma V^T$.

(i) The **pseudoinverse** of Σ is the $N \times M$ matrix Σ^+ obtained by taking reciprocals of the nonzero diagonal entries of Σ, leaving any 0s on the diagonal of Σ as they are, and taking the transpose.

(ii) The **pseudoinverse** of the matrix A, also called the **Moore–Penrose inverse**, is given by

$$A^+ = V\Sigma^+U^T = (1/\sigma_1)\mathbf{v}_1\mathbf{u}_1^T + \cdots + (1/\sigma_r)\mathbf{v}_r\mathbf{u}_r^T .$$

(iii) For any compatible vector $\mathbf{b}$, the **Moore–Penrose solution** to the normal equation $A^TA\mathbf{x} = A^T\mathbf{b}$ is

$$\begin{aligned} \mathbf{x}^+ &= A^+\mathbf{b} \\ &= \left(V\Sigma^+U^T\right)\mathbf{b} \\ &= (1/\sigma_1)(\mathbf{u}_1^T\mathbf{b})\mathbf{v}_1 + \cdots + (1/\sigma_r)(\mathbf{u}_r^T\mathbf{b})\mathbf{v}_r . \end{aligned} \tag{10.13}$$

Example 10.11 Take $A = \begin{bmatrix} 1 & 1 & 0 \\ 1 & 1 & 0 \\ 1 & 0 & 1 \\ 1 & 0 & 1 \end{bmatrix}$ and $\mathbf{b} = \begin{bmatrix} 1 & 3 & 8 & 2 \end{bmatrix}^T$, as in Example 10.9.

Following Example 9.5, the SVD is $A = U\Sigma V^T$, where

$$U = \begin{bmatrix} 1/2 & -1/2 & 1/\sqrt{2} & 0 \\ 1/2 & -1/2 & -1/\sqrt{2} & 0 \\ 1/2 & 1/2 & 0 & 1/\sqrt{2} \\ 1/2 & 1/2 & 0 & -1/\sqrt{2} \end{bmatrix}, \quad \Sigma = \begin{bmatrix} \sqrt{6} & 0 & 0 \\ 0 & \sqrt{2} & 0 \\ 0 & 0 & 0 \\ 0 & 0 & 0 \end{bmatrix},$$

$$\text{and } V = \begin{bmatrix} 2/\sqrt{6} & 0 & -1/\sqrt{3} \\ 1/\sqrt{6} & -1/\sqrt{2} & 1/\sqrt{3} \\ 1/\sqrt{6} & 1/\sqrt{2} & 1/\sqrt{3} \end{bmatrix}.$$

The pseudoinverses are

$$\Sigma^+ = \begin{bmatrix} 1/\sqrt{6} & 0 & 0 & 0 \\ 0 & 1/\sqrt{2} & 0 & 0 \\ 0 & 0 & 0 & 0 \end{bmatrix},$$

$$\text{and } A^+ = V\Sigma^+U^T = \begin{bmatrix} 1/6 & 1/6 & 1/6 & 1/6 \\ 1/3 & 1/3 & -1/6 & -1/6 \\ -1/6 & -1/6 & 1/3 & 1/3 \end{bmatrix}.$$

The Moore–Penrose solution is now

$$\mathbf{x}^+ = A^+\mathbf{b} = \begin{bmatrix} 1/6 & 1/6 & 1/6 & 1/6 \\ 1/3 & 1/3 & -1/6 & -1/6 \\ -1/6 & -1/6 & 1/3 & 1/3 \end{bmatrix} \begin{bmatrix} 1 \\ 3 \\ 8 \\ 2 \end{bmatrix} = \begin{bmatrix} 7/3 \\ -1/3 \\ 8/3 \end{bmatrix},$$

as we found before.

For a problem as small as the one in Examples 10.9 and 10.11, all of this machinery may seem overwhelming. Isn't it hard to find the eigenvalues of a matrix larger than 2×2? So isn't the singular value decomposition hard to compute? It didn't seem that bad to solve the normal equation using the familiar method of Gaussian elimination. These are reasonable concerns. We must keep in mind that the matrices that arise in many real-world applications may be quite large. Solving the normal equation using Gaussian elimination could get unwieldy, and finding the solution of minimum norm may be an onerous task. So there are potential computational difficulties whether we use Gaussian elimination or the pseudoinverse.

10.7 Approximate Least Squares Solutions

The formula (10.13) for the Moore–Penrose solution to the normal equation requires the reciprocals of the singular values of the coefficient matrix. This can be problematic when some of the singular values are near 0 or when the ratio of the largest and smallest positive singular values, called the *condition number* of the matrix, is big. Then, the reciprocals of the small values assume an outsized role in the pseudoinverse. One way to tackle this problem is to truncate the singular value decomposition and use a reduced rank approximation in its place. We make a judgment about which singular values are "too small" and cut those terms out. We replace A with its best rank k approximation $A_k = \sum_{j=1}^{k} \sigma_j \mathbf{u}_j \mathbf{v}_j^T$. The approximate solution to the normal equation $A^T A\mathbf{x} = A^T\mathbf{b}$ is then

$$\mathbf{x}_k^+ = (1/\sigma_1)(\mathbf{u}_1^T\mathbf{b})\mathbf{v}_1 + \cdots + (1/\sigma_k)(\mathbf{u}_k^T\mathbf{b})\mathbf{v}_k = \sum_{j=1}^{k} (1/\sigma_j)(\mathbf{u}_j^T\mathbf{b})\mathbf{v}_j\,.$$

Example 10.12 Let $B = \begin{bmatrix} 1 & 1 & 0 & 0.05 \\ 1 & 1 & 0 & -0.05 \\ 1 & 0 & 1 & 0.05 \\ 1 & 0 & 1 & -0.05 \end{bmatrix}$. The singular values are $\sqrt{6}$, $\sqrt{2}$, 0.1, and 0. The condition number is $\sqrt{6}/(0.1) \approx 24.5$. The full pseudoinverse is

$$B^+ = \begin{bmatrix} 1/6 & 1/6 & 1/6 & 1/6 \\ 1/3 & 1/3 & -1/6 & -1/6 \\ -1/6 & -1/6 & 1/3 & 1/3 \\ 5 & -5 & 5 & -5 \end{bmatrix}.$$

Taking $\mathbf{b} = \begin{bmatrix} 1 \; 3 \; 8 \; 2 \end{bmatrix}^T$, the Moore–Penrose solution to the normal equation $B^T B\mathbf{x} = B^T\mathbf{b}$ is $\mathbf{x}^+ = B^+\mathbf{b} = \begin{bmatrix} 7/3 \; -1/3 \; 8/3 \; 20 \end{bmatrix}^T$. The norm of $\mathbf{x}^+$ is dominated by the last coordinate of 20. By contrast, the rank 2 truncated SVD of B is

$$B_2 = \sqrt{6}\mathbf{u}_1\mathbf{v}_1^T + \sqrt{2}\mathbf{u}_2\mathbf{v}_2^T = \begin{bmatrix} 1 & 1 & 0 & 0 \\ 1 & 1 & 0 & 0 \\ 1 & 0 & 1 & 0 \\ 1 & 0 & 1 & 0 \end{bmatrix}.$$

This is the same as B except the fourth column is now all 0s. The pseudoinverse B_2^+ is the same as B^+ except that the bottom row is now all 0s. The approximate Moore–Penrose solution is

$$\mathbf{x}_2^+ = B_2^+\mathbf{b} = \begin{bmatrix} 7/3 & -1/3 & 8/3 & 0 \end{bmatrix}^T .$$

Comparing $B^T B\mathbf{x}_2^+$ to $B^T\mathbf{b}$, we have

$$||B^T B\mathbf{x}_2^+ - B^T\mathbf{b}|| = ||\begin{bmatrix} 0 & 0 & 0 & 0.2 \end{bmatrix}^T|| = 0.2 ,$$

so this approximation is pretty good. Omitting the smallest nonzero singular value did not cost much.

Tikhonov regularization is a different approach to approximately solving a normal equation when the condition number of the coefficient matrix is large. Where truncating the SVD removes the smallest singular values from the picture, Tikhonov regularization uses damping factors to modify all of the singular values and reduce the condition number. Specifically, we select a positive number $\alpha > 0$ as the desired lower bound on the positive singular values. We then dampen each positive entry $1/\sigma_j$ in the pseudoinverse Σ^+ by a factor of $\sigma_j^2/(\sigma_j^2 + \alpha^2)$. Let $\Sigma^\sharp$ denote the resulting matrix. That is, the diagonal entries of $\Sigma^\sharp$ are given by

$$\Sigma^\sharp{}_{j,j} = \begin{cases} \frac{1}{\sigma_j}\left(\frac{\sigma_j^2}{\sigma_j^2+\alpha^2}\right) = \frac{\sigma_j}{\sigma_j^2+\alpha^2} & \text{if } \sigma_j > 0, \\ 0 & \text{if } \sigma_j = 0, \end{cases} \tag{10.14}$$

Now use

$$\mathbf{x}^\sharp = V\Sigma^\sharp U^T\mathbf{b} = \sum_{j:\sigma_j>0} \frac{1}{\sigma_j}\left(\frac{\sigma_j^2}{\sigma_j^2+\alpha^2}\right)\left(\mathbf{u}_j^T\mathbf{b}\right)\mathbf{v}_j \tag{10.15}$$

as an approximate solution to the normal equation $A^T A\mathbf{x} = A^T\mathbf{b}$. The vector $\mathbf{x}^\sharp$ does not actually satisfy this normal equation, but it does solve a related minimization problem that we discuss below.

The matrix $V\Sigma^\sharp U^T$ must be the pseudoinverse of *some* matrix that we call $\widehat{A}$ for now. The vector $\mathbf{x}^\sharp$ in (10.15) is then a solution to the normal equation $\widehat{A}^T\widehat{A}\mathbf{x} = \widehat{A}^T\mathbf{b}$. That also means that $\mathbf{x}^\sharp$ minimizes the quantity $||\widehat{A}\mathbf{x} - \mathbf{b}||$, *not* $||A\mathbf{x} - \mathbf{b}||$. In fact, we can write down exactly what $\widehat{A}$ is in relation to A. When A is an $M \times N$ matrix, then $\widehat{A}$ is the $(M+N) \times N$ matrix

$$\widehat{A} = \begin{bmatrix} A \\ \alpha\mathcal{I}_N \end{bmatrix},$$

where $\mathcal{I}_N$ is the $N \times N$ identity matrix. This gives us

$$\widehat{A}^T\widehat{A} = \begin{bmatrix} A^T & \alpha\mathcal{I} \end{bmatrix} \begin{bmatrix} A \\ \alpha\mathcal{I} \end{bmatrix} = A^T A + \alpha^2\mathcal{I}.$$

Given the eigenvalue decomposition $A^T A = V\Lambda V^T$, we now get $\widehat{A}^T\widehat{A} = \left(A^T A + \alpha^2\mathcal{I}\right) = V(\Lambda + \alpha^2\mathcal{I})V^T$, since $VV^T = \mathcal{I}$. The matrix $(\Lambda + \alpha^2\mathcal{I})$ is invertible, since its diagonal entries are at least as big as α^2. Thus, $\widehat{A}^T\widehat{A}$ is also invertible, and the pseudoinverse of $\widehat{A}$ is given by

$$\begin{aligned}
\widehat{A}^+ &= (\widehat{A}^T\widehat{A})^{-1}\widehat{A}^T \\
&= V(\Lambda + \alpha^2\mathcal{I})^{-1}V^T \cdot \left[A^T \middle| \alpha\mathcal{I} \right] \\
&= V(\Lambda + \alpha^2\mathcal{I})^{-1}V^T \cdot \left[V\Sigma^T U^T \middle| \alpha\mathcal{I} \right] \\
&= \left[V(\Lambda + \alpha^2\mathcal{I})^{-1}\Sigma^T U^T \middle| \alpha V(\Lambda + \alpha^2\mathcal{I})^{-1}V^T \right].
\end{aligned}$$

The diagonal matrix $(\Lambda + \alpha^2\mathcal{I})^{-1}\Sigma^T$ has diagonal entries $\sigma_j/(\sigma_j^2 + \alpha^2)$. When $\sigma_j > 0$, this is the same as $\frac{1}{\sigma_j}\left(\frac{\sigma_j^2}{\sigma_j^2+\alpha^2}\right)$, as in the matrix $\Sigma^\sharp$ defined above. The pseudoinverse of $\widehat{A}$ is

$$\widehat{A}^+ = \begin{bmatrix} V\Sigma^\sharp U^T & \alpha(\widehat{A}^T\widehat{A})^{-1} \end{bmatrix}.$$

This is an $N \times (M + N)$ matrix. We are really interested in the restriction of this matrix to vectors in $\mathbb{R}^M$. So, returning to the normal equation, suppose $\mathbf{b}$ is any vector in $\mathbb{R}^M$, and set $\mathbf{b}^\sharp = \begin{bmatrix} \mathbf{b} \\ \mathbf{0} \end{bmatrix}$, where N 0s have been added below the coordinates of $\mathbf{b}$. Notice that $\widehat{A}^T\mathbf{b}^\sharp = A^T\mathbf{b}$. The solution $\mathbf{x}^\sharp$ satisfies the normal equation $\widehat{A}^T\widehat{A}\mathbf{x} = \widehat{A}^T\mathbf{b}^\sharp$. Hence, $\mathbf{x}^\sharp$ solves the minimization problem

$$\begin{aligned}
\min_{\mathbf{x}} ||\widehat{A}\mathbf{x} - \mathbf{b}^\sharp||^2 &= \min_{\mathbf{x}} \left|\left| \begin{bmatrix} A\mathbf{x} \\ \alpha\mathbf{x} \end{bmatrix} - \begin{bmatrix} \mathbf{b} \\ \mathbf{0} \end{bmatrix} \right|\right|^2 \\
&= \min_{\mathbf{x}} \left|\left| \begin{bmatrix} A\mathbf{x} - \mathbf{b} \\ \alpha\mathbf{x} \end{bmatrix} \right|\right|^2 \\
&= \min_{\mathbf{x}} \left\{ ||A\mathbf{x} - \mathbf{b}||^2 + \alpha^2||\mathbf{x}||^2 \right\}. \qquad (10.16)
\end{aligned}$$

The last equality follows from the Pythagorean theorem. The extra term $\alpha^2||\mathbf{x}||^2$ adds a penalty that is larger when $||\mathbf{x}||$ is large. In particular, we are willing to accept a larger value of $||A\mathbf{x} - \mathbf{b}||$ if it means we can use a smaller $\mathbf{x}$ that reduces the value in (10.16).

Example 10.13 The matrix B in Example 10.12 has singular values $\sqrt{6}$, $\sqrt{2}$, 0.1, and 0 and condition number $10 \cdot \sqrt{6} \approx 24.5$. Set $\alpha = 1/\sqrt{10}$, so $\alpha^2 = 0.1$. Using formula (10.14), we have $\Sigma^\sharp$ given by

$$\begin{bmatrix} \frac{1}{\sqrt{6}}\left(\frac{6}{6.1}\right) & 0 & 0 & 0 \\ 0 & \frac{1}{\sqrt{2}}\left(\frac{2}{2.1}\right) & 0 & 0 \\ 0 & 0 & 10\left(\frac{0.01}{0.11}\right) & 0 \\ 0 & 0 & 0 & 0 \end{bmatrix} \approx \begin{bmatrix} 0.4016 & 0 & 0 & 0 \\ 0 & 0.6734 & 0 & 0 \\ 0 & 0 & 0.909 & 0 \\ 0 & 0 & 0 & 0 \end{bmatrix}.$$

Take $\mathbf{b} = \begin{bmatrix} 1 & 3 & 8 & 2 \end{bmatrix}^T$. Using the factors U and V from the singular value decomposition of B, we compute

$$\mathbf{x}^\sharp = V\Sigma^\sharp U^T\mathbf{b} \approx \begin{bmatrix} 2.295 & -0.281 & 2.561 & 1.818 \end{bmatrix}^T.$$

This is an approximate solution to the normal equation $B^T B\mathbf{x} = B^T\mathbf{b}$. The regularization has a smoothing effect in that the coordinates of $\mathbf{x}^\sharp$ are closer together than those of the Moore–Penrose solution, which we found to be $\mathbf{x}^+ = B^+\mathbf{b} = \begin{bmatrix} 7/3 & -1/3 & 8/3 & 20 \end{bmatrix}^T$. In this case, we have $||B\mathbf{x}^+ - \mathbf{b}|| = 4$ while $||B\mathbf{x}^\sharp - \mathbf{b}|| \approx 4.398$. But this is no longer the quantity being minimized. Instead, we have

$$||B\mathbf{x}^+ - \mathbf{b}||^2 + (0.1)\cdot||\mathbf{x}^+||^2 \approx 57.2667\text{ , while}$$
$$||B\mathbf{x}^\sharp - \mathbf{b}||^2 + (0.1)\cdot||\mathbf{x}^\sharp||^2 \approx 20.868\,.$$

So, the vector $\mathbf{x}^\sharp$ approximately satisfies the normal equation and has a much smaller norm than the Moore–Penrose vector.

Remark In Tikhonov regularization, the value of the parameter α is chosen by the user. A large value of α brings the components of the solution quite close together. The resulting solution is said to be be *over-smoothed*. At the other extreme, choosing $\alpha = 0$ leads to the Moore–Penrose solution. Similarly, with the truncated SVD, using too few terms omits important features of A. With too many terms, we risk obscuring the forest by highlighting every tree.

Remark We have considered several approaches to finding approximate solutions to $A\mathbf{x} = \mathbf{b}$. In every case, we arrive at a vector which is a sum of terms of the form $\phi_j\frac{1}{\sigma_j}\left(\mathbf{u}_j^T\mathbf{b}\right)\mathbf{v}_j$, for some choice of weights $\{\phi_j\}$. For the Moore–Penrose solution, $\mathbf{x}^+$, we have $\phi_j = 1$ for all j. In the truncated SVD, ϕ_j is either equal to 1, when j is small, or 0, when j is large. Tikhonov regularization uses the weights $\phi_j = \left(\frac{\sigma_j^2}{\sigma_j^2+\alpha^2}\right)$. A more general theory of *spectral filtering* allows for other choices of weights. See [10].

10.8 Exercises

1. Let $A = \begin{bmatrix} 1 & 1 \\ 1 & 2 \\ 1 & 3 \\ 1 & 4 \end{bmatrix}$, as in Example 10.2.

   (a) Compute the eigenvalue decomposition of $A^T A$.
   (b) Compute the singular values of A.
   (c) Compute $\mathbf{u}_j = (1/\sigma_j) A \mathbf{v}_j$ for each nonzero singular value σ_j.
   (d) Write down the *thin SVD* of A.
   (e) Compute the *full SVD* of A.

2. Let $B = \begin{bmatrix} 1 & 1 & -1 \\ 1 & 2 & 0 \\ 1 & 3 & 1 \\ 1 & 4 & 2 \end{bmatrix}$, as in Example 10.3.

   (a) Compute the eigenvalue decomposition of $B^T B$.
   (b) Compute the singular values of B.
   (c) Compute $\mathbf{u}_j = (1/\sigma_j) B \mathbf{v}_j$ for each nonzero singular value σ_j.
   (d) Write down the *thin SVD* of B.
   (e) Compute the *full SVD* of B.

3. For each of the following matrices: (i) determine the singular values; (ii) write down the thin SVD; (iii) write down the full SVD; and (iv) write down an orthonormal basis for each of the four fundamental subspaces associated with the matrix. Use a computer to help you.

   (a) $A = \begin{bmatrix} 1 & 1 \\ 2 & -1 \\ 3 & 2 \end{bmatrix}$.

   (b) $B = \begin{bmatrix} 1 & 1 & 2 \\ 2 & -1 & 1 \\ 3 & 2 & 5 \end{bmatrix}$.

   (c) $X = \begin{bmatrix} 1 & 1 \\ 1 & -1 \\ 1 & -1 \\ 1 & 1 \end{bmatrix}$.

   (d) $Y = \begin{bmatrix} 1 & 1 & 1 \\ 1 & -1 & 2 \\ 1 & -1 & 3 \\ 1 & 1 & 1 \end{bmatrix}$.

4. A digital image of a plus sign (+) can be formed using

$$A = \begin{bmatrix} 0\,0\,1\,1\,0\,0 \\ 0\,0\,1\,1\,0\,0 \\ 1\,1\,1\,1\,1\,1 \\ 1\,1\,1\,1\,1\,1 \\ 0\,0\,1\,1\,0\,0 \\ 0\,0\,1\,1\,0\,0 \end{bmatrix}.$$

(a) Use a computer to verify that $\sigma_1 = 4$ and $\sigma_2 = 2$ are the nonzero singular values of A.

(b) Verify that the singular vectors corresponding to $\sigma_1 = 4$ satisfy

$$\mathbf{u}_1 = \mathbf{v}_1 = \frac{\sqrt{3}}{6} \cdot \begin{bmatrix} 1 \\ 1 \\ 2 \\ 2 \\ 1 \\ 1 \end{bmatrix}.$$

(c) Verify that the singular vectors corresponding to $\sigma_2 = 2$ satisfy

$$\mathbf{u}_2 = -\mathbf{v}_2 = \frac{\sqrt{6}}{6} \cdot \begin{bmatrix} -1 \\ -1 \\ 1 \\ 1 \\ -1 \\ -1 \end{bmatrix}.$$

(d) Thus, the outer product expansion of A is given by

$$\sigma_1 \mathbf{u}_1 \mathbf{v}_1^T + \sigma_2 \mathbf{u}_2 \mathbf{v}_2^T .$$

Compute these summands separately. Verify that their sum is A. (See Figure 10.4.)

5. A digital image of a small checkerboard can be formed using

$$A = \begin{bmatrix} 1\,1\,0\,0\,1\,1 \\ 1\,1\,0\,0\,1\,1 \\ 0\,0\,1\,1\,0\,0 \\ 0\,0\,1\,1\,0\,0 \\ 1\,1\,0\,0\,1\,1 \\ 1\,1\,0\,0\,1\,1 \end{bmatrix}.$$

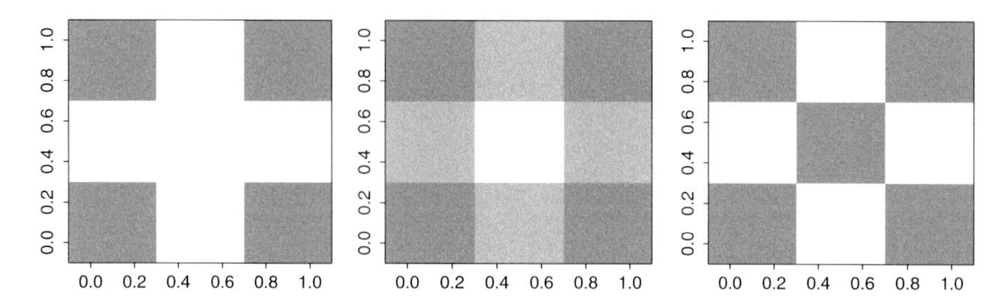

Fig. 10.4 An image of a plus sign + and the two nonzero summands in the outer product expansion. See Exercise #4

(a) Use a compute to verify that $\sigma_1 = 4$ and $\sigma_2 = 2$ are the nonzero singular values of A.

(b) Verify that the singular vectors corresponding to $\sigma_1 = 4$ satisfy

$$\mathbf{u}_1 = \mathbf{v}_1 = \frac{1}{2} \cdot \begin{bmatrix} 1 \\ 1 \\ 0 \\ 0 \\ 1 \\ 1 \end{bmatrix} .$$

(c) Verify that the singular vectors corresponding to $\sigma_2 = 2$ satisfy

$$\mathbf{u}_2 = \mathbf{v}_2 = \frac{\sqrt{2}}{2} \cdot \begin{bmatrix} 0 \\ 0 \\ 1 \\ 1 \\ 0 \\ 0 \end{bmatrix} .$$

(d) Thus, the outer product expansion of A is given by

$$\sigma_1 \mathbf{u}_1 \mathbf{v}_1^T + \sigma_2 \mathbf{u}_2 \mathbf{v}_2^T .$$

Compute these summands separately. Verify that their sum is A. (See Figure 10.5.)

6. Refer to the term–document matrix L shown in Table 1.5. Use a computer to do these exercises.

(a) Compute the singular value decomposition $L = U \Sigma V^T$.

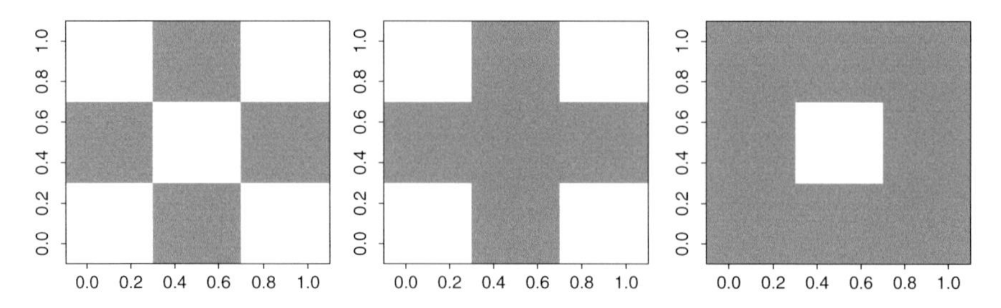

Fig. 10.5 An image of a checkerboard and the two nonzero summands in the outer product expansion. See Exercise #5

(b) Compute $L_2 = U_2 \Sigma_2 V_2^T$, the best rank 2 approximation to L.
(c) Use formula (10.9) to calculate the cosines of the angles $\widehat{\theta_j}$ between the column vectors of L_2 and each given query vector $\mathbf{q}$.

(i) $\mathbf{q} = \begin{bmatrix} 0\ 1\ 1\ 0\ 0 \end{bmatrix}^T$ = geometry + probability.
(ii) $\mathbf{q} = \begin{bmatrix} 0\ 1\ 0\ 0\ 1 \end{bmatrix}^T$ = geometry + symmetry.
(iii) $\mathbf{q} = \begin{bmatrix} 1\ 0\ 0\ 1\ 0 \end{bmatrix}^T$ = dimension + statistics.

(d) Use formula (10.10) to calculate the cosines of the angles θ_j^* between the column vectors of $\Sigma_2 V_2^T$ and the modified query vector $\widehat{\mathbf{q}} = U_2^T \mathbf{q}$, for each of the following:

(i) $\mathbf{q} = \begin{bmatrix} 0\ 1\ 1\ 0\ 0 \end{bmatrix}^T$ = geometry + probability.
(ii) $\mathbf{q} = \begin{bmatrix} 0\ 1\ 0\ 0\ 1 \end{bmatrix}^T$ = geometry + symmetry.
(iii) $\mathbf{q} = \begin{bmatrix} 1\ 0\ 0\ 1\ 0 \end{bmatrix}^T$ = dimension + statistics.

7. Refer to the term–document matrix L shown in Table 1.5. Use a computer to do these exercises.

(a) Compute the matrix $\Sigma_2 U_2^T$, where $L_2 = U_2 \Sigma_2 V_2^T$ is the best rank 2 approximation to L.
(b) Form the matrix W by normalizing the column vectors of $\Sigma_2 U_2^T$ to be unit vectors.
(c) Compute $W^T W$ to obtain the values of $\cos(\widehat{\omega}_{i,\,j})$, as in formula (10.11).
(d) Identify those pairs of terms for which the value of $\cos(\widehat{\omega}_{i,\,j})$ is

(i) greater than 0.5;
(ii) greater than 0.7;
(iii) greater than 0.8.

8. Let $A = \begin{bmatrix} 1 & 1 & 0 \\ 1 & 1 & 0 \\ 1 & 0 & 1 \\ 1 & 0 & 1 \end{bmatrix}$, as in Example 10.9.

   (a) Compute the singular value decomposition of A.
   (b) Compute the pseudoinverse $A^+ = V\Sigma^+U^T$, as in Definition 10.10.
   (c) With $\mathbf{b} = \begin{bmatrix} 1 & 3 & 8 & 2 \end{bmatrix}^T$, compute the Moore-Penrose solution to $A^T A\mathbf{x} = A^T\mathbf{b}$.

9. Let $B = \begin{bmatrix} 1 & 1 & 0 & 0.05 \\ 1 & 1 & 0 & -0.05 \\ 1 & 0 & 1 & 0.05 \\ 1 & 0 & 1 & -0.05 \end{bmatrix}$, as in Example 10.12.

   (a) Compute the singular value decomposition of B.
   (b) Compute the pseudoinverse of $B^+ = V\Sigma^+U^T$, as in Definition 10.10.
   (c) With $\mathbf{b} = \begin{bmatrix} 1 & 3 & 8 & 2 \end{bmatrix}^T$, compute the Moore-Penrose solution to $B^T B\mathbf{x} = B^T\mathbf{b}$.

10. Let $B = \begin{bmatrix} 1 & 1 & 0 & 0.05 \\ 1 & 1 & 0 & -0.05 \\ 1 & 0 & 1 & 0.05 \\ 1 & 0 & 1 & -0.05 \end{bmatrix}$ and $\mathbf{b} = \begin{bmatrix} 1 \\ 3 \\ 8 \\ 2 \end{bmatrix}$.

    (a) Take $\alpha = 0.5$. Using Tikhonov regularization, compute the approximate solution $\mathbf{x}^\sharp$ to $B\mathbf{x} = \mathbf{b}$.
    (b) Compute $||B\mathbf{x}^\sharp - \mathbf{b}||^2 + \alpha^2||\mathbf{x}^\sharp||^2$ for the answer to part (a).
    (c) Compare this to $||B\mathbf{x}^+ - \mathbf{b}||^2 + \alpha^2||\mathbf{x}^+||^2$ for the Moore-Penrose solution $\mathbf{x}^+$.
    (d) Repeat steps (a)–(c) with $\alpha = 0.2$.

11. *(Proof Problem)* Explain the following fact: *If a matrix is positive semidefinite, then its eigenvalue decomposition and its SVD are the same.* Is this still true if we replace *positive semidefinite* with *symmetric*? Explain.
12. *(Proof Problem)*

    (a) Show that $Null(A^T A) = Null(A)$, for every matrix A. (*Hint:* If $A^T A\mathbf{x} = \mathbf{0}$, then $||A\mathbf{x}||^2 = 0$.)
    (b) Show that the normal equation $A^T A\mathbf{x} = A^T\mathbf{b}$ has a *unique* solution orthogonal to $Null(A)$. (*Hint:* Show that the difference between two such solutions must be both orthogonal to $Null(A)$ and in $Null(A) = Null(A^T A)$.)

10.9 Projects

Project 10.1 (SVD and Image Compression)

1. Using a computer, create a matrix with numerical entries between 0 and 1. For example, the image in Figure 10.1 has the matrix A given in (3.1).
2. Create the grayscale image corresponding to this matrix. Example 10.5 might be helpful.
3. Compute the singular value decomposition of the image matrix.
4. Looking at the singular values, make a judgment about how many of them—k, say—are the most significant. Form the rank k truncated SVD.
5. Create the grayscale image corresponding to the truncated SVD. Compare this to the original image.
6. Look at the truncated SVD for a few different choices of rank k. Also, one can form individual layers $\sigma_j \mathbf{u}_j \mathbf{v}_j^T$ and add the corresponding matrices together.
7. Open a digital image from the library of a computing environment. Use truncations of the singular value decomposition to build reduced rank approximations of the image. See Example 10.5 for guidance.

Project 10.2 (Latent Semantic Indexing)
Let L denote the term–document matrix from Table 1.3.

1. Compute $L_3 = U_3 \Sigma_3 V_3^T$, the best rank 3 approximation to L.
2. Create a query vector using terms 3, 8, 12, and 18. That is,

$$\mathbf{q} = \begin{bmatrix} 0\,0\,1\,0\,0\,0\,0\,1\,0\,0\,0\,1\,0\,0\,0\,0\,0\,1 \end{bmatrix}^T .$$

3. Use formula (10.9) to calculate the cosines of the angles $\widehat{\theta}_j$ between the column vectors of L_3 and the query vector $\mathbf{q}$.

   - Using a threshold of *cosine* > 0.5, determine which documents are relevant to the query.
   - Using a threshold of *cosine* > 0.7, determine which documents are relevant to the query.

4. Use formula (10.10) to calculate the cosines of the angles θ_j^* between the column vectors of $\Sigma_3 V_3^T$ and the modified query $\widehat{\mathbf{q}} = U_3^T \mathbf{q}$.

   - Using a threshold of *cosine* > 0.5, determine which documents are relevant to the query.
   - Using a threshold of *cosine* > 0.7, determine which documents are relevant to the query.

5. Create two additional queries of your own and repeat the analysis for each one.
6. Apply formula (10.11) to compare the rows of L_3.
7. Verify the claims made in Example 10.7 about the number of comparable pairs of terms.

8. Determine the number of pairs of terms that compare favorably when the threshold used for the *cosine* of the corresponding angle is
   - *cosine* > 0.5;
   - *cosine* > 0.65;
   - *cosine* > 0.8.

Project 10.3 (Factor Analysis for Pitchers in Major League Baseball) The following matrix (A) shows several pitching statistics for the 2020 season of Major League Baseball for five top pitchers in the American League. The pitchers are listed in descending order of the evaluative statistic *WAR* (wins above replacement).

	ERA	WHIP	SO9	HR9	IP
Shane Bieber (WAR = 3.3)	1.63	0.866	14.2	0.8	77.1
Gerrit Cole (WAR = 2.2)	2.84	0.959	11.6	1.9	73.0
Kenta Maeda (WAR = 1.5)	2.70	0.750	10.0	1.3	66.2
Marco Gonzales (WAR = 1.5)	3.10	0.947	8.3	1.0	69.2
Zack Greinke (WAR = 1.2)	4.03	1.134	9.0	0.8	67.0

1. Compute the mean-adjusted matrix A_0 with column means of 0 in each column.
2. Compute the *variance/covariance matrix* $(0.25) \cdot A_0^T A_0$.
3. Compute the matrices V and Λ in the eigenvalue decomposition of $(0.25) \cdot A_0^T A_0$.
4. What percent of the overall variability in the data is accounted for by the first principal component direction $\mathbf{v}_1$?
5. Compute the row sums of $(A\mathbf{v}_1)\mathbf{v}_1^T$ to determine a new ranking of the five pitchers based on their scores in the first principal component direction.
6. Repeat the above analysis for a different example of your own choosing.

Bibliography

1. Axler, Sheldon, *Linear Algebra Done Right*, Springer, New York, 2015.
2. Berry, Michael W., and Murray Browne, *Understanding Search Engines: Mathematical Modeling and Text Retrieval*, 2nd edition, SIAM, Philadelphia, 2005.
3. Brouwer, A. E., and W. H. Haemers, *Spectra of Graphs*, Springer, New York, 2012.
4. Eldén, Lars, *Matrix Methods in Data Mining and Pattern Recognition*, SIAM, Philadelphia, 2007.
5. Epstein, Charles L., *Introduction to the Mathematics of Medical Imaging*, 2nd edition, SIAM, Philadelphia, 2008.
6. Estrada, E., and D. J. Higham, *Network properties revealed through matrix functions*, SIAM Review, Vol. 52, No. 4 (Dec., 2010), pp. 696–714.
7. Feeman, Timothy G., *The Mathematics of Medical Imaging: A Beginner's Guide*, 2nd ed., Springer, New York, 2015.
8. Fiedler, Miroslav, *Algebraic connectivity of graphs*, Czechoslovak Math. J., Vol. 23 (1973), pp. 298–305.
9. Gujarati, Damodar N., *Basic Econometrics*, 4th ed., McGraw–Hill, New York, 2003.
10. Hansen, Per Christian, James G. Nagy, and Dianne P. O'Leary, *Deblurring Images: Matrices, Spectra, and Filtering*, SIAM, Philadelphia, 2006.
11. Isaacson, Dean L., and Richard W. Madsen, *Markov Chains: Theory and Applications*, John Wiley & Sons, New York, 1976.
12. Jolliffe, I., *Principal Component Analysis*, John Wiley & Sons, New York, 2002.
13. Kalman, Dan, *A Singularly Valuable Decomposition: The SVD of a Matrix*, The College Mathematics Journal, Vol. 27, No. 1 (Jan., 1996), pp. 2–23.
14. Langville, Amy N., and Carl D. Meyer, *Google's PageRank and Beyond: The Science of Search Engine Rankings*, Princeton University Press, Princeton, 2006.
15. Langville, Amy N., and Carl D. Meyer, *Who's #1? The Science of Rating and Ranking*, Princeton University Press, Princeton, 2012.
16. Larsen, Richard J., and Morris L. Marx, *An Introduction to Mathematical Statistics and its Applications*, 2nd edition, Prentice Hall, Englewood Cliffs, 1986.
17. Noble, B., and J. W. Daniel, *Applied Linear Algebra*, 3rd ed., Prentice–Hall, Englewood Cliffs, 1988.
18. Noble, Safiya Umoja, *Algorithms of Oppression: How Search Engines Reinforce Racism*, New York University Press, New York, 2018.
19. Daniel A. Spielman, *Graphs, vectors, and matrices*, Bulletin of the AMS, Vol. 54, No. 1 (Jan., 2017), pp. 45–61.

T. G. Feeman, *Applied Linear Algebra and Matrix Methods*,
Springer Undergraduate Texts in Mathematics and Technology,
https://doi.org/10.1007/978-3-031-39562-8

20. Strang, Gilbert, *Linear Algebra and its Applications*, 4th edition, Cengage Learning, Boston, 2006.
21. Trefethen, Lloyd N., and David Bau III, *Numerical Linear Algebra*, SIAM, Philadelphia, 1997.
22. United States Bureau of Economic Analysis; URLs: https://www.bea.gov/industry/input-output-accounts-data; https://apps.bea.gov/iTable/index_industry_io.cfm These are official websites of the United States government. Accessed 05 January 2022.

Index

T. G. Feeman, *Applied Linear Algebra and Matrix Methods*, Springer Undergraduate Texts in Mathematics and Technology,
https://doi.org/10.1007/978-3-031-39562-8

ed in the United States
aker & Taylor Publisher Services